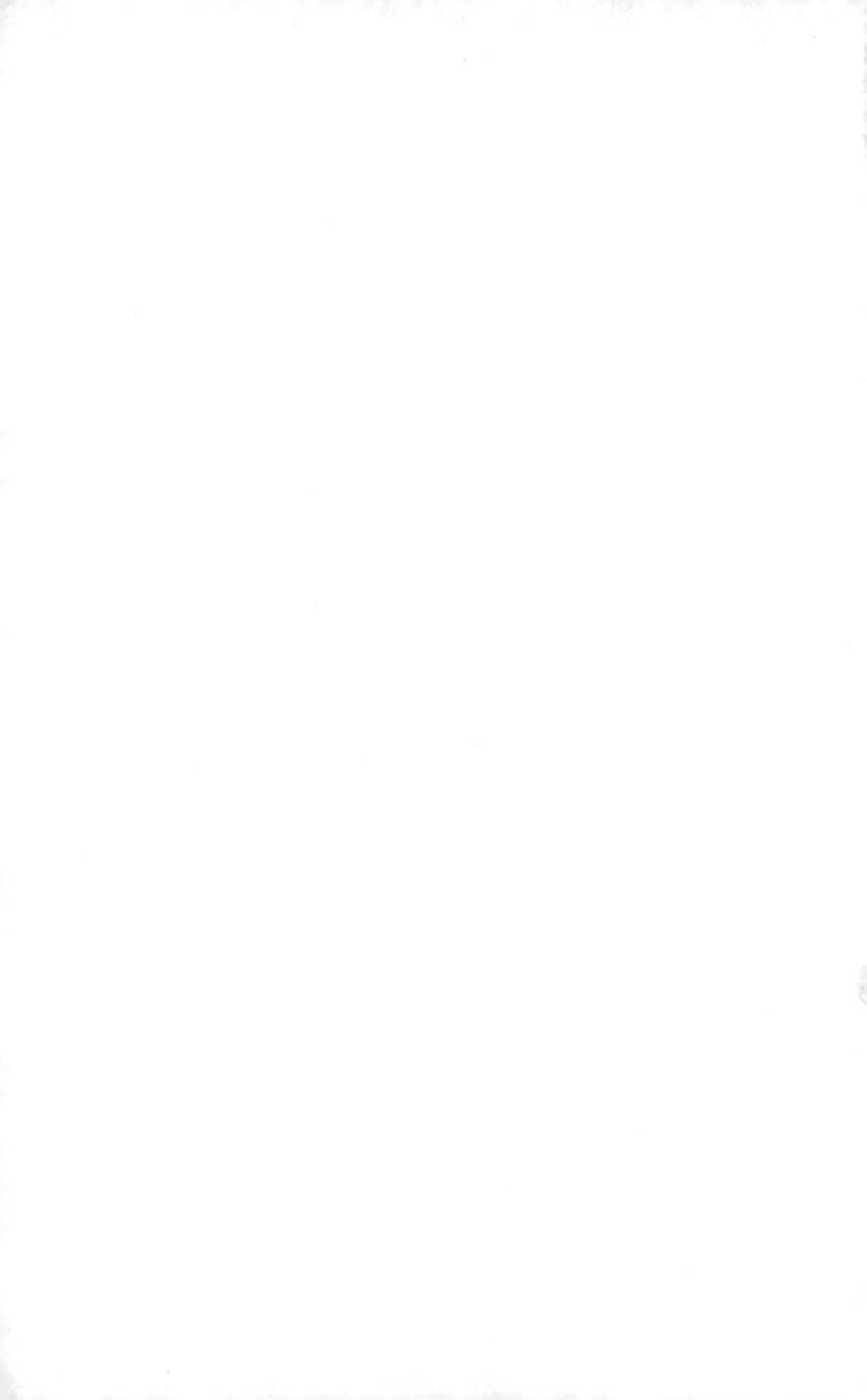

EARTH MATTERS ON STAGE

Earth Matters on Stage: Ecology and Environment in American Theater tells the story of how American theater has shaped popular understandings of the environment throughout the twentieth century as it argues for theater's potential power in the age of climate change. Using cultural and environmental history, seven chapters interrogate key moments in American theater and American environmentalism over the course of the twentieth century in the United States. It focuses, in particular, on how drama has represented environmental injustice and how inequality has become part of the American environmental landscape.

As the first book-length ecocritical study of American theater, *Earth Matters* examines both familiar dramas and lesser-known grassroots plays in an effort to show that theater can be a powerful force for social change from frontier drama of the late nineteenth century to the ecotheater movement. This book argues that theater has always and already been part of the history of environmental ideas and action in the United States.

Earth Matters also maps the rise of an ecocritical thought and ecotheater practice – what the author calls *ecodramaturgy* – showing how theater has informed environmental perceptions and policies. Through key plays and productions, it identifies strategies for artists who want their work to contribute to cultural transformation in the face of climate change.

Theresa J. May is the author of *Salmon Is Everything: Community-based Theatre in the Klamath Watershed*, the co-editor of *Readings in Performance and Ecology*, the co-author of *Greening Up Our Houses*, and the co-founder and artistic director of the EMOS Ecodrama Playwrights Festival. She is a professor of theater at the University of Oregon.

EARTH MATTERS ON STAGE

Ecology and Environment
in American Theater

Theresa J. May

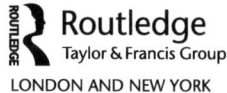
LONDON AND NEW YORK

First published 2021
by Routledge
2 Park Square, Milton Park, Abingdon, Oxon OX14 4RN

and by Routledge
52 Vanderbilt Avenue, New York, NY 10017

Routledge is an imprint of the Taylor & Francis Group, an informa business

© 2022 Theresa J. May

The right of Theresa J. May to be identified as author of this work has been asserted by her in accordance with sections 77 and 78 of the Copyright, Designs and Patents Act 1988.

All rights reserved. No part of this book may be reprinted or reproduced or utilized in any form or by any electronic, mechanical, or other means, now known or hereafter invented, including photocopying and recording, or in any information storage or retrieval system, without permission in writing from the publishers.

Trademark notice: Product or corporate names may be trademarks or registered trademarks, and are used only for identification and explanation without intent to infringe.

British Library Cataloguing-in-Publication Data
A catalogue record for this book is available from the British Library

Library of Congress Cataloging-in-Publication Data
A catalog record for this book has been requested

ISBN: 978-0-367-46464-6 (hbk)
ISBN: 978-0-367-46462-2 (pbk)
ISBN: 978-1-003-02888-8 (ebk)

DOI: 10.4324/9781003028888

An earlier version of part of Chapter 4 was previously published in the *New England Theatre Journal* 14:63–76 and is republished here with permission.

An earlier version of part of Chapter 7 was previously published in the *Journal of American Drama* 29:2 (Spring 2017): 1–18 and is republished here with permission.

Cover Photo: *Burning Vision* by Marie Clements, produced by Rumble Productions in association with Urban Ink Productions. Photo by Tim Matheson; Marcus Hondro pictured.

Index by Mary Harper, Access Points Indexing, OR.

*With gratitude to my grandparents,
Joseph Parsons and Rose Marie Fogarty Parsons*

CONTENTS

Acknowledgments		ix
Preface: from ecotheater to ecodramaturgy		xi
	Introduction: where has theater been while the world's been falling apart?	1
1	Stories that kill	18
	Augustin Daly's *Horizon* and William F. Cody's *Wild West: The Drama of Civilization*	
2	The Sabine wilderness and Progressive conservation	57
	David Belasco's *The Girl of the Golden West* and William Vaughn Moody's *The Great Divide*	
3	Dynamos, dust, and discontent	87
	Eugene O'Neal's *Dynamo* and the Federal Theatre Project's Living Newspapers *Triple-A Plowed Under* and *Power*	
4	We know we belong to the land	120
	Rogers and Hammerstein's *Oklahoma!* and Arthur Miller's *Death of a Salesman*	

| 5 | (Re)claiming home and homelands | 159 |

Lorraine Hansberry's *A Raisin in the Sun*, Luis Valdez's *Bernabé*, and Sam Shepard's *Buried Child*

| 6 | Stories in the land/legacies in the body | 200 |

Robert Schenkkan's *The Kentucky Cycle*, Cherríe Moraga's *Heroes and Saints*, and Anne Galjour's *Alligator Tales*

| 7 | Kinship, community, and climate change | 238 |

Marie Clements's *Burning Vision* and Chantal Bilodeau's *Sila*

Epilogue: theater as a site of civic generosity   276

Index   *283*

ACKNOWLEDGMENTS

This book was written on the traditional homelands of the Duwamish people of Puget Sound, the Wiyot people of Humboldt Bay, and the Kalapuya people of the Willamette Valley – lands that inspired and nourished me, each with histories that unsettled and motivated me. I register my respect and gratitude, first and foremost, there.

A wise editor-friend once told me that "a book, like a child, takes a village." Certainly, if ecological perspectives teach us anything, it is that we do nothing alone. Every achievement arises from the sustaining power of relationships – with friends, family, colleagues, community, and environment. A litany of gratitude cannot possibly give voice to the love, support, and keen critical insights that I have received on this journey from dozens of colleagues, graduate and undergraduate students, friends, and family members – my heart is overfull.

This book grew in the good soil of three institutions and the diverse intellectual communities they house. Many colleagues gifted me with their willingness to talk about the ideas in this book and the challenges of writing it. At a time when the words *ecology* and *theater* were not often found in the same paragraph, Barry Witham and my Ph.D. committee at the University of Washington School of Drama encouraged me to embark on this venture; comrades Tina Redd and Ken Cerniglia provided insights through peer readings. At Humboldt State University, faculty and students alike engaged in fruitful conversations at the intersection of environmental and social justice pedagogy. Big thanks to Christina Accomondo, Jean O'Hara, John Meyer, Marlon Sherman, and many other colleagues in the Environmental and Community Master's Program, in which I had the privilege to teach. The University of Oregon's Indigenous Studies Research Colloquium and Center for the Study of Women in Society provided rich intellectual communities. The research of gifted colleagues – including Stacy Alaimo, Kirby Brown, Stephanie LeMenager, Kari Marie Norgaard, Laura Pulido, Jennifer

O'Neal, Analisa Taylor, Sarah Wald, Louise Westling (among many others) – has been an integral and enriching part of that good soil. An Oregon Humanities Center Fellowship allowed me to move this project forward at a crucial time, along with research support from the University of Oregon's College of Arts and Humanities. Particular thanks goes here to the Knight Library's Elizabeth Peterson for her kind assistance. I have appreciated the intellectual curiosity of Department of Theatre Arts colleagues Michael Najjar, Sara Freeman, and LaDonna Forsgren. I have also learned much from my graduate students – including Jean O'Hara, Miriam Kammer, Jonathon Taylor, Brian Cook, Damond Morris, Dan Carlgren, Theresa Dudeck, Natasha Kolosowsky, J.M. Bacon, Zeina Salame, Olga Sanchez Saltviet, Liz Fairchild, Josh Evans, and Anna Dubla-Barnett – whose enthusiasm and engagement in this green conversation has deepened my own thinking about theater practice and questions of representation.

I owe a particular debt to my fellow travelers in this subfield of ecodramaturgy for their intellectual generosity and solidarity, especially Wendy Arons, Una Chaudhuri, Downing Cless, Nelson Gray, and Sarah Standing, as well as members past and present of ASTR's Performance and Ecology Working Group. Downing Cless read an early draft and provided crucial suggestions, which were only exceeded by his indefatigable moral support. Nelson Gray gave insights that helped to shape the final chapter. Thanks to Linda Forest for the walkin'-and-talkin' breaks that cleared my head and reset my will. Arwen Spicer provided helpful suggestions on early drafts. Erin Moberg's crucial insights, spot-on suggestions, and attention to detail have been a gift indeed. Most recently, I have appreciated the enthusiasm of Routledge editor Laura Hussey and the entire Routledge production team.

Kinship is always at the heart of the values that hold us this side of the abyss. In the end, it has been my family and friends who have sustained me throughout. For your constant and kind encouragement, big-hearted thanks to Lorraine, Susan, Don, Dave, Lee, Melanie, Alex, Linda, Ed, Nicoli, Marta, Gordon, Lisa, and Adell, and to my parents and family members, especially Donna, Jim, Mary Lou, Bill, Barbara, and Shinan, who were never too busy to listen. Finally, this book would not have been possible, nor its writing sustainable, without the faith and encouragement of my husband, collaborator, and best friend, Larry Fried (and our faithful Milo), who helps me daily to remember that we are all part of a precious ecological community on this planet and in this place we call home.

Theresa J. May, Kalapuya illihi/Eugene, Oregon

PREFACE

From ecotheater to ecodramaturgy

I wrote this book because I wish I'd had one like it to read when I was a young artist-activist. Growing up during the civil rights movements of the 1960s, and coming of age during the second-wave environmental movement of the 1970s, I thought theater could change the world – and there was ample evidence for my optimism. Throughout the U.S. theater was part of protest movements, consciousness raising, and political action: El Teatro Campesino was an empowering arm of the United Farmworkers Movement; Amiri Baraka's Black Revolutionary Theater was rehearsing revolution; Bread and Puppet Theater was inspiring communities to activism (see, for example, Elam (1997). When I learned about the work of Polish theater director Jerzy Grotowski ([1968] 1975), who sought to "abolish the distance between actor and audience," I understood that this art form carried a power of immediacy and connection that was unique. Grotowski urged artists to recognize the primacy in theater of the "the closeness of the living organism," even in the face of new technologies that (perennially) threatened the very survival of the art. "The theatre must recognize its own limitations. If it cannot be richer than the cinema, then let it be poor. If it cannot be as lavish as television, let it be ascetic. If it cannot be a technical attraction, let it renounce all outward technique." Staging plays in which actors and audience were "within arm's reach" of one another and the audience member could "feel [the actor's] breathing," Grotowski approached each play as an encounter – not only between artist and text and text and audience, but also between individual and community (41). I was inspired by his assertion that the uniquely embodied, immediate, and communal aspects of theater were at the heart of its power and potential to affect social change.

In 1977, I participated in the Teatr Laboratorium's *Mountain Project*[1] in the Polish towns of Wrocław and Legnica (May 2005).[2] The Project took place in the countryside in and around the nearly thousand-year-old Grodziec castle (Zmysłowski and Burzynski 2015). What was compelling to me at the time was the subtle way in which the embodied experience of land helped to foster a sense of community among participants. In part, that experience forged in me a theatrical aesthetic by which form – including the embodied immediacy of audience/actor as community – seemed a potent force for social change and environmental awareness. I had to find a way to reconcile my creative ambitions with my social and ecological values, and to balance my passion for the great outdoors with a craft that kept me indoors, often in dark and musty spaces. Inspired by the work of Bread and Puppet, San Francisco Mime Troupe, El Teatro Campesino, and other groups that were using theater to leverage both economic and environmental justice, I started a fringe company with actors who were willing to muck about in the woods – for central to our emerging tenets was to leave the black box behind. If nature were not part of the theater, we reasoned, theater should go to nature. We made giant puppets, collected found objects, talked about audience participation, wrote "new myths," and occasionally took industry jobs to pay the rent when the grants did not come through.

After moving to Seattle in the mid-1980s, I founded Theatre in the Wild (TITW), whose mission was to use theater to inspire in our audiences a sense of connection with, and compassion for, the natural world. Through Washington State's Department of Ecology's Public Involvement and Education, TITW expanded the environmental education curriculum through school residencies in which students devised and performed original plays about their watersheds. We also devised site-specific public performances – immersion theater designed to connect audiences to the natural world and integrate stories into the landscape (May 1999).[3]

In 1991, my co-director Larry Fried initiated Act Green under the umbrella of TITW, a project to support theater artists in walking their environmental talk. In partnership with Seattle's Intiman Theatre, Act Green brought theater-makers from across the country together with policy-makers for a three-day conference, called *Theatre in an Ecological Age*. The conference posed this central question: how might theater better respond to the environmental crisis? Robert Schenkkan (1994) – whose play, *The Kentucky Cycle*, had just premiered at the Intiman Theatre – gave the keynote. Schenkkan indicted the U.S. frontier narrative as the root story that sanctioned, indeed mandated, environmental destruction. The idea that stories encapsulate ideologies from which policies and behaviors toward the environment emerge affected me greatly: it was, at least in part, what compelled me to return to the academy a few years later.

At the same conference, Molly Smith, then the Artistic Director of Perseverance Theatre in Juneau, Alaska, gave a second keynote about her company's commitment to place. Perseverance had staged Homer's *Odyssey* as a response to the 1989 *Exxon Valdez* oil spill off the Kenai Peninsula; the theater had also adapted Sophocles' *Antigone* through a Native Alaskan cultural lens. Cast members of *Yup'ik Antigone* spoke of what it meant to bend western theater traditions to speak through and about indigenous cultural and environmental concerns

(Denning 1984). This example of forwarding and foregrounding indigenous voices also shaped the direction of my own creative and scholarly work. The conference, however, was centrally focused not on stories, but on stuff – the material practice – of theater. Following the conference, Fried and I published *Greening Up Our Houses* (1994) with suggestions for how to reduce the environmental and human health impacts of theater making. Our desire to nurture new works – new stories – would continue to haunt and motivate us going forward.

The early 1990s was a watershed moment for ecotheater in the U.S. In a 1992 cover story in *American Theatre*, Lynn Jacobson gave a shout-out to regional and community-based companies that were putting environmental issues and ecological values at the center of their productions. In a special "ecological" issue of *Theater* released in 1994, guest editor Una Chaudhuri charged playwrights to take up the ecological themes on stage, and called on scholars to use ecology as a critical vantage point. In 1996, Downing Cless pointed out that grassroot theaters were already "greener" in ways that empowered their communities (79–82). Together, this scholarship began to break open a discursive space, a critical standpoint from which to view theater history, theory, and practice through the lens of ecological thought.

At the University of Washington, I was blessed with faculty who supported my desire to look at theater through the lens of ecology. I believed that Chaudhuri's (1994) analysis provided a framework for understanding the ecological themes and implications in *any and all* plays and performances, not merely plays expressly about environmental themes. Yet although I'd spent years doing arts-based watershed education and pollution prevention, I did not know the story of the land on which I lived, worked, and played. Despite my passionate sentiments about preserving wilderness spaces, my moral indignation at polluters, and my conviction that the environmental crisis rose out of long ingrained Western European intellectual frameworks that had shaped a capitalist-consumerist culture – I did not know how things came to be the way they are, nor how I might be implicated in the destruction around me.

Alongside theater history and critical theories, I devoured the works of environmental historians William Cronon, Vine Deloria, Jr., Carolyn Merchant, Richard White, and Donald Worster. I began to see how the stories in early American melodramas, or 1930s labor plays, were inscribed in the land around me. I began to see stories everywhere, stories that carry and concretize ideologies that give rise, in turn, to the policies that shape the land. Narratives rose up before my eyes in the form of giant gravity dams spread out in the pastoral vistas of the Seattle arboretum and on the streets of Seattle's Pioneer district. As I learned of the legacies and of settler colonialism, I also saw how theater in the U.S. had participated in the propagation of stories that shored up and defended its worst (and ongoing) impact on land and lives. I became more fully convinced that theater could be a force for healing, justice, and resilience.

In 2002, as assistant professor at Humboldt State University, I saw first-hand the ecological and cultural impacts of settler colonialism in the form of a devastating fish kill on the Klamath River. The death of over 70 thousand salmon on Yurok homelands on the lower Klamath River brought home to me not only the artificial separation of *nature* from *culture*, but also the subtle ways this binary thinking perpetuates environmental injustices. Working with Karuk, Yurok, and Hupa tribal members, students, and community members, I wrote *Salmon Is Everything* (2019), a community-based play that documents the cultural significance of that historic fish kill. The process of developing *Salmon Is Everything* was an exploration of the complex ways ecology and environment are materially intertwined with culture, identity, and sovereignty. That process of community collaboration changed me.[4] As I listened to and learned from my Native collaborators, I became keenly aware of the systems of power and privilege evidenced not only in the control of water on the Klamath River and in the economic priorities that privileged some communities while ignoring others, but also in who was included in policy solutions and whose narrative drove public debate. The process also changed my thinking about what theater can do as a site of civic engagement.

The convergence of ecologically conscious theater practice and ecocritical theater scholarship gained momentum in 2004 when Downing Cless and I started the Performance and Ecology Working Group for the American Society for Theatre Research, a project which has continued to grow and incubate new scholarship. Also in 2004, director Larry Fried and I started the EMOS (an acronym for "earth matters on stage") Ecodrama Playwrights Festival.[5] Hosted at Humboldt State University in partnership with the Dell'Arte Company, EMOS called on dramatists to "respond to the ecological crisis and explore new possibilities of being in relationship with the more-than-human world" (quoted from EMOS guidelines; see also Arons and May [2013]). We called for plays centrally focused on ecologies, environmental injustice, and sense of place. EMOS, now hosted by various institutions, is one of several national artistic initiatives focused on the environment crisis and climate change. In 2007, Wendy Arons invited me to write an article in a special "green" edition of *Theatre Topics*. In "Beyond Bambi," I offered a series of "Green Questions to Ask a Play" as a starting point for ecodramaturgy – the critical and historiographic examinations of plays and performances (2007). In order to put new artistic work in conversation with new scholarly work, EMOS 2009, hosted by the University of Oregon, added a symposium on performance and ecology. *Readings in Performance and Ecology* (Arons, May, 2012) emerged from that symposium.

What emerged as an outcropping of environmentally concerned theater artists (myself and so many others) in the 1990s is now a groundswell of creative innovation fueled by networks of media-savvy artists who are committed to theater becoming a significant force for global ecological transformation. This book represents a personal as well as a scholarly journey – from the sobering confrontation with the ways history lives in the land and reverberates as patterns of privilege, to the work of contemporary playwrights whose dramas (re)envision ecological relations. It stands alongside many other projects, both scholarly and artistic, that are part of what I've called ecodramaturgy – an unfolding environmental consciousness in North American theater. As we stand together at the threshold of the Anthropocene, I hope that it will provoke your thinking about what theater has, can, and must do.

Notes

1 The *Mountain Project* (1975– 1978) grew out of the "paratheatre" work of the Teatr Laboratorium (also known as the Polish Laboratory Theatre); it took place over several years and included multiple and cumulative phases, including *Night Vigil*, *The Way*, and *Mountain of Flame* (Kumiega 1985, 183–214.)
2 A copy of my journal from my *Mountain Project* can be found at the University of Washington Drama Library in Seattle, Washington.
3 Today, I would take issue with some of my own assertions and assumptions in that (1999) article, but I list it here because it describes TITW's site-specific work in some detail.
4 In "The Education of an Artist," I write in detail about my personal process of learning indigenous methodologies from Karuk, Yurok, and Hupa collaborators in the development of *Salmon Is Everything* (111–152). The volume that contains my essay and the play also contains first-person accounts by Native collaborators (see May 2017).
5 EMOS consists of a new play contest and symposium hosted by university/community partnerships across the country. EMOS events have included workshop productions of winning scripts, play readings, panels, critical papers, and performance workshops. Following EMOS 2004 at Humboldt State, the festival was hosted by the University of Oregon in 2009; Carnegie Mellon University in 2012; the University of Nevada-Reno in 2015; and the University of Alaska, Anchorage, in 2018.

References

Arons, Wendy, and Theresa May, eds. 2012. *Readings in Performance and Ecology*. New York: Palgrave Macmillan.
Arons, Wendy, and Theresa May. 2013. "Ecodramaturgy and/of Contemporary Women Playwrights." In *Contemporary Women Playwrights*, edited by Lesley Ferris and Penny Farfan, 181–196. New York: Palgrave MacMillan.
Chaudhuri, Una. 1994. "'There Must Be A Lot of Fish in that Lake': Toward an Ecological Theater." *Theater* 25 (1): 23–31. DOI: ttps://doi-org.libproxy.uoregon.edu/10.1215/01610775-25-1-23.
Cless, Downing, 1992. "Eco-Theatre, USA: The Grassroots Is Greener." *TDR* 40 (2): 79–102. DOI: 10.2307/1146531. https://www-jstor-org.libproxy.uoregon.edu/stable/1146531
Denning, Jennifer. 1984. "Stage: *Yup'ik Antigone*." *New York Times*, June 29.
Elam, Harry Justin. 1997. *Taking It to the Streets: The Social Protest Theater of Luis Valdez and Amiri Baraka*. Ann Arbor: University of Michigan Press.
Fried, Larry K., and Theresa May. 1994. *Greening Up Our Houses: A Guide to a More Ecologically Sound Theatre*. New York: Drama Book Publishers.
Grotowski, Jerzy. (1968) 1975. *Towards a Poor Theatre*. London: Methuen.
Jacobsen, Lynn. 1992. "Green Theatre: Confessions of an Eco-reporter." *American Theatre* 8 (1): 17–25, 55.
Kumiega, Jennifer. 1997. "Laboratory Theatre/Grotowski/The Mountain Project." In *The Grotowski Sourcebook*, edited by Richard Schechner and Lisa Wolford, 231–247. London and New York: Routledge.

Kumiega, Jennifer. 1985. *The Theatre of Grotowski*. London: Methuen.
May, Theresa. 1999. "Bakhtin on Site: Chronotopes in Theatre in the Wild's *Dragon Island*." *On-Stage Studies* 22: 19–38.
May, Theresa. 2005. "Remembering the Mountain: Grotowski's Deep Ecology." In *Performing Nature: Explorations in Ecology and the Arts,* edited by Gabriella Giannachi and Stewart Nigel, 345–359. Bern, Switzerland: Peter Lang.
May, Theresa. 2007. "Beyond Bambi: Toward a Dangerous Ecocriticism in Theatre Studies." *Theatre Topics* 17 (2): 95–110. DOI: 10.1353/tt.2008.0001.
May, Theresa. 2019. *Salmon Is Everything: Community-Based Theatre in the Klamath Watershed*. 2nd ed. Corvallis, OR: Oregon State University Press.
Schenkkan, Robert. 1994. *The Kentucky Cycle*. New York: Dramatists Play Service.
Zmysłowski, Jacek, and Tadeusz Burzynski. 2015. "On the Opposite Pole of the Mundane." In *Voices from Within: Grotowski's Polish Collaborators*, edited by Paul Allain and Grezgorz Ziolkowski, 100–105. London: Polish Theatre Perspectives/TAPAC Theatre and Performance Across Cultures.

INTRODUCTION

Where has theater been while the world's been falling apart?

Present disasters, like present debates, have deep roots in our collective past. Consequently, sound activism requires knowledge of history, just as history told in ways that bend toward justice is itself a kind of activism. Many theater artists, and I count myself among them, hope to contribute to positive social and environmental transformation through their work. But we cannot proceed without understanding the ways in which theater has already participated in shaping social behavior and national policies. American theater has represented some of the central environmental debates of the last century. At times, it has intervened to strengthen democratic and ecological values, but many plays and productions championed U.S. environmental imperialism and were complicit in the project of plunder.

The need for a history of American theater written through an environmental lens was clear when, in the context of a lively discussion about plays thematically related to the climate crisis, one of my students threw up their hands and exclaimed, "OMG! Where has theater *been* while the world's been falling apart?!" While the student's incredulity arose from an activist impulse, it was in fact a request – a demand, really – for an accounting. Traditional theater history and literature courses had not revealed the many ways that theater *has* engaged the environmental crises of the twentieth century. In the uncertain future that is now unfolding, this book aims to lay a foundation for artists and scholars who seek to put their shoulders to the wheel of what Una Chaudhuri (1994) has called an ecological "transvaluation so profound as to be nearly unimaginable at present" (24).

In the chapters that follow, I take a wide-angle historical perspective, together with detailed analyses of plays and productions, in order to map how theater has responded at key moments in U.S. environmental history. How has theater disseminated ecological ideas or stimulated new perceptions of the natural world?

How has it helped to perpetuate ecological violence? How might theater be a tool for *de*colonizing movements? Finally, how might theater participate in the transformation of conscience and consciousness so desperately needed now?

In each chapter I consider emblematic plays from turning points in U.S. popular understandings of our relatedness to the environment, namely, the closure of the frontier, the wilderness preservation movement, the New Deal conservation era, the rise of the postwar consumer culture, the environmental and civil rights movements of the 1960s and 1970s, the environmental justice movement of the 1990s, and, finally, the increasing social and political awareness of climate change. Some of these plays have enjoyed critical success that has made them part of the canon; others rose out of grassroots activism. Some represent the deeply rooted "American" stories about the land; others critique those master narratives, telling new stories infused with the ecological values of interdependence and justice. Whether mainstream or grassroots, these plays, performances, and the stories they transmit have become part of not only *what* we think but also *how* we think, and how we understand our place in and kinship with the land.

At the heart of this braided history and play analysis is an assertion that the human imagination is an ecological force and that our stories have social and ecological consequences. The stories we tell touch the land. Storytelling becomes policy-making as stories told and enacted inform values and ideologies, which, in turn, shape individual and collective behaviors. In this way, theatrical representation participates in shaping perceptions, desires, behaviors, and policies toward the land and its biotic communities. As a theater-maker, activist, educator, and scholar, I also am centrally concerned with how the dynamics of racism, sexism, and economic inequity intersect with environmental concerns. To underscore how theater might activate an eco-civic imagination, throughout this book I elucidate the ways by which theater has represented the complex interweaving of ecological degradation and human oppression. Ultimately, this is a book that argues for theater's potential to assist in mending broken relations (with the land and others) and to inspire ecologically responsible action.

Theater as civic practice

This book conceives of theater as a site of civic discourse that has influenced and reflected society's relatedness to the land, and that might help us compassionately navigate the social changes that have occurred and will occur as a result of climate change. Plays are blueprints for live performance that require collective co-imagining by people who have come together in time and place. It is necessarily immediate, embodied, and communal – attributes that have everything to do with how theater makes meaning and how it might make a difference. As characters, places, and temporalities are brought to life on stage, a give-and-take occurs between the imaginations, sensibilities, and visceral experiences of the performers and the audience. This alchemy of embodied communal feeling, visceral response, and shared risk reminds us that no matter how abstract or virtual our interactions may be outside the theater space, we nonetheless inhabit the world as embodied organisms. At the same time, theater's inherent reciprocity encourages dialogue not only between performers and audience but also between the event of the performance and the larger sociopolitical milieu in which it takes place. In an age when digital, virtual, and remote realities dominate our everyday experience, it is useful to take a moment to consider these defining qualities of theater.

Theater has long served as a site of civic practice and a forum for civic action; through plays, diverse societies have explored questions of free will, social conscience, community obligation, moral leadership, and the social and ecological consequences of hubris. This is civic power, as inherent in theatrical *form* as it is in any particular narrative content. Theater's form begins with an invitation. You may have heard the axiom that theater starts with the question *What if?* Theater depends on the willing suspension of disbelief by which audiences and actors set aside rigidly guarded worldviews about self, world, and other. The prologue of Shakespeare's *Henry V* makes this invitation clear when the Chorus asks the audience to "entertain conjecture of a time" (1.1, 1). For a few hours, audience and performers are expected to summon the generosity to collectively conjure the world of the play – a world that may appear fantastical or even revolutionary. In this task, both audience and performers are responsible for bringing the play to life in the present time and in a shared place of the stage. This fundamental contract – the willingness to collectively engage in fiction for the purpose of bearing witness and finding meaning – makes theater a vital civic tool. Theater cultivates democratic values and strengthens the civic muscles of tolerance, empathy, and self-reflection.

Applied and community-based theater practitioners tend to argue for theater's efficacy as a tool for social change and civic engagement (see, e.g., Cohen-Cruz [2005, 2010]). I argue, however, that the idea of civic engagement provides a framework for understanding how theater functions more generally. As a communal act, theater hones our capacity to be in a relationship – a sense of connection that is foundational to ecological health and democracy. Theater's immediacy requires us to be attentive and responsive to others. As stories play out in real time and physical space, theater invites us to *live into* the world of the play in order to examine together the consequences of human actions. In doing so we experience the ways we are unexpectedly connected and implicated. The ruse of the play – the *What if?* – allows for open-minded, playful consideration of possibilities actualized as sensorial experience so that we might taste and feel those possibilities and glimpse the wisdom that otherwise comes from lived experience. Theater thus exercises our capacity to listen, to acknowledge worlds of experience different from our own, to simultaneously hold multiple and conflicting viewpoints as plausible and real, and to *give audience* to values that may conflict with our own. In this way, theater exercises the imaginative elasticity necessary to safeguard democracy, justice, and compassion as climate change unfolds.

In the chapters that follow I examine the democratic value of theater's unique capacity to gather people together in community. Throughout, I argue that because theater is a practice of collaborative imagination and collective conjuring, it has ample capacity to intervene in the large master narratives that have helped perpetuate environmental injustice. By telling untold stories theater can reclaim histories that have been erased or distorted, unmask and expose the impact of ideologies, assert new voices, and carve out space for those that have been silenced or marginalized. These stories can build relationships, crack outdated ideologies, open new possibilities, envision futures, and help to (re)shape the social, political, and ecological landscapes of our lives. In the present historical moment, theater artists have an opportunity to tell stories and explore forms that actively practice compassion and demand justice – stories that are visionary, generative, and healing. For theater artists in particular, I pose this question: In the face of the shattering facts of climate change and other human-caused planetary biocides, *what if* the skills that you possess, the stories that you tell, and the forms through which you tell them could help to save lives, prevent suffering, heal destruction, reclaim worlds, and transform what it means to be a human animal in a diverse ecological community? *What if?*

Ecodramaturgy as methodology

Ecodramaturgy is theater praxis that centers ecological relations by foregrounding as permeable and fluid the socially-constructed boundaries between nature and culture, human and nonhuman, individual and community. It encompasses both artistic work (making theater) and critical work (history, dramaturgy, and criticism) in three interwoven endeavors: (1) examining the often invisible environmental message of a play or production, making its ecological ideologies and implications visible; (2) using theater as a methodology to approach contemporary environmental problems (writing, devising, and producing new plays that engage environmental issues and themes); and (3) examining how theater as a material craft creates its own ecological footprint and works both to reduce waste and invent new approaches to material practice.[1] As a critical lens, ecodramaturgy examines the role of theater in the face of rising ecological crises, foregrounding the material ecologies represented on stage.[2]

Although environmental issues have concerned dramatists throughout the twentieth century, ecodramaturgy as a critical discourse can perhaps be traced to the summer of 1994, in which a landmark issue of *Theater* was dedicated to ecology as a vantage point for theater studies. Editor Erica Munk (1994) wrote that "our playwrights' silence on the environment as a political issue and our critics' neglect of the ecological implications of theatrical form are rather astonishing" (5). Guest editor Una Chaudhuri (1994) likewise envisioned an ecological theater that would be possible only if and when we recognize ecological themes as more than metaphorical comments on the human condition. Theater's human-centric bias, she argued in that piece, stems from an ideology that shares common cause with Euro-American industrial civilization's celebration of artifice as an expression of human superiority over and separation from nature. Ultimately, Chaudhuri's analysis not only identified a probable cause for mainstream theater's perceived silence on ecological issues; it also suggested a framework for viewing and conceptualizing the ecological themes and implications in *any and all* plays and performances. Munk likewise urged investigations of "the way ecologies – physical, perceptual, imagined – shape dramatic forms" and suggested that "we stand at the edge of a vast, open field of histories to be rewritten, styles to rediscuss, contexts to reperceive" (5). Since that time, artists and scholars have responded in myriad ways, and their work is part of the many-handed project that I call ecodramaturgy.

As a component of theatrical production, ecodramaturgy helps to decipher the meaning a play might have had in the past and what it might mean to a contemporary audience, thus informing a decision about how or whether to produce it today. However, merely overlaying green themes on narratives and forms that still sow the seeds and structures of oppression is not enough. Theater artists must continue to ask: What is the history of the land we are representing on stage? How was the idea of nature or environment understood in the historical moment represented in the play? How does the play represent the consequences of those ideas as they impact people and land? How do those ideas and representations resonate differently today? Questions about the social, political, economic, and ecological context of a play or production shed light on a play's potential for meaning-making and expose the ways a play might inadvertently reinscribe environmental imperialism through the repetition of familiar tropes and narratives. Such questions form the axis of the chapters that follow.

My argument in this book is that theater matters precisely because the stories that we have told as a nation have *already* had material-ecological impact. Theater has long served as a means through which human beings have questioned and examined their relationships with the natural world.[3] Indeed, the environment, the land, and ecological paradigms of thought have always been present in the drama and on the stages of this nation; in this way, theater has helped to produce today's environmental realities. U.S. theater has participated in spinning the stories that forwarded the ideologies and practices of Manifest Destiny, white supremacy, and extractive capitalism. But it has also provided potent means to intervene in longstanding destructive narratives. As an activist framework, ecodramaturgy acknowledges this history *in order to* take responsibility for the destruction of land and culture that some plays have supported, as well as to forward theater's participation in the *decolonization* of people and land. These chapters examine the ways in which theater in the United States has participated in bringing us to this perilous moment and invite readers to consider the stories they have encountered in the theater or in life. As part of U.S. cultural history, those stories have had material consequences for the environments in which we all live, work, play, and worship.

Nature, in a word

In the preface to this book, I mention my early passion for *the great outdoors*. How did I learn to use this phrase to describe the same *wilderness* that terrified my settler ancestors? The phrase carries layers of personal as well as sociopolitical and environmental history. For starters, it signals my privilege as a person who has access to and is assured a measure of safety in *outdoor* places. Furthermore, what might be the association between those *outdoors* and the nationalism that is implied in the descriptor *great*? None of these questions coursed through my mind when, at age 22, I hiked the length of the John Muir Trail alone. Yet, my very ability to venture forth, even the awe that I felt in the high country, were products of my sociopolitical position. What I called *nature* was in every way a product of culture.

All the terms that we use to name or describe the natural world carry layers of historical and cultural meanings. The words themselves encapsulate stories and histories buried in the land. Terms such as *wilderness, natural resource, landscape,* and *ecosystem,* carry socioeconomic implications and embedded understandings of power that give agency to certain persons or institutions and divest it from others. The point here is that none of these are stable or self-evident terms. Science, likewise, is of its historical moment. The science of ecology developed (and continues to develop) over decades, and its conceptual framework is not independent of the political and economic pressures in which it was forged. In the chapters that follow, I use *ecosystem* when referring to the interconnectedness of living and geologic components in a specific locale (e.g., the ecosystem of the prairies or the ecosystem of the inner city). I use the term *land* to mean the entire field of presence that includes plants, animals, and inorganic geologic features such as waterways, as well as human beings. I have tended to use the term *natural world* when referring to those plants and animals with which humans share the planet and the term *landscape* when it is important to emphasize the visual and consumptive behaviors that have often governed U.S. society's relationship to the land. In discussing the way the natural world was conceptualized during a certain period I use the terms common at that time, such as *nature* or *wilderness.*[4] Throughout this book, all these terms (and more) are to be understood as constructed notions open to critique and unpacking and are not called out in quotations.

The title of this book poses a similar challenge in its use of the word *American,* and I have similarly dispensed with quotation marks in this case – but not with the awareness of the contested and changing meanings of the term. While *American* has many, and sometimes dubious, connotations, it is a term that must be engaged with in discussing the history of what has been and still is known as American drama. As the descriptor of national identity for those who live within the political boundaries of the United States, *American* has been used as a tool of exclusion as well as belonging. Ideas about what is and is not American, as well as who does and does not belong, are much contested by residents of the United States and by residents of the Americas. In the United States, who or what is American has often been defined in contrast to who or what is *not* American, resulting in the exclusion of sovereign indigenous nations as well as millions of other residents of two continents. In the chapters that follow, I use the term *American* to critique such problematic connotations but also to invoke the term's complex history, which no synonym can replicate. I use the term *United States* (or *U.S.*) when discussing national policies, programs, or events that have specifically to do with the operations of the nation-state. I use the term *American* when the artists or historical figures under consideration have used that term to describe themselves and when their work has been invested in defining, protecting, or challenging that identity. I have also used it to describe mainstream U.S. cultural contexts and affinities that are or have been understood, particularly by residents of the United States, as American – and there, its meaning has changed over time. The chapters that follow deal with how theater has participated in that discourse.

American is an identity associated with power – imperial, white supremacist, and economic – for many of the artists and historic figures in the early chapters; it is an identity that is called into question and preempted by works and artists discussed in the later chapters. The plays discussed in the first six chapters were authored by U.S. citizens – many were celebrated as canonical examples of American drama – and these participated in the accumulation of meaning and identity around the term. Others, which I discuss in the final chapter of this book, challenge that hegemony through works that blur the borders of belonging. These plays reclaim, redefine, and redraw what American means – and what it means to be American – in hemispheric terms. They stand on different American ground, one with both indigenous and colonial histories and an indigenous present; and one that asserts shared ecologies, languages, and cultures to be as significant, and perhaps more significant, than national citizenship. Indeed, the plays at the center of Chapter 7 were written by women who are Canadian citizens, but their identities (as First Nations/Métis and Québécois) are more complex and fluid.

Finally, as a U.S. citizen of settler descent, the stories and ideologies contained in many of the plays I discuss have also been deeply ingrained within me as a part of my own cultural inheritance. In each chapter, I wrestle with ideologies, assumptions, representations, and master narratives in which I myself am implicated. With this in mind, the chapters that follow are my own meditations – personal as well as scholarly – on how theater has not only been part of the problem but also how it might contribute to civic and ecological justice and healing.

Chapter summary: from cowboys to climate change

No story has proven as useful to the U.S. nor as destructive to the environment as that of the frontier. As mythos, the frontier continues to inform who Americans think they should be.[5] As an ethos – the guiding values that inform actions – frontier ideology is operative everywhere around us, playing out in individual lives, businesses, families, and communities. Frontier ideology is still reinforced on stage in U.S. theater today, while its violence continues to be felt by people and land in the twenty-first century. The recent Broadway musical *Bloody Bloody Andrew Jackson* is a case in point, with its cavalier use of Jackson as a hero and stereotypical representation of Native peoples and history (Friedman and Timbers 2010). As individuals, we may reject frontier values and directives, yet as a society we are no more postfrontier than we are postracial, in part because of the interconnection of both discourses and corresponding, institutionalized practices. In Chapter 1, I consider how Augustin Daly's *Horizon* (produced in 1871) and William F. "Buffalo Bill" Cody's *Wild West: The Drama of Civilization* (1886) propagated what J. M. Bacon calls the "ecological violence" of settler colonialism (2019, 59–69). Both Daly's play and Cody's extravaganza reveal how dominant Anglo narratives justified U.S. military occupation of indigenous lands, promoted resource extraction from western lands by eastern capital, and normalized white supremacy and the extermination of people and animals (see, e.g., Opie [1998], 2). I argue that a critical, self-reflexive awareness about how we represent, discuss, and frame history is crucial for ecodramaturgy to guard against propagating the very values we hope to dismantle.

In the early twentieth century, as the nation sought to cope with the diverse ecologies of a single continent, a vast majority of Euro-Americans saw the land through a viewfinder of stories – such as the Judeo-Christian Garden of Eden – that justified ongoing Anglo-U.S. expansion. Meanwhile, Gifford Pinchot's (1947) idea of conservation for use sanctioned ongoing extractive capitalism, albeit through a gloved managerial hand. The mutually interlocking stories of the frontier and the biblical garden worked on and in the imaginations of early conservationists, as well as the politicians, capitalists, and citizens they hoped to sway. In Chapter 2, I examine how David Belasco's *Girl of the Golden West* (1929 [produced in 1905]) and William Vaughn Moody's *The Great Divide* (1906) reflected the discourse of the early twentieth-century conservation and preservation movements, which characterized progress as a reclamation of the biblical garden (see, e.g., Merchant [1994]). At the time of its 1906 production, Moody's story of a woman who agrees to marry an outlaw to save her life posed questions of national significance. How will this marriage, which begins with an act of violent and violatory conquest, survive to nourish family and home? As a metaphor for Euro-Americans' relationship with the land, *Divide* provides a complicated engagement with – if not answer to – this quandary, along with a good measure of denial. Both works reflect a dichotomous framework that categorizes land as utilitarian or scenic – binary thinking at the heart of the early wilderness movement that shapes national environmental policy today.

As the United States caught up with Europe's industrialization, rivers, forests, and mineral deposits became the raw materials of industrial capitalism and the basis for the nation's growing international power. Progressives of the 1920s envisioned nature itself as an "organic machine,"[6] while many dramatists (including Elmer Rice, Sophie Treadwell, and Eugene O'Neill) were concerned with how technology was idolized.[7] I begin Chapter 3 by exploring these tensions through an analysis of O'Neill's *Dynamo* (1929), a curious and troubling play about a young man infatuated with the power and potential of hydroelectric power. By the early 1930s, the ecological disaster of the Dust Bowl had illustrated the cost of poorly managed resources. During Franklin Roosevelt's New Deal, conservation-for-use gained credibility as the nation's governing framework for land and resource management. Beginning in 1935, with unemployment at 25 percent, the Works Progress Administration (WPA) began to put people back to work building national infrastructure, revitalizing eroded land, controlling rivers and waterways, and building national parks. Under the WPA, the Federal Theatre Project (FTP) not only employed out-of-work artists but also sought to use the theater to educate citizens. In Chapter 3, I consider how FTP's *Triple-A Plowed Under* (produced in 1936) popularized New Deal conservation policies and economic programs by arguing that the welfare of society and land are bound together.[8] *Triple-A* foreshadowed the New Deal Soil Conservation Act of 1937, while it also advocated the more socialist position of political solidarity between workers and farmers in the form of a Farmer-Labor party.

In 1942, at a turning point in World War II, New York's Theatre Guild's hit musical *Oklahoma!* (1942) revealed cultural perceptions of land and environment saturated with racial and jingoist overtones. Chapter 4 examines how *Oklahoma!* repurposed the frontier narrative to champion the industrial development of the west and the mounting antiradical/anticommunist sentiment of the period during and after the war. As a musical adaptation of *Green Grow the Lilacs* by Cherokee playwright Lynn Riggs (1931), Oscar Hammerstein's version tells the story of Oklahoma settler-pioneers and statehood in 1906 without mention of the region's indigenous presence or history. In his introduction to Riggs's play, Jace Weaver (2003) describes *Lilacs* as "the story lived by Lynn Riggs's relatives, and by all people who migrate in desperation, in search of a better life, building their new home with nothing but the land and their own hands" (3). What was a multiracial, multiethnic community in Riggs's play becomes, in *Oklahoma!*, principally white. I argue that Rogers and Hammerstein's adaptation of *Lilacs* helped to justify the federal termination of tribes and absorption of their lands during the 1950s (see, e.g., Dunbar-Ortiz [2014], 173–175). Meanwhile, the economic fantasies celebrated in *Oklahoma!* give way to tragedy in Arthur Miller's 1948 *Death of a Salesman* (1971), which opened on Broadway only six months after *Oklahoma!* closed. Miller's play sent a prescient warning about the mental and spiritual, as well as environmental, impacts of consumer capitalism. Considered together, these plays reveal the continuing cognitive dissonance in the nation's relationship with the land.

The second-wave environmental movement grew out of both the antinuclear movement of the 1950s and the publication of Rachel Carson's (1962) *Silent Spring*, which details the effects of chemical toxins on both humans and nonhumans. As consumerism was reframed as an expression of patriotism, scientists warned of the dangers posed by the many new chemicals that saturated every aspect of modern life. Despite the impact of those chemicals on human health and habitat, the environmental movement of the 1960s and 1970s was steeped in the mythologies and values of wilderness preservation and thus missed the connection it might have made to social justice. Meanwhile, theater artists in the 1960s and 1970s were making clear connections between environmental and social problems. In Chapter 5, I argue that during the civil rights movements of the 1950s through the 1970s, theater concerned with social justice prefigured the central issues of the environmental justice movement that would come two decades later. In *A Raisin in the Sun* (1994 [produced in 1959]), for example, Lorraine Hansberry is concerned with the human price paid for the so-called American dream. In her story of the Younger family, Hansberry illuminates the human health impacts of racism and poverty on women, children, and families – issues that would be at the center of environmental activism into the twenty-first century. Theater was also a central activating force for the *Movimiento*, helping to engender a sense of pride in shared culture, language, history, and stewardship of the land among Chicano/as.[9] The work of El Teatro Campesino, for example, contributed to ecological understandings rooted in histories of mestizo/a presence on the land and in the ancestral homelands of Aztlán. Yet the second-wave environmental movement and the civil rights movements of this period seemed to run on separate rails due, in part, to a milieu that defined nature as something apart from everyday human affairs. The environmental movement of the 1970s was primarily a conversation among whites about recreational places long ingrained with white and masculine privileges that were so much a part of the wilderness tradition. Nevertheless, on stage the embodied art of theater reclaimed both urban and rural environments as sites of habitation where human rights and nature's rights had been violated by a system that regarded land and labor as mere commodities.

Throughout the 1980s, female activists in communities of color gained visibility as they resisted the siting of landfills, incinerators, and toxic waste sites near homes and schools. Then, in 1991, the People of Color Environmental Leadership Summit redefined "environment as the places where people live, work, play and worship," initiating a culture shift in environmental discourse in the United States (see, e.g., Adamson, Evans and Stein [2002]; and Sandler and Pessullo [2007]). Environmentalism, activists argued, must attend to, and redress, the way that women, children, communities of color, and those living in poverty have been more severely impacted by the shadow side of consumer capitalism (see, e.g., Di Chiro [1996], 298–320).[10] In Chapter 6, I turn to consider plays that deal directly with environmental justice concerns, from the impact of resource extraction on the Cumberland Plateau in Robert Schenkkan's (1992) *The Kentucky Cycle* (1994) to the long-term effect of agricultural pesticides in Cherríe Moraga's 1993 *Heroes and Saints* (pub. 1994). Taken together, these two works and their productions reveal trends and cautions. Critics at the time called out the way in which Schenkkan's play recycled frontier stereotypes, and seemed to trade on the stories of Kentuckians, simplifying lived experiences of communities for dramatic effect. *The Kentucky Cycle*'s ultimate failure on Broadway demonstrated a tension between mainstream theater's appetite for universal stories and stories particular to specific communities and places. Building on Cless's (1996) analysis, I discuss *Heroes and Saints* with particular attention to the embodied representation of environmental injustice as the lived experience of women and children. Driving questions for this chapter include: How can theater increasingly empower narratives that demonstrate the interdependence of people and nature that lies at the heart of environmental justice? Moreover, how can settler-descendant dramatists take care to not replicate the very patterns of colonization that they hope to critique?

In 2001, the National Aeronautics and Space Administration (NASA) scientist James Hansen met with then Vice President Richard Cheney to brief the administration on a matter Hansen believed was central to national security – namely, the earth was warming at rates that far exceeded normal geologic fluctuations, and it appeared that humans were the cause.[11] Throughout the late twentieth century, the scientific community warned officials and the public about the effects of increasing carbon emissions on the earth's atmosphere. Indeed, scientific evidence had been mounting since the 1950s. Meanwhile, growing public awareness and firsthand experience of climate change in the first decade of the twenty-first century has contributed to an increased sense of urgency. Yet mainstream U.S. media and popular discourse often characterize the climate crisis in terms of disparate and unexpected cataclysmic events (e.g., hurricanes, wildfires, Antarctic ice cracking). Contemporary dramatists, meanwhile, connect human and nonhuman stories to the long-term causes of climate change through multivocal, multitemporal, transnational, and transspecies stories.

In Chapter 6, I consider theater at the millennium that has focused on the ecological implications of globalized economies and climate change. The runaway power of transnational corporations has caused ecological havoc for people, places, and biotic communities, and has fueled transnational environmental justice activism. Indigenous peoples and developing nations throughout the world have argued that the culpability and risks of climate change are not equally shared by all peoples or nations of the world. Moreover, research shows that the burden of impact from climate change will fall more heavily on the most vulnerable in society, and on indigenous communities in regions at higher risk, such as the Arctic (see, e.g., Watt-Cloutier [2015]; and Nagel [2016]). These new issues and understandings have inspired theater-makers to expose environmental and cultural imperialism in the age of climate change by amplifying the voices of those places and peoples who have been silenced, ignored, or who are at greater risk. In the final chapter, I examine climate change theater more broadly, with a close reading of two plays that employ innovative artistic forms.

Demonstrating the complex, far-reaching continuance of settler colonialism's ecological violence, First Nations playwright Marie Clements's *Burning Vision*

(2003) traces the mining of uranium on Dene land in the northern Canadian territories and its transnational impacts on lives and land. First produced in 2001, *Burning Vision* connects the dots between the exposure of Dene workers to radiation poisoning and the Japanese deaths in Hiroshima and Nagasaki, where the bombs developed from the Dene uranium were ultimately used. Clements's play also marks precisely the connection between the nuclear age, the birth of the Anthropocene, and environmental injustice.[12] Using a ceremonial structure, *Burning Vision* ruptures separations across time and space, collapses past and present, and fuses human and nonhuman life into a single fabric of ecological and embodied relationship.

As the millennium brought increasing warning by the Intergovernmental Panel on Climate Change, dozens of plays about climate change began to hit the U.S., Canadian, and British stages, among them *Sila* by Québécois dramatist Chantal Bilodeau (2014). The first of Bilodeau's Arctic Cycle, *Sila* calls attention to the climate justice implications of resource extraction and geopolitical economic pressures by centering Inuit traditional ecological knowledge in which humans and animals share both kinship and culture. Both *Sila* and *Burning Vision* echo Inuit activist Sheila Watt-Cloutier's (2015) argument that the degradation of the environment is a violation of human rights.[13] Social health issues such as teen suicide, depression, alcoholism, and loss of cultural traditions, Cloutier argues, must be understood as symptoms of climate change. Both plays suggest that part of theater's function in the age of climate change is grief work. In "Climate Change as the Work of Mourning," Ashlee Cunsolo Willox (2012) posits that "grief and mourning have the unique potential to expand and transform the discursive spaces around climate change to include not only the lives of people who are grieving because of the changes, but also to value what is being altered, degraded, and harmed as something mournable" (141). The questions that should compel our artistic will moving forward include: How might theater and performance help us to remain present to the loss that will occur as climate change continues? And perhaps more importantly, how might theater work to resist ongoing ecological violence through enacted histories of decolonization?

Taking a stand where we stand

The environmental and climate crises of the twentieth and twenty-first centuries are crises of identity and relationship, concerned at once with our most basic material needs as well as our most abstract notions of who we conceive ourselves to be in the web of life. At the very nexus of problems that concern governments or neighborhoods, the central questions become: How do we live and behave in relation to one another and to the land in a way that sustains life, land, community, and justice? How can we take responsibility for the stories that have helped to perpetuate over a century of destruction on a global scale? How can theater respond to the consequences of those narratives without perpetuating them? What, if anything, is recoverable in stories that we have inherited? How can theater transform narratives of exploitation into stories of habitation? How can theater empower stories of interdependence and amplify the voices of those who have been most impacted by environmental crisis? And in all of these, how can we make the best use of the ways of knowing that are at the heart of theatrical practice – embodied exploration, story sharing, communal creation, imaginative experimentation, and the palpable immediacy of being together?

Theater can help us to remember and re-member our relationship with the land and to consider the permeable boundaries between human life and the environment. In doing this, theater can help us to examine our own ecological identities: Where do we draw our boundaries of skin and kin? How permeable or fixed are our own notions of self, culture, and humanness? This most ephemeral of art forms can help us imagine into and embody ways of being human consistent with ecological knowledge and ecological sensibilities. Theater's fluid, mutable, and palpable way of knowing is useful not only to acknowledge what *is* happening but also to envision multiple and generous possible futures. If the art of theater is our homeplace, then we are presented now, as at past historical crossroads, with the opportunity to *take a stand from where we stand*. From here, we have a unique opportunity to tell new stories and to apply the searing, sensuous edge of our critical and creative practices with humility and courage as we live into the questions of the future. Surely the world needs theater's *What if?* now more than ever.

Now is a powerful time for stories. What artists do matters; how we represent gender, ethnic, and racial identities matters; how we represent animals, food, and lands also matters. The systems of oppression, domination, and exploitation that commodify the labor of women or people of color, for example, stand on the presumptive ground that land and natural resources exist for the purpose of wealth accumulation. As I underscore in the pages that follow, what we consider to be our most cherished notions of identity, and the material culture that issues from that self-sense, are constructed from the fabric of our stories. Stories are, in this sense, ecological forces that inscribe both the land and our bodies. We belong – to the earth and to one another – but with this kinship comes a reciprocal and sacred trust. The environmental crisis is an invitation not only to develop new behaviors but also to tell new stories that reflect our ecological reciprocity with the planet, with the land we share and have single-handedly decimated. Emerging from within these dire and often hopeless awakenings and moments of consciousness-raising, theater can offer a source of *new* stories that reconfigure who we understand ourselves to be within the circle of life of the earth. Theater, too, can be a kind of nourishment for our species and for the nonhuman communities that share this home planet with us. This book is for young artists – like the student I mentioned earlier and the many like them – who possess a passionate awareness of both the power of theater and the perilous conditions on earth. To those students – incredulous, just fired-up artists that remind me of myself back then: your work

14 Introduction

matters. Your work will carry us forward. And finally, this book is for anyone who has faith in the power of stories and the role of the arts to illuminate, inspire, and actualize an ecologically just and interdependent future.

Notes

1 This third strand (the ecological footprint of the material craft of theater) is not part of my study here, as its technical demands deserve separate treatment. For the interested reader, however, many resources on environmentally friendly theater-craft do exist. See, for example, Garret (2012), Fried and May (1994), and Rossol (1991). See also the *Center for Sustainable Practice in the Arts* (CSPA), which publishes a quarterly journal on the topic and maintains a blog with up-to-date resources. See also the Broadway Green Alliance.
2 The ecodramaturgy lens has gone by various monikers in recent decades, including green dramaturgy, environmental dramaturgy, and ecological theater. As a starting point for ecodramaturgical, critical and historiographic examinations of plays and performance, see my "Green Questions to Ask a Play" (May 2007, 110). See also Arons (2012), Chaudhuri (1994), Cless (1996, 2010), and Lavery (2018).
3 Downing Cless (2010) argues as much in *Ecology and Environment in European Drama*, in which he recasts the history of European theater as a conversation about Europeans' shifting values and understandings of the natural world. See also Arons and May (2013) for more regarding ecodramaturgy's intersection with other critical frameworks.
4 For a more in-depth discussion of the shifting meanings of wilderness, see Roderick Nash's *Wilderness and the American Mind* ([1973] 2001). See also Cronon (1994) for more on this topic.
5 In "The Adventures of the Frontier in the Twentieth Century," environmental historian Patricia Nelson Limerick (1994) argues that the frontier has become a ubiquitous cultural tool for making meaning and interpreting events. Deeply ingrained in the U.S. collective imaginary, Limerick describes the frontier as "the fly paper of our mental world [because] it attaches itself to everything" (94).
6 Use of the phrase "organic machine" was ubiquitous among early twentieth-century engineers, managers, and conservationists. See, for example, Richard White's *The Organic Machine* (1995), which offers a history of hydropower development on the Columbia River, and Donald Worster's *Rivers of Empire* (1992), which details the extent to which rivers, and nature itself, have been conceived in mechanical terms.
7 The critique of technology and mechanization in the drama of this period has been a popular topic among scholars. See, for example, Charles Thorpe (2009) and Dennis G. Jerz (2002). Scholars have also drawn connections between *Machinal*'s implicit critique of technology and violence against women. See, for example, Jennifer J. Parent (1982), Miriam López Rodríguez (2011) and Merrill Schleier (2005).
8 A decade later, Aldo Leopold (1949) would articulate this concept as society's "contract with the land," and he then expanded upon the ethical implications of such a contract. His ideas were later applied to describe the atmosphere itself as a commons to which society has an ethical obligation.
9 *El Movimiento* refers to the larger Chicano/a movement in the U.S. Southwest that gained momentum in the 1960s and 1970s and continues today in the twenty-first century across the United States. This sociopolitical movement has included political, social, and cultural activism; cultural and literary production; and community and global consciousness-raising. See, for example, Laura Pulido (1996), as well as Yolanda Broyles-González's (2004) pioneering work on the history and widely held misconceptions concerning El Teatro Campesino's creation and evolution within the broader Chicano movement and U.S. civil rights movements.
10 As the environmental justice movement exposed the disproportional impact of environmental degradation on workers, mothers and children, and communities of color, it also asserted a human place *in*, rather than apart from, the natural world. This new

footing for environmental thought is more fully explored in Giovanna Di Chiro's (1996) "Nature as Community."
11 Hansen (2011) details his attempts to discuss climate change with the Bush administration in *Storms of My Grandchildren*. See also Andrew C. Revkin's (2006) *New York Times* article on Hansen's report of attempted silencing by NASA.
12 Stratigraphers differ about the start of dramatic change in the earth's climate, but many place its beginnings at the time of colonialization. At the very least, the advent of the nuclear age, Jeremy Davies (2016) argues, marks the start of the Anthropocene's Great Acceleration.
13 Watt-Cloutier was nominated for the 2007 Nobel Peace Prize for her work (although she did not win the award).

References

Adamson, Joni, Mei Mei Evans, and Rachel Stein, eds. 2002. *The Environmental Justice Reader: Politics, Poetics and Pedagogy*. Tucson, AZ: University of Arizona Press.
Arent, Arthur. 1938. "Triple-A Plowed Under." In *Federal Theatre Plays*, edited by Pierre de Rohan, 1–57, Vol. 2. New York: Random House.
Arons, Wendy. 2012. "Queer Ecology/Contemporary Plays." *Theatre Journal* 64 (4): 565–582. www.jstor.org/stable/41819890
Arons, Wendy, and Theresa May, eds. 2012. *Readings in Performance and Ecology*. New York: Palgrave Macmillan.
Arons, Wendy, and Theresa May. 2013. "Ecodramaturgy in/and Contemporary Women's Playwriting." In *Contemporary Women Playwrights*, edited by Lesley Ferris and Penny Farfan, 181–196. New York: Palgrave Macmillan.
Bacon, J.M. 2019. "Settler Colonialism as Eco-social Structure and the Production of Colonial Ecological Violence." *Environmental Sociology* 5 (1): 59–69. https://doi.org/10.1080/23251042.2018.1474725
Belasco, David. 1929. "The Girl of the Golden West." In *Six Plays by David Belasco*, edited by Montrose J. Moses, 310–403. Boston, MA: Little, Brown and Company.
Bilodeau, Chantal. 2014. *Sila*. Vancouver, BC: Talonbooks.
Broyles-González, Yolanda. 2004. *El Teatro Campesino: Theater in the Chicano Movement*. Austin, TX: University of Texas Press.
Carson, Rachel. 1962. *Silent Spring*. Boston, MA: Houghton Mifflin; Cambridge, MA: Riverside Press.
Chaudhuri, Una. 1994. "'There Must Be A Lot of Fish in that Lake': Toward an Ecological Theater." *Theater* 25 (1): 23–31. https://doi-org.libproxy.uoregon.edu/10.1215/01610775-25-1-23
Clements, Marie. 2003. *Burning Vision*. Vancouver, BC: Talonbooks.
Cless, Downing. 1996. "Eco-Theatre, USA: The Grassroots Is Greener." *TDR* 40 (2): 79–102. https://doi.org/10.2307/1146531. www-jstor-org.libproxy.uoregon.edu/stable/1146531
Cless, Downing. 2006. "Ecologically Conjuring Doctor Faustus." *Journal of Dramatic Theory and Criticism* 20 (2): 145–167.
Cless, Downing. 2010. *Ecology and Environment in European Drama*. New York: Routledge.
Cohen-Cruz, Jan. 2005. *Local Acts Community-based Performance in the United States*. New Brunswick, NJ: Rutgers University Press.
Cohen-Cruz, Jan. 2010. *Engaging Performance: Theatre as Call and Response*. New York: Routledge.
Cronon, William, ed. 1994. "The Trouble with Wilderness: Or, Getting Back to the Wrong Nature." In *Uncommon Ground: Rethinking the Human Place in Nature*, 69–90. New York: W.W. Norton & Company.

Daly, Augustin. 1978. *Horizon*. In *Six Great American Plays*, edited by Allan G. Halline, 335–375. New York: Modern Library.

Davies, Jeremy. 2016. *The Birth of the Anthropocene*. Oakland, CA: University of California Press.

Di Chiro, Giovanna. 1996. "Nature as Community: The Convergence of Environment and Social Justice." In *Uncommon Ground: Rethinking the Human Place in Nature*, edited by William Cronon, 298–320. New York: W.W. Norton & Company.

Dunbar-Ortiz, Roxanne. 2014. *An Indigenous Peoples' History of the United States*. Boston, MA: Beacon Press.

Fried, Larry K., and Theresa May. 1994. *Greening Up Our Houses: A Guide to a More Ecologically Sound Theatre*. New York: Drama Book.

Friedman, Michael, and Alex Timbers. 2010. *Bloody Bloody Andrew Jackson*. New York: Music Theater International.

Garret, Ian. 2012. "Theatrical Production's Carbon Footprint." In *Readings in Performance and Ecology*, edited by Wendy Arons and Theresa May, 201–210. New York: Palgrave Macmillan.

Gottlieb, Robert. 2005. *Forcing the Spring: Transformation of the American Environmental Movement*. San Francisco, CA: Island Press.

Hansberry, Lorraine. 1994. *A Raisin in the Sun*. New York: Vintage/Random House, Inc.

Hansen, James E. 2011. *Storms of My Grandchildren: The Truth about the Coming Climate Catastrophe and Our Last Chance to Save Humanity*. London and New York: Bloomsbury.

Jerz, Dennis G. 2002. "The Experimental Seduction of Mechanistic Modernism in Eugene O'Neill's 'Dynamo' and the Federal Theatre Project's 'Altars of Steel'." *Interdisciplinary Science Reviews* 2 (3): 1–9. https://doi.org/10.1179/030801801802225005608

Lavery, Carl. 2018. *Performance and Ecology: What Can Theatre Do?* Milton Park, Abingdon, Oxon and New York: Routledge.

Leopold, Aldo. 1949. "The Land Ethic." In *Sand Country Almanac and Sketches Here and There*, 201–226. Oxford: Oxford University Press.

Limerick, Patricia Nelson. 1994. "The Adventures of the Frontier in the Twentieth Century." In *The Frontier in American Culture an Exhibition at the Newberry Library, August 26, 1994–January 7, 1995*, edited by Richard White, Patricia Nelson Limerick, James Grossman, and Newberry Library, 69–100. Chicago and Berkeley, CA: University of California Press.

López Rodríguez, Miriam. 2011. "New Critical Approaches to *Machinal*: Sophie Treadwell's Response to Structural Violence." In *Violence in American Drama: Essays on Its Staging, Meanings and Effects*, edited by Alfonso Ceballos Muñoz, Ramón Espejo Romero and Bernardo Muños Martínez, 72–84. Jefferson, NC: McFarland.

May, Theresa J. 2005. "Greening the Theatre: Taking Ecocriticism from Page to Stage." *Journal of Interdisciplinary Studies* 7 (1): 84–103. www-jstor-org.libproxy.uoregon.edu/stable/41209932

May, Theresa J. 2007. "Beyond Bambi: Toward a Dangerous Ecocriticism in Theatre Studies." *Theatre Topics* 17 (2): 95–110. https://doi.org/10.1353/tt.2008.0001

Merchant, Carolyn. 1994. "Reinventing Eden: Western Culture as a Recovery Narrative." In *Uncommon Ground: Rethinking the Human Place in Nature*, edited by William Cronon, 132–159. New York and London: W.W. Norton & Company.

Miller, Arthur. 1971. *Death of a Salesman*. In *The Portable Arthur Miller*, edited by Harold Clurman, 3–133. New York: Viking Press.

Moody, William Vaughn. 1906. *The Great Divide*. New York: Houghton Mifflin Company.

Moraga, Cherríe. 1994. "Heroes and Saints." In *Heroes and Saints & Other Plays*, 85–149. Albuquerque, NM: West End Press.
Munk, Erica, ed. 1994. "A Beginning and an End." *Theater* 25 (1): 5. https://doi-org.libproxy.uoregon.edu/10.1215/01610775-25-1-5
Nagel, Joane. 2016. *Gender and Climate Change: Impacts, Science, Policy.* New York: Routledge.
Nash, Roderick. 2001. *Wilderness and the American Mind*, 4th ed. New Haven, CT and London: Yale University Press.
O'Neill, Eugene. 1929. *Dynamo*. New York: H. Liveright.
Opie, John. 1998. *Nature's Nation: An Environmental History of the United States*. New York: Harcourt Brace College Publishers.
Parent, Jennifer J. 1982. "Arthur Hopkins' Production of Sophie Treadwell's 'Machinal'." *The Drama Review* 26 (1): 87–100. https://doi.org/10.2307/1145447
Pinchot, Gifford. [1947] 1987. *Breaking New Ground*. Washington, DC: Island Press.
Pulido, Laura. 1996. *Environmentalism and Economic Justice: Two Chicano Struggles in the Southwest*. Tucson, AZ: University of Arizona Press.
Revkin, Andrew C. 2006. "Climate Scientist Says NASA Tried to Silence Him." *New York Times*, January 29.
Rice, Elmer. 1929. *The Adding Machine: A Play in Seven Scenes*. New York, Los Angeles and London: S. French Ltd.
Riggs, Lynn. 1931. *Green Grow the Lilacs: A Play*. New York: Samuel French.
Rogers, Richard, and Oscar Hammerstein. 1942. *Oklahoma!* New York: Random House.
Rossol, Monona. 1991. *Stage Fright – Health & Safety in the Theater*. New York: Saint Paul, MN: Allworth Press.
Sandler, Ronald, and Phaedra C. Pessullo. 2007. *Environmental Justice and Environmentalism: The Social Justice Challenge to the Environmental Movement*. Cambridge, MA: MIT Press.
Schenkkan, Robert. 1994. *The Kentucky Cycle*. New York: Dramatists Play Service.
Schleier Merrill. 2005. "The Skyscraper, Gender, and Mental Life: Sophie Treadwell's Play *Machinal* of 1928." In *The American Skyscraper: Cultural Histories*, edited by Roberta Moudry, 234–254. Cambridge: Cambridge University Press.
Taylor, Diana. 2003. *The Archive & the Repertoire: Performing Cultural Memory in the Americas*. Durham, NC: Duke University Press.
Thorpe, Charles. 2009. "Alienation as Death: Technology, Capital, and the Degradation of Everyday Life in Elmer Rice's *The Adding Machine*." *Science as Culture* 18 (3): 261–279. https://doi.org/10.1080/09505430903123032
Valdéz, Luis. 1990. *Early Works: Actos, Bérnabe, and Pensamiento Serpentino*. Houston, TX: Arte Público Press.
Watt-Cloutier, Sheila. 2015. *The Right to Be Cold*. Minneapolis, MN: University of Minnesota Press.
Weaver, Jace. 2003. "Introduction to *Green Grow the Lilacs*." In *Cherokee Night and Other Plays*, edited by Lynn Riggs, 3. Norman, OK: University of Oklahoma Press.
White, Richard. 1995. *The Organic Machine*. New York: Hill and Wang.
Willox, Ashlee Cunsolo. 2012. "Climate Change as the Work of Mourning." *Ethics and Environment* 17 (2): 137–164. https://doi.org/10.2979/ethicsenviro.17.2.137
Worster, Donald. [1985] 1992. *Rivers of Empire: Water, Aridity, and the Growth of the American West*. Reprint, Oxford: Oxford University Press.

1
STORIES THAT KILL

Augustin Daly's *Horizon* and William F. Cody's *Wild West: The Drama of Civilization*

In 2010, the Public Theater of New York produced a punk-rock sketch-comedy musical called *Bloody Bloody Andrew Jackson*. Critics praised its irreverent bravado and "sexy pants" take on the seventh president of the United States (Levine 2010). In one musical number, white actors, dressed in stereotypical deerskin dresses, sang and danced to the (explicitly racist) vaudeville tune "Ten Little Indians" (see, e.g., Jennings [2012]). *Politico*'s D.M. Levine (2010) describes *Jackson* as a musical

> full of death and destruction and bow-and-arrow fights. There's murder of natives and soldiers and betrayal and insults. But it's all gag. This bleak, tragic history is filtered through a steam punk lens of red neon lights and modern iconography . . . all designed to foster an atmosphere of absurdity that makes it okay for people to delight in this bleak stuff.

Playing fast and loose with history, the musical distorted Jackson's life, in effect obscuring (even justifying) his policy of ethnic cleansing, executed through the Indian Removal Act of 1830.

Jackson was the Public's second-largest grossing production at the time and moved quickly to Broadway; over the next several years, touring and regional productions took place in cities across the United States (Brantley 2009).[1] Native artists associated with the Public were shocked and offended by the production's cavalier representation of indigenous lives and history, including the fictionalized and historically inaccurate portrayal of the murder of Jackson's family by Native Americans. Ironically, the Public had a Native Advisory Committee, but *Jackson*'s creative team had not consulted them (Yazzie 2014). How is it that the story of the Jacksonian era can be mined, fictionalized, and exploited without the inclusion of the viewpoints of real people whose bodies continue to

DOI: 10.4324/9781003028888-2

be impacted by that history? As touring and regional productions of the work continued to receive rave reviews, outrage and protests mounted among Native communities and allies across the United States (Levine 2010).

Defenders of the production protested that the work was satire, dismissing critics and detractors as too sensitive. Meanwhile, in a *New York Times* piece, Ben Brantley (2009) observed that *Jackson* reflects an "abiding strain in the American character." Whether Brantley was hinting at the production's potential appeal to white nationalists or, rather, to the unfortunate persistence of "bad-boy" heroes in U.S. culture is hard to decipher. Brantley also points out that *Jackson* is like a person who can "say what he really wants to [say] only by pretending that he doesn't mean it." As the American story of origin, the frontier narrative on which the musical capitalized has become so ingrained that even New York artists have expressed surprise upon learning that indigenous North Americans have not "vanished" and that they and others are offended by "vignettes in which historic acts of violence are rendered as deadpan slapstick." Such disregard for the unintended consequences of artistic work suggests ignorance or naiveté on the part of *Jackson*'s creative team or an intentional disregard for those who continue to experience the effects of history in the present day. The disconnect between the musical's smashing success according to critics and the protests against the production's content and aesthetic points to a cognitive dissonance that is not new in American popular culture or theater. Brantley's sense that the musical "feels anything but distant" stems from the way it redeploys the familiar vocabulary of the American frontier story as it folds the past into the present while promoting ongoing ideologies of white supremacy, heroism, and environmental domination.

Theater has long played an important role in telling stories that propagate the ideology of Manifest Destiny, the doctrine of discovery on which that ideology has stood, and its implicit claim of white entitlement to land, which precipitated and sanctioned historic patterns of environmental destruction on this continent that continues to accelerate today. When the racial stereotypes that populated the nineteenth-century American stage still inspire laughter in 2015, then those stories, images, and tropes deserve closer scrutiny. Likewise, the role of theater in perpetuating narratives designed to exclude the lives and experiences of some while preserving entitlement and privilege for others must be examined precisely because the consequences of those stories are ongoing. If we – and here I mean white Anglo artists like I am – are troubled by epidemic sexual assault on college campuses or excessive use of force in our nation's streets, then we must examine our part in perpetuating the old stories at the root of those present effects. The frontier narrative, even deployed in fun or fantasy, continues to condone the genocidal and ecocide project of U.S. expansion from first contact into the twenty-first century.

While the whole of U.S. history is implicated in this trajectory, this chapter deals with a key moment at the end of the nineteenth century, when the U.S. Census Bureau "closed" the frontier as the Anglo-identified nation sought to reposition as an international economic power. In 1893, Chicago's Columbia

Exposition celebrated a triumph of Anglo-Saxonism, with the United States positioned as the apex of its civilizing mission. Like a kind of Manifest Destiny theme park, the fairground's "White City" bragged pseudo-classical architecture and outdoor carnivalesque fun that included Ferris's great wheel that spun into the sky, taking spectators up-up-up over Lake Michigan to look down on the splendor of the city. Exposition spaces provided patrons with opportunities to view indigenous peoples and flora and fauna from throughout the colonized world, all on display as a testament to civilization's evolutionary path: a narrative writ large in flesh and bone. The public could also witness the latest industrial accomplishments and ponder the marvels of modern mechanical invention, just as they could attend lectures by noted scientific and historical experts.

It was here that the up-and-coming historian Frederick Jackson Turner delivered "The Significance of the Frontier in American History," or "Frontier Thesis," in which he described a national relationship with the land and shaped a collective imaginary about U.S. identity. The Turner thesis asserted who belonged and who did not and claimed a central place for the United States in the world as an emerging industrial capitalist nation (Turner 1956, 2–4). In 1890, the U.S. census had recently declared that the "frontier was closed," a conclusion based on average population densities in the western states. For an audience assimilating this loss, Turner's contention that the frontier lived on in the American *mind* had a strong attraction. The familiar personae of the "bad-boy hero" – who is buoyant, optimistic, irreverent, roughhewn, of the people, lawless by necessity, brutal, expedient, sexy, and, in short, the quintessential American male, who sang and danced at the center of *Bloody Bloody Andrew Jackson* – is a visage straight out of Turner's thesis. Americans, Turner theorized, had been "compelled to adapt" to forces of wilderness, an encounter with the environment that propels the pioneer in reverse along a social evolutionary path to a "savage state," where he "shouts the war cry and takes the scalp in orthodox Indian fashion" (2; see also Smith [1950], 150–160).

Turner (1956) asserted that the pioneer's savagery is understood as the building blocks of a uniquely American identity. As he explained it in 1893, the quintessential American was gregarious, optimistic, lawless when necessity required, with an aptitude for and expectation of violence. American character, Turner theorized, arose out of the settlers'/pioneers' embodied encounter with the environment, in which he included indigenous resistance, fierce terrain, inclement weather, and wild animals (the elements theatricalized in the works discussed later). Turner posited that the critical role of the frontier in this process of transformation was predicated on its vast expanse of "free land" (1). There, Turner theorized, American pioneers honed their new identity as distinct from their European origins by regressing and then regenerating through an encounter with the wilderness. Turner asserted that "[l]ong before the pioneer farmer appeared on the scene, primitive Indian life had passed away" (6). Rhetorically erasing thousands of diverse communities and cultures, he emptied the continent and conflated its indigenous human inhabitants with the wilderness (see, e.g., White [1994]). While the white pioneer can advance again from "the primitive economic and political condition of the frontier into the complexity of city life," Turner maintained that indigenous peoples are fixed in the "primitive" state, conflated with the land itself, merely one of the savage forces with which the pioneer must contend. "The wilderness masters the colonist. It finds him a European in dress, industries, tools, modes of travel and thoughts. It takes him from the railroad car and puts him in the birch canoe. It strips off the garments of civilization," Turner argued (4). The American character could be distinguished from its European ancestry as a new species of man. Historians since then have pondered how a thesis of so little originality captured the Zeitgeist of how Americans understood themselves for, as Theodore Roosevelt mused (and as the discussion that follows shows), Turner had simply gathered together a lot of ideas floating around. Generations of Native and non-Native historians have actively rejected this framework for American history, demonstrating how it gave rhetorical support to imperialism and mapped its corrosive effect on popular imagination (see, e.g., Hofstadter [1944]; White [1993, 1994]; Limerick [1994]; Cronon [1996]; and Deloria [2004]).

The story Turner (1956) told was not new. These ideas had long been reflected in popular culture, propagated by stage plays and Wild West shows. The story's elements – the pioneer setting out in a covered wagon, fighting the wilderness against all odds, clearing forest to build a small cabin, and violent encounters with indigenous warriors – were already familiar tropes in news reports, dime novels, melodramas, and visual art of the nineteenth century. In more recent decades, frontier iconography, and the implicit values it carries, has continued to saturate nearly every sector of mainstream U.S. culture (Limerick 1994). We have only to look at the figures of speech that saturate rhetoric used in some sectors of U.S. society to discuss sexual assault, or to defend white supremacists or corporate privilege, to see that everyday speech is haunted by the ubiquitous stories and images of leather stockings, log cabins, pioneers and covered wagons, cowboys and Indians, stagecoach holdups, gunfights, and endangered or compromised women, all encircled in a frightening yet sublime wilderness. Reinscribed and maintained through symbolic language, the frontier story is indicative of deeply ingrained ways of thinking and being – what environmental historian Patricia Nelson Limerick calls the "fly paper" of the [Anglo] American mind (94).[2] Everyday euphemisms invoke frontieresque masculinity as casually as weather, as invisibly as wallpaper: a professional who pushes the edges of the status quo is a "pioneer," "cowboy," or "maverick"; a businessperson "wrangles a deal"; in life's challenges, one can "play the cards you're dealt," "stake your claim," "stick to your guns," "play it fast and loose," or "get outta Dodge" and "head for the hills" when "the chips are down." These ubiquitous and seemingly harmless tropes contain implicit performances of gender, racial, and cultural privilege. They also imply an extractive relationship with the land and its ecological communities.

In what follows, I argue that the complex story-knot known collectively as the frontier has mythologized U.S. environmental imperialism as a "natural" evolutionary progression of Anglo-Saxon civilization, one that in turn shapes social norms, propagates structures of power and privilege, and informs the institutions and policies that affect shared ecological systems. Environmental sociologist J.M. Bacon (2019) argues that settler colonialism constitutes the foundational and precipitating origins of ecological destruction on a mass scale on this continent. Colonial ecological violence manifested first and foremost as intentional human-led annihilation of human cultures, communities, and nations, and then in the destruction of ecosystems and bioregions and, ultimately, global earth systems such as climate. As a theater-maker, as well as a citizen, I am particularly interested in how theater helped to propagate the ideas – as stories – that resulted in the long-term effects of settler colonialism on the ecologies of the continent. This chapter argues that theater has been complicit in sanctioning the frontier as an *ecological ethos* from which the destruction of people and lands ensues.

Thomas King writes, "The truth about stories is that that's all we are" (2005, 8). We live by and within stories. "Stories are wondrous things," King continues, and "they are dangerous" (9). Stories have marked the land and lives. As an artist of settler descent, I hope to encourage both increased cognizance of and responsibility for the power of the stories theater tells and has told. Engaging in a deeper, more comprehensive understanding of the stories that have shaped mainstream U.S. culture thus becomes an integral part of the work entailed in apprehending environmental and environmental justice problems and solutions. Overall, this chapter aims to foster a frontier literacy with an eye toward promoting 'ownership' of these narratives, in such a way that artists (and others) might mindfully critique them with an awareness of their socioecological implications, as well as the power and privilege they inscribe and protect. By exposing the working elements of the frontier narrative, I hope that artists and readers of settler descent – myself included – might become more sensitive and discerning of the reach of this story, particularly in recognizing how its consistently reapplied and recirculated tropes continue to sanction injury of land and peoples.

Setting the stage

Westward expansion was aided and executed by showmen as well as soldiers and settlers. In the nineteenth century, theaters were not only places of entertainment but also centers of community in which people met and mingled and spheres of influence intersected. Information about the "Far West" came to easterners, in part, through the supposedly true-to-life stories and personages that appeared on the melodrama stage. Theater played a vital role in popularizing ideas through stories and images about the west and in translating an ideology about western lands into stories that explained those territories to the east coast public. As American empire-building pushed farther west, frontier plays that claimed true-to-life escapades helped to justify and popularize ideologies of expansionism and imperialism. As ideology took on flesh through the visceral experience of live performance, theater became a weapon fired at the bodies, cultures, communities, and ecosystems of the land west of the Mississippi (White 1994, 10).

This chapter examines the connection between the social violence and the environmental violence in two emblematic productions – Augustin Daly's *Horizon* (1978; see also Hall [2001], 38–43), which opened at New York's Olympic Theatre in 1871, and William F. Cody's *The Wild West/Drama of Civilization*, produced in 1886 and directed by Steele MacKaye at Madison Square Garden.[3] Together, they reveal how cultural performances helped to advance American imperialism. *Horizon* appeared as the U.S. government pursued an aggressive policy towards indigenous nations of the plains, and critics praised *Horizon* for its realistic depiction of frontier life, including hostile Indians, which aided in the construction of a rationale for U.S. military occupation of indigenous lands.[4] MacKaye's direction of Cody's *Wild West* laid claim to veracity through a spectacle of bodies that included Native performers and military personnel, as it lionized acts of warfare against indigenous populations, including the state-sponsored killing of millions of bison in the 1880s (White 1993, 212–222). While Cody began his circus-like *Wild West Show* in 1883, it was his collaboration with well-known New York director-producer Steele MacKaye in 1886 that shaped a cohesive, viscerally scenic narrative (Blackstone 1986). The show's reenactment of battles with Native performers billed as "real" Indian warriors in a simulated wilderness, and the climactic symbolic martyrdom of General George Armstrong Custer, indelibly shaped Anglo-American's long-term collective imaginary in ways that are still felt today.[5]

Cody had already earned celebrity status as a military scout and guide for elite hunting parties (similar to the one parodied in *Horizon*'s subplot of the British nobleman Mr. Smith). After Cody's much-publicized hunt with the Grand Duke of Russia, the well-known theater producer Ned Buntline encouraged Cody to come to Chicago to play himself on stage. Thus, Cody began a successful ten-year stage career in which he managed and starred with his own melodrama company.[6] He spent summers with the U.S. Calvary fighting indigenous resistance warriors and leading hunting parties, while he continued to play in melodrama on and off through 1886. Cody claimed to be presenting his own experiences gleaned from years as a rider for the Pony Express and as a scout for the U.S. Cavalry. He had led expeditions into remote regions, fought hand to hand with resistance warriors, and (so he claimed) negotiated peace with tribal chiefs. Season by season, Cody increasingly transferred and translated his military exploits on the plains into stage vis-à-vis shows touting veracity and realism. He often arrived at actual battles in the costume he would ordinarily wear for a performance. After playing the winter season on stage in the summer of 1876, for example, Cody returned west and worked as a scout for the Fifth Cavalry; here again, the real and the represented frontiers blurred further as the players in each endeavored to be more like the other, prescient of contemporary reality television programs. Aware of the members of the press who were also present to cover the anticipated conflicts, Cody wore his costume onto the battlefield. At Hat Creek, Cody slew the Cheyenne warrior Yellow Hair (White 1994, 34). The victory received wide publicity, and Cody capitalized on his actions by staging the event the following season in a work he called *The Red Right Hand* or *Buffalo Bill's First Scalp for Custer* (see, e.g., Russell [1960], 16); Deloria [2004], 53–94; and Buss [2011]). The violence and sheer cynicism of such a publicity stunt on Cody's part epitomize theater as a weapon of colonization on the plains.[7]

In the messy interplay between the frontier of the stage and lived experience on the plains, we can observe how powerfully and profoundly the embodied art of theater gives flesh to ideas as it also affects social understandings of human and human–nature relationships. *Horizon* and *The Wild West* are palpable examples of ways that theatrical production implicitly and explicitly endorsed state-sponsored violence and corporate entitlement on the plains. Given that the frontier ideology is a story that still lives among and within us, we can examine these performances in their historical context to help us understand better – if not fully resolve – contemporary theater's fraught relationship with this tenacious narrative.

Daly's stage play and Cody's extravaganza both gather meaning from the backdrop of westward expansion of the United States during the nineteenth century. U.S. territory tripled with the Louisiana Purchase in 1830, followed by the 1848 annexation of one million acres of Mexico, including lands that are now commonly known as the U.S. states of Texas, New Mexico, Arizona, California, and parts of Colorado and Utah. Discovery of gold in California in 1849, as well as silver and copper throughout the U.S. Southwest, along with the rise of the timber and salmon industries in the Pacific Northwest continued to propel settlers, speculators, and the railroads westward after the Civil War. Railroads, which had to raise the funds for their own construction ventures, as well as politicians and others who sought to benefit from increased resource extraction, reflected a concerted zeal for a common vision to unite eastern and western North America under the banner of a single nation with growing international prowess, namely, the United States of America.

As the U.S. federal government mounted an increasingly aggressive campaign on the plains to seize the territory and vanquish the Nez Perce and the Sioux in the north and the Comanche and Apache in the southwest, public debate on what the mainstream U.S. press called "the Indian problem" fell into two general schools of thought (White 1993, 85–118).[8] Those who advocated a "peace policy" believed in the gradual assimilation of indigenous North Americans into mainstream society. Others advocated a military solution – referred to in the press as "extermination" – and constituting ethnic cleansing that Roxanne Dunbar-Ortiz (2015) observes falls within the definition of genocide set out in 1948 by the United Nations. Those who favored assimilation (understood as the peaceful solution) proposed teaching the indigenous people to farm, thus forcing them along the path prescribed by Thomas Jefferson toward civil society. Assimilationists seemed ignorant, however, of the fact that many tribes already tended and cultivated herbaceous plants, maintained healthy animal populations through knowledgeable hunting practices, and in many other ways utilized and shaped their environments. Those who advocated for the forceful removal of indigenous tribal communities found further evidence of inferiority in the failure of such ventures. These (genocidal, profiling, tokenizing) stereotypes – also represented in the works discussed in this chapter – reflect white ignorance of the different work rhythms of the hunting/gathering cultures of the plains. While farmers get up and go to work each day, hunters do their work according to seasonal cycles and animal migration patterns. But neither position – forced assimilation or forced removal – gave consideration or voice to the sovereign rights and desires of tribes. Public officials increasingly ignored the terms of treaties that had previously been made with indigenous nations since the arrival of Europeans. As the history of boarding schools shows, assimilation is yet another form of violence and violation (see, e.g., Archuleta, Child, Lomawaima, and Heard Museum [2000]; and Gram [2016], 251–273). Both strategies, however, were rooted in the contemporaneous and interlocking big ideas of Anglo-Saxonism and Darwin's theory of evolution, which informed the framing, reception, and interpretation of unfolding conflicts on the Great Plains, and provided a rationale for imperialism by justifying genocidal ethnic cleansing as part of what was rationalized as the evolutionary progress of civilization.[9] Those ideas also saturate Daly's *Horizon*, which indicted assimilationists' naiveté as it spun a tale that supported military aggression, and Cody's *Wild West*, which similarly asserted the success of U.S. military while purporting to demonstrate assimilation in the employment of indigenous talent.

In the circular logic of Anglo-Saxonism, industrial capitalism is the engine of progress and synonymous with civilization, and, because industry required raw materials, those resources rightfully belonged to those who would use them to advance civilization. Within the expanding U.S. territories, the presumed cultural superiority of Anglo-Saxon culture translated into a sense of entitlement. A growing nationalist vision of industrial and economic dominance in the world characterized those who saw themselves as emissaries of U.S. Anglo-Saxonism. Economic supremacy was predicated on opening the continent and its raw materials – furs, fish, timber, minerals, and water. Those who held capital assets, almost entirely white men, had a right to claim, extract, transport, and process anything that served the advancement of Anglo civilization, including other human beings.

Similarly, Darwin's (1859) compressive narrative about the relationship between living creatures (including humans) and their environments fits easily with the ideology of Manifest Destiny. Darwin theorized that as species cope with their environments over generations, changes or adaptations occur that are sometimes so profound that a new species emerges. His theory underscored the profound reciprocity between living creatures and their environment as an interdependent mutually unfolding web of relatedness. It was received, however, within an already-existing intellectual milieu of Anglo-Saxonism. Consequently, ideas of reciprocity and interdependence were subsumed by social thinkers who instead favored the theory's themes of competition and hierarchical dominance. Historian Richard Hofstadter further observes that Darwin's theories were a convenient fit for the

Anglo-Saxonism of the nineteenth century, providing a scientific rationale for white supremacy and the exclusionary privilege it enforced ([1944] 1992, 170–200). Popular circulation of Darwin's ideas, delayed in the United States until after the Civil War, began to shape public perceptions of indigenous–settler conflicts.

Herbert Spencer's "social Darwinism," a spurious application of Darwin's theories to human social and economic relations, seemed to provide scientific justification for genocidal U.S. policies (Hofstadter [1944] 1992, 31–50). Nationalists used Spencer's framework to rationalize the ethnic cleansing on the plains and to explain social unrest in cities by characterizing state-sponsored violence against people and land as part of the "natural" evolutionary progress of society toward civilization. Daly's *Horizon*, as we will see, exemplifies this kind of Darwinian bravado – an ideological amalgam that idealized and naturalized white Christian nationalism.[10]

Fierce and uncontainable, the land west of the Mississippi River seemed to demand a new conceptual model for understanding the natural world, and Darwin's thesis provided a fitting new paradigm. Survivalist interpretations of Darwin's theories emphasized the environment as a collection of forces against which men had to struggle, ultimately win, and then dominate, and control became a convenient narrative supporting regional expansion and increased resource extraction that would make the United States a global imperial power. The construction of nature as a kind of brute force resonated with Euro-Americans, who were facing environments that sometimes seemed malevolent (see, e.g., Smith [1950], 4). Like the biblical myth of the Garden of Eden (which I consider more closely in the following chapter), Anglo-Saxonist interpretations of Darwin's theory put Anglo men at the center and apex of a natural order (Hofstadter [1944] 1992). The idea of a man's "war" with nature made sense to many westward-moving settlers who endured unexplainable weather patterns and strange geologic formations, as well as unwelcoming deserts, impassable mountains, and other unfamiliar ecosystems (see, e.g. Stegner [1992]).

Horizon – weaponizing the stage

By the end of the nineteenth century, U.S. dramatists were struggling to free themselves from European aesthetics. Many such playwrights believed that theater could and should play a role in shaping national identity and destiny by giving expression to what they perceived as a uniquely American experience. While late nineteenth century critics often used the term *Native drama* to describe an emerging uniquely American dramaturgy, they were not referring to dramas written by indigenous authors. On the contrary, these critics used the term to describe plays written by white, Anglo men from the United States, works which could stand shoulder to shoulder with the dramatic canon of Europe (see, e.g., Felheim [1956], 67).

Thus, frontier dramas like *Horizon* and performance spectacles like the *Wild West* shows provided an embodied and visceral experience (if simulated and vicarious) that bolstered the burgeoning U.S. empire. Populated with stereotypically "savage" Indians, corrupt railroad barons, helpless women, courageous cavalrymen, and the lawless populace of frontier towns, *Horizon* quickly became a model for the frontier drama genre. Critics praised *Horizon* for its treatment of matters of "national interest and significance," a play meant "not only to amuse but to instruct."[11] Those main "instructive" themes included: presenting indigenous characters instead of vicious and vengeful foes and thus dispelling the prior white literary visage of the noble savage, invoking the expansion of the transcontinental railroad the instrument of progress, asserting settler concepts of property, affirming the separation of racial groupings, and subsuming indigenous resistance to occupation as part and parcel of the wilderness. I contend that *Horizon* aided in naturalizing an increasingly aggressive military occupation necessary and unproblematic step in the advancement of the U.S. nation-state. Two decades before Turner's frontier thesis, *Horizon* exemplifies how theater functioned as a kind of weapon providing rhetorical and imaginary support to the U.S. military occupation and ethnic cleansing that took place on the plains after the Civil War. Moreover, *Horizon*'s rhetorical/visual prongs show how social and ecological violence are interlocked in the project of settler colonialism, with ongoing consequences for people and the land.

The cotton economy in the southern United States had collapsed as a result of the Civil War, along with the institution of slavery on which it depended. In the economic, social, and environmental chaos that followed the Civil War, that economy was slowly rebuilt on the same foundation of racism and coercive economic systems, including sharecropping, forced labor, and servitude (Blackmon 2009). Meanwhile, in the west, the railroad industry recruited and hired Chinese and Irish workers at low pay to build the railroads in exchange for very low wages and poor working conditions. Settlers of all identities went west seeking opportunities they did not have in eastern cities. Many hoped to secure land for farming, find work, or start an enterprise that allowed them to pay off debts or support poorer family members; others went to escape persecution or prosecution. Yet most found themselves on the short end of the laissez-faire capitalism that presided throughout the territories.[12] *Horizon*'s cast of characters includes an assortment of racial and gender stereotypes that typified popular prejudices of the day and would surely offend contemporary readers. A snapshot of the racist ideologies espoused by the play and presumed palatable to its listening audience include Wannemucka, a "civilized Indian"; Wahcotah, a "friendly Indian"; Onata, a "prairie princess,"; Cephas, a "Fifteenth Amendment"; and "The Heathen Chinee," along with a dozen others who fill out the popular stereotypes of outlaws, vigilante townsfolk, Irish immigrants, and fragile white women (Daly 1978, 335). After the Civil War, these constructions worked to accomplish the ongoing project of colonialization by characterizing groups of people – whose labor on the land profited white landowners, venture capitalists, and extractive industries – in demeaning and dehumanizing ways.

The politics of embodied representation are complicated to be sure; yet when white actors play characters of color as imagined and stereotypified by whites, characters whose real-world counterparts suffer(ed) exploitation, dispossession, and death to benefit whites, including audience members, such representation asserts white entitlement to both lands and bodies. As was typical in theater of the era, white actors played all characters by wearing black-, brown-, red-, or yellow-face as necessary (see, e.g. Young [2013]). Arthur Hobson Quinn (1927) also notes that the "Heathen Chinese" was a popular stock character of the period, drawn from Bret Harte and Mark Twain's *Ah Sin: A Dramatic Work* (109). Indeed, theatrical makeup used by professional actors at the time came with various base colors named for racial phenotypes for the many identities that a professional white actor might play over the course of a career.[13] *Horizon* thus performs white entitlement explicitly as white actors dawned coded constructions of Indians, as well as other characters representing groups of people whose bodies were mined for their labor in the project of extractive capitalism that drove settlement of the west. (Is it any wonder, then, that white actors dressed in buckskin, dancing to "Ten Little Indians" in *Bloody Bloody Andrew Jackson* was offensive?)

A brief plot summary here will help orient the discussion that follows. A five-act play, *Horizon* is a full evening's entertainment that takes the audience on a wild if vicarious ride into the "Far West." The story begins in an upper-class New York home where a young West Point graduate, Captain Alleyn Van Dorp, teams up with Sundown Rowse, a blustering politician and railroad tycoon. Rowse has secured a large railroad land grant in the western territories and requires a military escort to tour his new holdings. With Alleyn as their military escort, Rowse and his entourage – including his debutante daughter (Columbia) and a flamboyant British nobleman (Mr. Smith) – set out to encounter the myriad dangers of the Far West. In the town of Rogue's Rest (Act II), the New York elites meet local ruffians, outlaws, vigilantes, Indians, and Med, a young mysterious white woman known as "the flower of the plains" (335). Rowse must contend, it seems, not only with the dangers of wilderness foes but also with a lawless world in which vigilantes exact rough-hewn justice on local outlaws. In theater performances of this period, the curtain dropped at the end of each act, giving the audience time to reflect on what just occurred, to talk with persons sitting nearby, or to walk about the theater for a smoke, a brandy, or a bit of gossip. These interludes allowed the play's narrative, images, and ideas to settle into the imagination and to whet the audience's anticipation of what comes next. In Act III, Rowse and party head upriver to explore the further reaches of his land grant. As the wilderness deepens around them, they are (predictably) attacked by Indians, taken prisoner, and held captive. Acts IV and V fan fears through a series of scenes in which the hapless easterners are safe only behind the walls of a military fort. Meanwhile Alleyn and his regime leave to search for the warriors who attacked their boat. While the soldiers are gone, the fort is attacked, and Med and the other women are taken prisoner. Act V opens in a ravine where the Indians hold their captives; Rowse and later Mr. Smith are also brought in at knifepoint; meanwhile, Med is taken ill, crying out for Alleyn to return. At the close of Act V, however, Captain Alleyn and the U.S. Cavalry rescue the party of whites and vanquish the Indians, but not without the help of the play's two unsavory outlaws, Wolf and Loder. While the play clearly pokes fun at the naiveté of the elite characters, it delivers a real-world warning: the dangers are real, the military intervention is necessary, and woe is the person who ventures west unprepared!

Themes that work to legitimize military occupations in the west are evident in the lively opening dialogue, which touches on Indian fighting, bison hunting, and the general lawlessness of the frontier. Captain Alleyn quips about the dangers of "ferocious Indians" and introduces his benefactor as a man who "owns a slice of every Territory in the West" (343), thus centering the notion of property as one of the play's ideological assumptions. Sundown Rowse has Manifest Destiny in his bones, a Darwinian entrepreneur with big plans, he fancies himself as a kind of missionary "carrying civilization into the Far West." Mrs. Van Dorp, who is preoccupied by the "real" dangers of the west, charges Alleyn with finding her long-lost daughter, who was stolen by her husband and taken west years before.[14] Meanwhile, Mr. Smith, a naïve "English nobleman," has romantic visions and is "anxious to see the great west. . . . The aboriginal red men and the real original white settlers," and hopes to "hunt buffalo and bison" – both apparently mythical beasts to him.[15] Rowse reminds them of the purpose of their adventure, "Well, after I've located my railroad –" but his daughter cuts him off before he can invite Mr. Smith to shoot from the comfort of a palace car, brandy snifter in one hand, Winchester rifle in the other. Columbia chimes in that "[i]t's like a grand picnic. We'll go over the prairies on wild horses and camp out in the woods" (347). After the war, the U.S. government unofficially encouraged the railroads to promote bison hunting as a high sport for the elite. (A similar military tactic was used by Northerners who destroyed Southern food sources during the Civil War.) With vicious irony, Mr. Smith's touristic hopes work hand in glove with the central military tactic used against the Plains tribes: destroying the bison herds on which sustenance and culture depended.

Throughout the 1870s, U.S. state-sponsored annihilation of the vast herds of bison was facilitated by the railroad industry. Euro-American hunters shot bison from the comfort of their railcar or hired guides to lead hunting parties. While some of the meat was harvested to feed the railroad construction crews, and the skins were harvested for export to Europe, the sport hunters left most of the slain bison to rot, unused and wasted. Indigenous subsistence hunting was part of the maintenance of healthy herds and ecological balance on the plains since time immemorial (Cajete 2000, 168–174). Herds that had moved by the millions over the plains like vast hooved clouds – sustaining tribal life and culture and representing an ecological relationship between people and the land that had characterized the continent since time immemorial – were destroyed in under a decade. Skinned bison bodies were piled 20-feet high in some instances. Piles of flesh and bone were visual demonstrations of the interlocked U.S. project of genocide/ecocide on the plains. Richard White (1993) writes that

> the southern hunt peaked between 1872 and 1874. By 1875 the southern herd was virtually gone; by 1883, the northern herd had vanished. The slaughter was so thorough and so quick that not even the hunters could believe what they had done. In the fall of 1883, many outfitted themselves as usual. But there was nothing to hunt except piles of bones bleaching in the sun and wind.
>
> *(218–219)*

Daly's humorous treatment makes light of this real-world damage to the interconnectivity between lives, land, ecology, and cultural practice. As white hunting parties decimated bison populations and portioned off the hunting territories of plains tribes by rail and barbed-wire fencing, the plains' ecosystems deteriorated. Today, the combined populations of bison, bighorn sheep, elk, and pronghorn antelopes in the American West amount to less than one percent of original estimates (Worster 1992, 45–46). The buffalo were among the many animal populations of North America (otter, whale, salmon, elk, wolf, coyote, mountain lion) that would be decimated through state-sponsored slaughter either for economic gain directly or to ensure and protect the project of colonialization not only by attempting to destroy people's sustenance but also their cultural fabric. White observes that

> the disappearance of the buffalo marked the final blow to the old life. In the words of the Crow warrior Two Leggings: "Nothing happened after that. We just lived. There were no more war parties, no capturing horses from the Piegan and the Sioux, no buffalo to hunt. There is nothing more to tell."
>
> *(220)*

After the Civil War, Northern military personnel – particularly officers who had made names for themselves in the Civil War – remained in the public eye by taking their regiments west, where they built new careers by battling Native resistance fighters. In historically analogous ways, *Horizon*'s implicit endorsement of military occupation of tribal lands comes through in the character of Captain Alleyn, who typifies the western cavalry hero of the era, including George Armstrong Custer, already much in the public eye at the time. Young Alleyn believes "a soldier should fight," and Rowse has commissioned him for just that:

MRS. VAN DORP: Allow me to thank you, Mr. Rowse, most warmly, for the kind interest you have taken in my son. . . .

ROWSE: Don't mention it, ma'am. It wasn't much. I had a cousin wanted the commission, but he didn't like to go and fight the Indians. Your son jumped at the offer. My cousin backed down, asked me if I thought he was a chicken to go for the Chickasaws, and told me, I might go myself and keno the Kiutes.

MRS. VAN DORP. Alleyn is very courageous, and believes a soldier ought to fight.

ROWSE: He's a trump. I appreciate pluck. I come from a fighting family. They were the first settlers of Kansas.

(346–347)

Here, both Mrs. Van Dorp and Rowse consider willingness to fight indigenous people – and forcibly dispossess them of their ancestral land – to be a manly virtue. Rowse, in particular, clearly prefers Alleyn's aggressive "pluck" to his cousin's implicit cowardice. While the claim to being "first settlers" erased indigenous residents, the racist and demeaning mockery of tribal names trivializes indigenous claims to the land, exemplifying the play's endorsement of the alliance between the U.S. military and the interests of eastern railroad investors who aimed to acquire that land. Such theatrical propaganda validated and vindicated audiences far removed from conflicts on the plains, many of whom were no doubt closely aligned with railroad interests and stood to profit from the conquest of the west (White 1993, 247).[16]

Railroading land and lives

The completion of the transcontinental railroad in 1869 highlighted a kind of performance of power that enabled a neocolonial relationship between the urban industrial east and the undeveloped west. The railroads ultimately functioned as arteries through which the harvest of western resources could flow to eastern industrial centers, and towns that sprang up along the railroad lines were nodes in a chain of commerce where laborers unloaded commodities from wagons onto railcars headed north and east to St. Louis, Chicago, and eastern cities. As its iron signature stretched across the continent, the locomotive became the central symbol of the westward advance of American civilization in landscape paintings and literary imagery (see, e.g., Marx [1964]).

It came as no surprise, then, that Daly chose to position a railroad baron at the center of *Horizon*, invoking at once the powerful symbol of a unified nation and the economic machine that came to embody the idea of progress. *Horizon* satirizes the corruption rampant in the railroad industry of the period, even as it endorses the economic and cultural expansion of the U.S. Alleyn claims that Rowse "owns immense tracks of *public lands*, granted him by Congress" (343, emphasis mine). Meanwhile, Rowse brags: "I'm going to prospect for the first hundred miles of the Fort Jackson and Big Run branch of the Union Pacific Railroad, chartered by Act of Congress and subsidized with twenty thousand acres, well adapted for farms and settlements" (347). Of course, what Alleyn calls "public lands" were anything but and did not belong to Congress to allocate.

In the 1850s, Congress granted land to a railroad company to build a connecting spur in the state of Illinois, thus setting a precedent for government to subsidize railroad construction. Throughout the 1860s and 1870s, western railroad developers lobbied Congress for increasingly larger parcels of public land until the size of the grants included 20-mile stretches on either side of the proposed tracks. According to grange movement advocate Edward Winslow Martin, the railroads owned ten percent of U.S. lands by 1873 (Winslow Martin 1873, 4). Railroad companies found themselves in a catch-22 each time a new line was proposed. Eastern investors were reluctant to put their money into a scheme that proposed to build lines in the "lawless" western wilderness when there were easier ways to make money. White (1993) observes that in the highly competitive financial market of the period, railroad companies "promised high returns on their securities in much the same manner that 'junk bond' promoters of the 1980s promised high returns on risky investments" (247). Faced with financing a capital venture in which profits were always down the line, railroads gave land grants as collateral to investors, not only ensuring the ready cash for labor and supplies to build the lines but also promoting the growth of settlements and resource extraction projects that would provide the railroad ongoing business in the future (248). Fraud, bribery, and corruption, White observes, ran rampant as speculators, politicians, and bankers lobbied Congress for additional land (248–249). Alleyn's description of Rowse as "an honest man" (343) in *Horizon*'s Act I was designed to get a laugh because New York audiences would have been well aware of this corrupt underbelly of the railroad finance market. Rowse the profiteer plans to use the land that Congress granted him, he boasts, not only to build a railroad but also to begin mining operations on the (fictional) Big Run River. By selling or leasing the more fertile land downriver to farmers, he plans to establish a town in which merchants will sell goods shipped west on his railroad to the miners and farmers indentured to him. In turn, his railroad would profit a second time by transporting the minerals and grain produced by that land and labor for east coast markets. Railroad corruption fueled long-standing grievances by farmers in the West and South against the urban industrial power centers of the East. Railroads extracted high prices for grain storage and shipping, even as farmers were tested by ongoing drought and too many years of farming land that was not well suited to water-intensive crops, and railroads extracted high prices for grain storage and shipping (see, e.g., Winslow Martin [1873]). These conditions precipitated the nationwide railroad strike in 1877 and fueled the rise of the Populist Party in western states.

Horizon demonstrates the ongoing cycles of erasure that the frontier narrative also sanctions, in which the loss of indigenous lives, communities, and cultures paved the way for the erasure of the white settlements, towns, and localities; all easily disappeared in the face of the progress that Rowse and the railroad represent (see, e.g., Slotkin [1973]). Rowse is intent on giving the land a new identity – his own – as he brings the full weight of his private dreams to bear on that land. *Horizon* reflects the ways naming functions as a tool of imperialism, laying claim to proprietary rights that erase prior human bonds with the land. Discussing the map of his land grant with Mrs. Van Dorp, Rowse uses the personal pronoun *I* instead of "my property" or "railroad land," when he remarks,

> Here's the map. . . . *I* run from here on the west bank of the Big Run River down to Dogs' Ears, that's the name of another settlement, then out to All Gone, that's an Indian camp, and then to Hollo Bill, that's a traders' settlement. Queer names, ain't they?
>
> *(347)*

Here Daly seems to acknowledge tensions that exist between the dominant narrative of white expansion and the call of conscience that mourned the violent subjugation of lands and peoples. Once the railroad is built, Rowse envisions renaming the settlement "Rowseville," thus marking the land with his own image and the civilizing forces he believes he represents. Turner's ([1921] 1956) notion of "free land" depended on similar cycles of erasure in which colonizing forces – whether early settlers or later industrialists – emptied the land and refilled it with their own ambition (1–2).

But in Act II, the land is anything but vacant. In Rogue's Rest, Rowse finds that he presides not so much over a plot of property but a web of relationships that precede him. Rowse meets this populace as they are conducting a vigilante expulsion of two unsavory characters, Wolf, a town drunk, and Loder, a gunslinging outlaw. When Rowse announces, "I'm the owner of this here settlement, the landlord of the premises, and proprietor generally. I won't have any mob law, and no Vigilance Committees, and no riots, and no games of that sort on my land," the stage directions tell us that the locals "draw knives, pistols" in response (350). By making Rowse the butt of this joke, the play makes two important ideological points: First, it reiterates the necessity of military presence on the plains to bring order to people who are undergoing the immersive process of "re-wilding" that Turner would later articulate. Second, even as Daly satirizes Rowse's naïveté, the scene also asserts the notion of property as a signifier of civilization. For Rowse, the land is composed of marketable "plots," which the railroad companies can exchange as they would with any other commodity. However, even as *Horizon* satirizes Rowse's self-aggrandizing in this scene, it also underscores the need for civic order – something in short supply in Rogue's Rest. Military occupation, the play shows, is not so necessary for safeguarding property rights of settlers (who, in this play, have developed their own systems of vigilante governance) but rather those of the railroad and other corporate land grabbers.

One of the reasons the Indian Bureau cites for its policy failures during this period was the "inability" of indigenous residents to grasp the idea of private property, a concept that is historically specific and ideologically aligned with European/Anglo-Saxon early capitalism. Antiassimilationists saw the Natives' less possessive relationship with the land as one more example of the cultural differences that was cited as evidence of white supremacy (Slotkin 1985, 301–324). Land traders typified by Daly's Rowse – who failed to understand the indigenous practice of shared homelands – swindled indigenous people.[17] The idea of private property, like the railroad, paved the way for extractive capitalism (including industrial agriculture) and would become a rationalization for delegitimizing Native land and water rights for generations to come (see, e.g., Dunbar-Ortiz [2015], 34–39, 205–208, 218–236]).

Yet even as *Horizon* argues for the necessity of military partnership for projects like Rowseville, the vigilante court scene seems to undercut eastern ideals of how civil society should proceed. Loder submits to the crowd's wishes and leaves town, growling, "I'm tired of this place. It's getting too civilized for me," and leaves for someplace "a little nearer the Horizon" (354). Wolf refuses (he cannot leave his daughter Med), and the crowd cries, "Hang him! Hang him!" At this moment, Rowse tries to diffuse the angry mob with a flourish of rationalist rhetoric and offers himself as an arbiter of justice:

> Fellow citizens: Let us not be irregular, let us not proceed to mob law . . . we are here proceeding according to law. Not the musty statutes of effete systems and oligarchies of the Old World, but the *natural* law implanted in the bosoms of man since our common ancestors were washed, wrung out and hung up to dry by the universal flood.
>
> *(356, emphasis mine)*

With logic that emerges from the very violence it seeks to control, Rowse is purposeful in his distinction between European law and custom (as typified by Mr. Smith), and an American democracy rooted in the land – "natural law" – that (as Turner would later claim) emerges from the special circumstances of men's encounters with "savage" lands and peoples. Furthermore, the biblical allusion (to the flood) presumes religious affiliation of the mob onstage (as well as the theater audience), as it expresses the assumed solidarity among Christianity, the nation-state, and populist democracy.

Staging a credible threat

In *Horizon*, the assertion of white economic models comes, unsurprisingly, at the expense of the indigenous characters. In Act II, Wannemucka challenges Rowse to a game of poker for his gold watch and chain. Rowse laments in reply: "If you had offered to scalp me, you red rascal, I might have forgiven you. But Poker! That knocks the romance." The play underscores the message that indigenous presence is a threat to Anglo-Saxon settlement by demonizing Native characters, whether or not they assimilate. As Loder chides further, "[t]hat's civilization, my friend. The noble savage was in his native state, he went for the hair of your head. Now . . . he carries the weapons of enlightenment and goes for the money in your pocket" (352). The joke points to the double bind in which Native characters were "savage" if they followed their own culture but dishonest if they joined Euro-American economic culture.

Earlier in the century, a genre of literature and an iconography of the "Noble savage" characterized the devastating impacts of colonization on the continent's indigenous inhabitants as regrettable but inevitable symptoms of progress of the civilization. James Fenimore Cooper's "Leatherstocking" tales (1985), written between 1823 and 1841, along with Henry Wadsworth Longfellow's *Hiawatha* (1855), depict indigenous peoples as peacefully fading away into the past. On stage, where a dramatic plot required high-stakes action, however, Indian characters were often shown being pressed into violence by circumstance. In John Augustus Stone's 1829 play *Metamora or, the Last of the Wampanoags* (1941), for example, the old chief is fated to witness the destruction of his people and his inevitable death understood as a prerequisite to Anglo-Saxon dominance. The requisite disappearance of indigenous characters is not only evident in the plot of *Metamora* but was also embodied and actualized theatrically by the well-known white actor Edwin Forrest, who played the character of Metamora in red-face. First staged in 1829 at the Park Theatre in New York City, *Metamora* coincided with and provided rhetorical support for Andrew Jackson's Indian Removal Act signed into law in 1830. As such, the play's debut also coincided with the forced removal and relocation of tens of thousands of Native people from their homelands in the U.S. South, as well as other territories east of the Mississippi, to newly designated "Indian territories" in the West (Sayre 2005, 117–125).

Daly replaced the noble savage with the visage of a fierce and cunning foe – a threatening enemy that should be taken seriously. Hobson Quinn (1927) notes that Daly was praised as one of "the first of the modern realists in American playwriting" (12). The *New York Times* (1871) review likewise affirms that Daly's "Indians . . . and United States soldiers [who] are more like *real* Indians and soldiers than any yet brought before us on the boards." *Horizon*'s Act I, as we saw earlier, satirizes Columbia and the buffoonish Mr. Smith's naïve desire to "speak to him as his paleface brother. I've read the Leatherstocking stories" (347). These and numerous other references in dialogue poke fun at the literary constructions made famous by Cooper, Longfellow, and others. The genre taxonomy of realism tends to silence other perspectives, in this case the romanticism of the early nineteenth century, in order to forward an ideological narrative that had empire as its goal. In Act III of *Horizon*, the easterners meet the Widow (perhaps the widow of a French fur trader), who warns them not to trust the Indians, although Columbia remains enamored:

COLUMBIA: I used to think I'd like to have an Indian brave fall in love with me – so romantic.

...

COLUMBIA: I mean a real noble savage, not a dirty common Indian.
ALLEYN: They're all the same.
COLUMBIA: Somebody like Fenimore Cooper's braves.
WIDOW: Coopers, is it? Faix [sic] all trades is alike, they're all a dirty pack, and coopers is no better nor any of 'em.

(365)

While the Widow is coded as an illiterate buffoon who mistakes Fenimore Cooper for a man who is a cooper (a barrel maker) by trade, the Widow's view alone might be dismissed, if not for Alleyn's agreement. Stereotypes gleaned from novels read in the quiet of a drawing room may seem on the surface to engender white sympathy, the literature that informed the Indian brave of Columbia's imagination did not argue in favor of indigenous rights to live and prosper on their own lands. However, the concept and visage of the so-called noble savage gave rhetorical support to white supremacy as surely as the binary other Columbia's quip above invokes.

In contrast, Daly's *Horizon* fed a new imaginary that saw white settlers as victims in the face of indigenous resistance. The story Daly tells operates in complete disregard for the realities of indigenous lives.[18] Only the presence of an orphan Indian child, who, in Act III, hangs on Rowse's pant leg, begging to go to "the big city," provides a clue to the tears in the fabric of tribal life as a result of colonialism and U.S. militarism. Throughout the scene, and no doubt to comic effect, Rowse tries to get rid of the child, calling him a "young mosquito" (365). As the *New York Times* review ("Horizon" 1871) notes, *Horizon* affirmed an Anglo-Saxonist/white supremacist understanding of the American project, and the play was consistent with the ongoing national trend to delegitimize Native resistance:

> Our people have pushed westward in a steadily advancing fringe that, through the golden discovery of Capt. Sutter, has been confronted by a smaller fringe from the east far sooner than the most sanguine observers of twenty-five years ago ever ventured to predict. The aborigines of the plains, thus caught, as it were, between two fires, are being 'improved' with terrible swiftness from off the face of the earth.

The preceding quotation's biting allusion to the rapidity of the ethnic cleansing taking place on the plains is chilling, as the unrelenting westward expansion advanced a culture dependent on resource extraction for capital gain. Whether a drama calls attention to the tragedy of the so-called "lost race," as in *Metamora* (1941), or demonizes Native people as savage yet cunning foes of civilization, as does *Horizon* and Frank Murdock's *Davy Crockett* (produced in 1873), these stories and stage plays were complicit in ethnic cleansing. Stories that masked and justified removal of indigenous peoples from their lands as natural, inevitable, and evolutionary forwarded colonization, and subtle permutations of these stories continue ongoing colonization of indigenous lands, perpetrating ecological and cultural harm.

One of the most potent foils in support of a military solution to the "Indian problem," Richard Slotkin (1985) notes, was the erroneous representation of Indians as "habitually devoted" to the "capture and rape of white women" (401). Echoing such rhetoric, Daly's Alleyn fumes at the suggestion that Wannemucka wants to capture Med. The cadet hopes to "use his sword" to slay the Indian and "make an example of him" to "amalgamationists" (a wordplay on "assimilationist") foolish enough to believe in the peaceful settlement of disputes between settlers and Indians (365). Thus *Horizon* compounds the perceived threat of indigenous presence by fueling white fear of racial mixing. In Act II, Wannemucka reveals that he desires Med and intends to "take her, or never go back to his tribe again" (352). The play's demonization is consistent with the nineteenth century's general racialization of sexual desire, and this becomes the pivotal threat around which the violence of *Horizon* builds. Meanwhile, the young Captain Alleyn has *also* fallen in love with Med. Female characters were often read as symbols for the land – wild, virginal, and pure – and this love triangle has everything to do with who controls the land.[19]

Horizon appeared as reports of violence between settlers and indigenous resistance fighters were increasingly touted as the rationale for land takings. Newspaper accounts of the period often described settlers as victims, while ignoring violent settler aggression. Yet even assimilationist condemnation of settler violence couches the violence on both sides as acts of individuals (e.g., a settler's provocations, Indian attacks in defense) rather than recognizing indigenous resistance to a *system* of land theft, broken treaties, stolen families, cultural desecration, subjugation, and abuse. Violent resistance to relocation, Slotkin (1985) observes, "became more potent as the maladministration of Indian affairs and provocative land seizures by whites continued" (320). Moreover, the vilification of indigenous resistance in the press and on stage as "savage red men" condoned and encouraged hate crimes by providing both an official and unofficial context of "war," which White (1993) explains justified individual violence, as well as collective violence in the form of military and vigilante aggression (337).

At the end of Act II, Wannemucka jumps on top of a shed, shoots Wolf, and cries, "Now Injun have white princess!" Loder, the outlaw, jumps to Med's defense: "You red devil! Come and take her!" (357). As the curtain falls, the play underscores – and not in a subtle way – that the greatest danger to the civilization that Rowse represents is not vigilantism (indeed, the outlaw is crucial to the security of the woman and the land!) but rather the implied violence of the Native characters. In this way, *Horizon* represents nascent American civil society as a unique amalgam in which a certain measure of "lawlessness" (including vigilante violence) is not only tolerated but valorized as long as a white man conducts it. Native "savagery" threatens American civilization, while white savagery, paradoxically, furthers it.

Standing (in) for wilderness

The visage of *Horizon*'s Indians as formidable enemies was consistent with the European symbolic conflation of indigenous people with nature and fundamental to the ethical dilemma of colonization because it worked to mask the moral implications of genocide. Native characters in frontier dramas of the period, whether romanticized or demonized, were often read as the personification of wilderness. The conflation of "nature" and "Native" is demonstrated in the frequent use of the trope "savage" to describe both. Characterizing the destruction of human lives, language, and cultures as necessary to and synonymous with conquering wilderness (which Christian settlers understood as a biblical directive) meant that both were expendable as it was exploitable in the project of nation building. *Horizon* accomplishes this dehumanizing equation not only through dialogue that degrades Native characters but even more powerfully through its visual staging of three attack scenes.

In Act III, Rowse's party floats downriver on a flatboat to explore his land grant, while Alleyn and the cavalry guard the boat by riding along the shore. Meanwhile, Wahcotah, the Native guide and tribesman of Wannemucka, has assured the cavalry that there is no danger of attack. This journey is accomplished on stage by scenery made to move across the stage while the actors stay in one place. The stage directions describe that first attack:

> A narrow bend in the Big Run River. From the bank on extreme left, a blasted and fallen tree trunk stretches over to right, dipping the water near the fourth groove. Wannemucka and three Indians are concealed on this tree. Wahcotah and another are in the water near left center. Other Indians concealed behind logs and trees.

As the personification of the dangerous wilderness,[20] "Wahcotah rises from the water," while Wannemucka and the other Indians drop from the trees onto the deck of the boat, hatchets drawn (364). The staging of the attack scenes works to embody the visual identification of *Horizon*'s Indians with such predatory animals and wilderness itself. In the visual logic of the performance, the warriors are a part of the wilderness, and thus the performance rationalizes the larger genocidal project of the United States as synonymous with conquering the wilderness and everything within it.

Access to resources unfettered by moral hesitation, ambiguity, or moral culpability was, and remains, essential to industrial capitalism. Turner's (1956) notion of "free land" (as the basis of American identity formation) is predicated on the erasure of indigenous habitation (1–4). This narrative and visual conflation of Native and nature espoused land vacant and available for Anglo-Saxon settlement and provided settlers a way to simultaneously mask and rationalize ethnic cleansing as a natural process, a mere necessary part of civilization's access to the land and its wealth. When combined with the Anglo-Saxonist imperative to, as Rowse boasts, "bring civilization to the Far West" (347), this early presumption and assertion of innocence is the ideology that now stands in the way of corporate culpability for environmental destruction or human health impacts.

At the top of Act IV, the party is no worse for the wear, but they have accumulated terrifying stories that they can now safely recount within the locked gates of a military outpost. At this point in the play, Alleyn and the cavalry leave Rowse and the women secured in the fort while they go out in search of their Indian attackers. Meanwhile, Wahcotah, the "friendly Indian" and guide, betrays the white travelers by reporting their location to his chief, Wannemucka. Then, while the soldiers are away, Wannemucka and his warriors attack the fort, capturing Rowse and the women, who scream and struggle as the curtain falls. Act V reveals a wilderness ravine in which Wannemucka's tribe holds the prisoners. Warriors dance around the white women, who squeal in horrified anticipation of being raped. It was colonization, however, that brought rape as a weapon to the continent. In *The Beginning and End of Rape* (2015), Sarah Deer argues that the use of rape can be credited to white colonizers who first introduced it as a practice of war (xiv-xv). Deer also asserts that "rape is a fundamental result of colonialism, a history [of] violence reaching back centuries" (x) and "inextricably linked to the way in which the United States developed and sustained a legal system that has usurped the sovereign authority of tribal nations" (xiv). In the previously described scene, *Horizon* performs a rhetorical flip in which whites – particularly white women – were repositioned as victims, providing a further rationale for aggressive military assault against plains tribes (see, e.g., White [1994], 6–65).

In the final, tense moments of the play, Rowse and the women become increasingly desperate, hoping that Alleyn and the U.S. Cavalry will return in the nick of time. This expectation powerfully embodies the play's clear stance that military occupation is the *only* practical solution to the "Indian problem" and the only choice that would ensure the safety of settlers and the railroad. However, Daly has one more twist in store for these New York theatergoers, a twist that delivers one of the play's most problematic "lessons." As the scene develops, Wannemucka sends Wahcotah, acting as a double agent, out on a reconnaissance mission to locate the soldiers. Meanwhile, Loder (the outlaw from Act II) has secretly followed Rowse's party downriver. Loder catches Wahcotah in the act of betrayal and kills him. He then steals Wahcotah's clothing and, thus disguised, Loder infiltrates the Indian camp, fooling even Chief Wannemucka. Then, in *Horizon*'s final moments, Loder enters "dressed in the disguise of Wahcotah, and [im]personating him" (374). Wannemucka, thinking this is Wahcotah, cries out: "Ha! You have slain White Panther!" A few moments later, as the audience hears the cavalry's drums in the distance, Loder casts off his disguise and shouts, "Indian! Stand back!" (375). At this moment, Wannemucka "springs toward him with uplifted knife. All the Indians spring upon their captives with a yell. Loder seizes Med and fires his rifle at Wannemucka, who falls. Only *then*, the stage directions indicate that "the ravine is filled with soldiers" (375). It's not the military that saves the day in *Horizon*; it's the outlaw. The scene not only foreshadows the popular romanticization of the outlaw in U.S. culture but also underscores something more troubling: the uneasy and unspoken solidarity between individuals who take matters into their own hands, whether at the U.S. borders or in the streets, to forward the cause of the nationalism.

Horizon not only endorses military occupation of western lands (where once the Indians are vanquished, the cavalry will continue to serve business interests); it proposes that the job of securing the continent cannot be accomplished without the help of the outlaw, who has cast off his civilized ways, name, and clothing. Loder goes by an Indian-type nickname, White Panther, which signals his ingenuity and cunning. His willingness to "go native," a process that Shari Huhndorf (2001) notes involves assuming the characteristics of the "savage" housed in the white imagination, Loder has moved along Turner's trajectory from "civilized" to "savage" (1–18, 51–62). Ironically, the stereotypical qualities assigned to Indians (e.g., deception, cunning, betrayal, willingness to kill) when transcribed onto a white body become markers of heroism. As a Turnerian example of the quintessential American, the visual performance of Loder the outlaw who assumes the persona of Wahcotah lays claim to a self-made indigeneity – Turner's ([1921] 1956) thesis in the flesh.

The implicit and uneasy partnership between the military (i.e., federal) forces, on one hand, and local vigilante/outlaw/ruffian cunning and impulsivity, on the other, was essential to the occupation of western lands, and this partnership has ongoing implications. *Horizon* earned praise for its "realistic" portrayal of western characteristics of lawlessness, mob rule, expediency, and disregard for civilized processes – understood through a Darwinian lens as a kind of naturally occurring disorder provoked by encounters with the wilderness. This idea – that a certain amount of lawlessness is requisite in the American character – also protected expedient eastern capitalists from incrimination. After all, such behavior was merely a *natural* part of frontierism (see, e.g., Hobson Quinn [1927], chaps. I and IV). The veneration of lawlessness (e.g., as heroic, as quintessentially American) serves the ideology and practices of white supremacists today, for example, as self-appointed border guards. An important part of the notion of what constitutes an American within this framework is the aptitude and willingness to function both within and outside of the law, within and outside of obligations, responsibility, and ethical consideration. Likewise, both individual and corporate disregard for government regulation (and the prioritizing of individual rights over the common good) leads to a culture in which the violation of environmental law is simply part of doing business, requiring frontieresque strategies including outsmarting, finding extra-legal work-arounds, or just plain ignoring existing law and regulations.

Horizon concludes with the revelation of Med's true identity as Mrs. Van Dorp's daughter and a child of eastern wealth. Captain Alleyn's proposal and the promise of their marriage implicitly represents the state's claim to the land itself. The nuptial ending (a familiar trope not only in frontier drama but in plays and films about the American West throughout the twentieth century) reflects the deeply held Anglo-Saxonist perspective that only white men could effectively carry out the imperative of civilization and get the land to actively produce the wealth it (passively) held.

In the year before *Horizon* opened (1871), Congress ended the treaty system that held tribes as sovereign nations with the right to negotiate access to their own land. This move signaled, among other shifts, the embracing of an ideology of entitlement without responsibility (Calloway 1996, 209–219). Congress's abandonment of the treaty system effectively promoted the violation and disregard of existing treaties with sovereign tribal nations, sanctioned increasing violence of settler land grabs, and contributed to devastating ecological changes to diverse human and nonhuman communities.

The following year brought the dedication of Yellowstone as the nation's first national park and a growing movement to preserve lands as monuments to national identity that will be discussed more closely in Chapter 2. However, it is important to note here that it, too, had everything to do with a Rowse-like project of claiming and naming for a privileged Anglo-American class. As monuments to the nation, the national parks were also predicated on the removal of indigenous inhabitants from "preserved" lands (Cronon 1996, 69–90). Daly's *Horizon* appeared in the five years *before* Custer's defeat at Greasy Grass/Little Big Horn in 1876, which further bolstered white America's support for military aggression on the plains. Alleyn, for his part, might be seen as Custer's prototype – the archetypal "boy hero" of the frontier, whose persona is reflected in early descriptions of Custer himself as the "Boy General" of the Civil War (Slotkin 1985, 376). This emblematic archetype would echo, transmute, and duplicate itself many times over, not only on the stage and in the stories of the late nineteenth century but also throughout the twentieth and twenty-first century. The lineage of this Turner-esque visage can be observed in various iterations such as Owen Wister's *The Virginian* ([1902] 1945), among countless others; the rough-hewn, arrogant, pragmatic, adolescent central character of *Bloody Bloody Andrew Jackson* (2010) is a recent example. These Turner-esque visages continue to propagate and rationalize and naturalize the socioecological violence of settler-colonialism.

Steele MacKaye's simulation: performing colonial ecological violence

In 1887, Congress passed the Dawes Allotment Act, under which the remaining lands held by tribes would be carved into plots, and assigned to individuals rather than being held collectively by tribes. What was left over – and here is the key that made it merely another land grab – would be sold by the federal government to settlers. The Dawes Allotment Act of 1887 would later divide even reservation land into single-family parcels in order to encourage Indians to farm. The policy made little sense to indigenous people of the plains, where life was dependent on hunting. The Dawes Act produced many benefits for whites, however, including allowing "surplus" reservation land (usually the most fertile) to be taken by white settlers or mining companies. Officials who opposed the "peaceful solution" argued that "the Indians did not have the same entitlement to property as the white man who sought to develop it, because they were incapable of using it productively and unwilling to tolerate its productive use by others" (Slotkin 1985, 324).[21] The government's cultural and ecological ignorance divided individuals from tribal groups and ripped at the fabric of Indian communities.

As assaults on tribal lands, lives, and sovereignty continued under the Dawes Act, "Wild West" shows emerged as a new performance genre that celebrated frontierism and helped transpose it into everyday American life. As horsemanship and marksmanship illustrated the various skills, risks, and exploits of frontier life, these circus-like spectacles argued that the task of vanquishing the land and its indigenous inhabitants had been accomplished. By 1885, more than 50 Wild West-type expositions were touring throughout the United States. William Cody is arguably the most iconic and remembered of the genre, claiming to represent an "authoritative," "realistic," and "authentic" reenactment of life on the Great Plains. Other "Wild West" performances included James B. Hickok (Wild Bill), Pawnee Bill, Captain Jack Crawford, Buck Taylor, Major Frank North, and Colonel Frederick T. Cummins. Many of these entertainers traversed back and forth from battlefield to footlights and eventually to civic leadership (Russell 1961, 17–33).

Beginning in the spring of 1883 (in the wake of the decimation of buffalo herds), Cody's *Wild West* toured until 1906. Its longevity was due in major part to Cody's collaboration with New York theater director/producer Steele MacKaye. Cody and his business partner, Nate Salsbury, wanted to set their *Wild West* apart from the countless others and so hired MacKaye (who recently had seen the early version of the *Wild West* on Staten Island in 1886) to reenvision and restage it. MacKaye pulled out all the stops, creating a simulation retitled the *Wild West: The Drama of Civilization* at Madison Square Garden that used all the theatrical effects and inventions for which he was well known (Percy MacKaye 1927). In his detailed description of the production, J.A. Sokalski (2007) documents how MacKaye's role as theatrical director transformed Cody's loose array of western stunts into a fully formed epic drama that ensured its longevity through 1906 (161–173). It was MacKaye who gave Cody's *Wild West* its climactic plot structure, definitive heroic characters, and indelible iconography. MacKaye coalesced the loose tropes and themes of frontierism into a perennial national narrative. *The Drama of Civilization* effectively utilized the stage as a weapon of Anglo dominance and dominion on the continent.

Envisioning himself as a theater producer in a position to author history, Steele MacKaye later characterized his collaboration with Bill Cody as a "history told by the actual creators of history" (Percy MacKaye 1927, 87) represented in "significant epochs that have emphasized the strides of Civilization into the wilderness, the conquest of barbaric nature by the supreme intelligence of our race . . . presented with a picturesque force" (88). MacKaye clearly understood the power of performance to shape national narratives and national identity; he understood that control over the representation of the past helps to shape the future. Intentionally blurring theater with reality, MacKaye devised an evolutionary scenario that, like Daly's *Horizon*, spelled out a vision of Manifest Destiny tableau by tableau. MacKaye gave the production a Darwinian "historic order," in which a series of pantomimes moved from "aboriginals" in their "forest primeval" through pioneer wagon trains, cowboys on the range, Indian attacks, prairie fires, mining camp, Pony Express, and Cavalry fort – all meant to depict the steady unrelenting march of MacKaye's civilizing forces, and all of which depended on military and paramilitary violence (87–88).

MacKaye's goal was "not only a new sensation, but a new Enlightenment" (Percy MacKaye 1927, 87), and he gave Cody's show a new subtitle, which MacKaye promised would be as "instructive as it was inspiring" (88). This new and improved *Wild West* would provide visceral experiences of wilderness, pioneer life, battles, and, not least, violent weather. MacKaye's vision constituted a public pedagogy that ingrained a narrative of white racial entitlement, indigenous defeat, environmental and industrial domination and exploitation, and national innocence – in short, precisely the ideology that supported the burgeoning extractive capitalism and industrial development of land west of the Mississippi.[22] MacKaye hired artist Matthew Morgan, who had recently completed a series of massive Civil War paintings, to develop panoramas of the primeval forest, the open prairies, Custer's battlefield, and the emblematic town of Deadwood on a giant cyclorama background "a half mile long and fifty feet high" (78). Morgan's panoramic wilderness transformed Madison Square Garden into a wilderness, which, together with MacKaye's dramaturgical influence, plot structure, and direction, gave flesh to ideology. Indeed, it was as if the characters in John Gast's painting *American Progress* (1872) had come to life in a monumental simulation.

Of all the realistic elements of Cody's *Wild West*, none is more astonishing and troubling than its objectification of Native persons and personalities.

In a kind of casting as coercion, Cody hired warriors, as well as women and children, through the Indian Bureau, including warriors arrested for participating in the Ghost Dance resistance movement.[23] White (1994) observes that, as a kind of demonstration of assimilation, Native performers were "imitating imitations of themselves" – playing roles assigned to them by whites (35). For an Anglo-American audience, there was no more convincing evidence of military/settler victory on the U.S. plains than the actual appearance of Native warriors, many of whom were prisoners of war. "Native performers authenticated the *Wild West* in ways that no mere cowboy could," writes Philip J. Deloria (2004, 60; see also 57–60). Cody's roster of Native stars read like a *Who's Who* of the embattled plains, including Sitting Bull, Black Elk, Chief Joseph, and Geronimo; however, their performances must be seen in the context of ongoing conflicts, such as the years of resistance over U.S. confiscation of Paha Sapa (the Black Hills) following the discovery of gold there in 1876 (Dunbar-Ortiz 2015, 188–189). For spectators, the embodied presence of Native performers in the arena were concrete evidence of land settled and railroads built – acreage, rivers, timber, ore deposits – somewhere in the west awaiting development. But the presence of famous indigenous personages in the *Wild West* and similar entertainments *also* sent a countervailing message, signaling that indigenous people had not "vanished" at all. Deloria argues that these performances also demonstrate Indians as modern, contemporary people, living through and into a collective future (3–14). Their presence, skill, and vitality constituted acts of resistance to the very project of conquest the production propagated.

The visceral, participatory quality of *The Drama of Civilization* animated the brutal nucleus of the frontier myth through spectacles in which induced terror became a form of entertainment. Bison, steers, and horses were a *Wild West* staple, but MacKaye also sought "real" wild animals to fill out and "animate" the simulated wilderness (Percy MacKaye 1927, 80; see also Sokalski 2007, 161–173). He harvested trees and other vegetation to dress the arena in primeval flora and borrowed a menagerie of captive animals from circus magnet Adam Forepaugh, including trained bears, moose, and a herd of antelope. Live animals coerced to dash "madly across the plains" before the specter of white civilization. "Real" Indians reenacting their defeat under the gaze of a primarily white audience, completed the picture of the land west of the Mississippi (Percy MacKaye 1927, 80). Reviewers note that "one of the most realistic prairie fires ever represented" frightened the animals into "a stampede of real horses, cattle, buffalo, elk and deer" (70). So comprehensive was MacKaye's commitment to a simulated reality that he insisted on raising the roof of Madison Square Garden to accommodate equipment for the special effects during the performance's finale in which selected audience members rode as passengers in the stagecoach during a scene titled "Attack on the Deadwood Stage." These high-stakes performance tricks provided a visceral experience for a few audience members and effectively raised the stakes for all – a vicarious reminder that "this could happen to you!"

spectators. Meanwhile, a "cyclone" devised by stage manager Nelson Waldron used four steam-powered ventilation machines and several tons of leaves (Sokalski 2007, 165–166). As if the very winds of change had blown into the arena, one reviewer describes the

> rush of air turned upon the camps of miners and troopers [that] lifted the tents from their fastenings, causing the flags to snap in the gale. Then when the storm was at its fury height, the leaves would be turned loose, sweeping the arena with terrific force.

The participating audience members inside the stagecoach would magically escape just before the scene finished.

> [With] light and cloud effects, the old Deadwood stage-coach striking a snag in the ravine and going to pieces, while the six mules fled on a dead run, with only the forward wheels, dragging the driver by the reins . . . [It] never failed to bring tremendous final curtain-calls.
>
> *(Percy MacKaye, 79)*

Scenes like the one described earlier put the audience in the role of survivor and potential heroes who had, like Cody, escaped a near-death experience and lived to tell the tale.

The reenactment of the Battle of Little Big Horn was one of Steele MacKaye's more important episodic additions to Cody's production and remained part of Cody's show until it closed in 1906. MacKaye capitalized not only on General George Armstrong Custer's cultural cache as martyr but also on Cody's own reputation as a white warrior. Cody biographer Sarah Blackstone (1986) notes that Cody used Custer's demise to reinscribe Euro-Americans' claim on the land:

> [T]he battle raged until the last soldier was killed. After a moment of silence in which the audience could take in the tableau, Cody would gallop into the arena at the head of his cowboy band, react to the battlefield scene, and doff his hat in respect for the dead. At this point he would be picked up by a spotlight, the arena lights would be lowered, and a projection of the words "Too late!" would be flashed on a screen behind him.
>
> *(20)*

The audience is left with the impression that if only Cody had arrived sooner, his martial prowess would have turned the tide for Custer. In reality, Custer's defeat by the Sioux in 1876 was of little military significance to the United States. But the Battle of Little Big Horn/Greasy Grass, occurred in the year of the first U.S. centennial, and the press quickly constructed its significance as a symbolic event – a martyrdom. As Slotkin (1985) has observed, the mythologizing of "Custer's last stand" found its way almost immediately into U.S. literature, melodrama, and pageantry as an obligatory climactic episode of the frontier story (407). This episode helped to "transform conquerors into victims" in order to justify increasing military force against Native resistance fighters (408).[24] It was a finale that played on fear, constructed Anglo-Americans as victims (and potential victims), and enlivened a sense of ongoing threat in a manner similar to the aphorism, "Remember the Alamo!" or "Never again!" In one swoop, the performance confirmed martyrdom and heroism on Custer, and, by Cody's example, charged the audience with vigilance. Absent from this lionization of Custer and Cody is any awareness of Native experiences. White agents of conquest are cast as martyrs, while the warriors whose land they invaded are dehumanized as adversaries with no rights, perspectives, or feelings that bear consideration. The pageantry of martyrdom, meant to assert a kind of blood-earned right to the land, would be used again and again to argue for Euro-American entitlement.

The *Wild West* toured Europe in the years to come, ingraining the equation between frontier ideology and U.S. identity in the imaginations of Europeans and their heads of state. According to Percy MacKaye (1927), prior to Cody's collaboration with his father, *Wild West* was "very sketchy and disjointed, wholly lacking in dramatic form – a production very much 'in the making' and an exposition 'practically devoid of imaginative or dramatic coherence'" (75–76; see also Blackstone 1988, 20). MacKaye's narrative arc remained at least partially intact until Cody's *Wild West* closed in 1906. The order of some of the scenes changed in subsequent performances, but the suggestion of a Darwinian evolutionary process from barbarous to civilized remained a constant as did the implicit justification of violence against indigenous people and nature.

No matter the version told – whether that of the pioneer battling savage wilderness or cowboys and cavalry fighting indigenous warriors – the arc of the frontier story follows a linear narrative that proceeds from "primitive" to "civilized." Expansion and domination ("taming") lead to development ("improving the land"), resulting in progress that is equated with profit ("prosperity"). This arc carries with it an operative ideology that sets the stage for extractive and industrial capitalism. The lives of frontier heroes like Cody tended to bend toward a similar end. Even as Cody performed a narrative of the Old West, he was fast becoming an architect of the New West. For Cody, the mythologized west was a commodity, a cash crop of images and stories that made him one of the most successful adventure capitalists of his day. The frontier as mythos informed and advanced an ethos that drove and defined the industrial development of the western United States. Buffalo Bill transitioned from roughrider to industrial capitalist, participating in the national project of taming, harnessing, and controlling the land in order to extract its resources. Between European tours, Cody pursued a number of investments and development plans for Big

Horn Basin, Wyoming, included irrigation projects, hotels, mining, ranching, and a railroad (Blackstone 1988, xix). He envisioned a massive irrigation project that would tap the land's potential as "prime ranching country" and attract settlement to what became Cody, Wyoming (reminiscent of Rowse's Rowseville). He mounted a campaign to construct three canals from the Shoshone River, pressuring his friends in government for funding for the canal, as well as railroad spurs between Cody, Wyoming, and Yellowstone National Park (14). Cody also organized the development of hot springs on the Shoshone River, claiming that "it is for the benefit of the whole public that these springs be given over to capital that will improve them and bring within the reach of the public their health giving qualities" (23). The spring's water, Cody argued, was going "to waste" and "should be shipped all over the country as soon as the railroad gets here. It will advertise the state" (21). He also had plans to locate a military academy near Cody, Wyoming. By 1910, there were two dams on the Shoshone – the Corbett and the Buffalo Bill – completed by the ironically named Shoshone Reclamation Project (20–21).

The power of performance

Turner formalized an idea of American *as a performance* of certain character traits and (perhaps inadvertently) encouraged Euro-Americans on the cusp of the twentieth century to imitate the fictions of western dime novels, melodramas, frontier plays, and *Wild West* shows as a way to demonstrate their certain identity. The embodied frontier of the stage also informed how Anglo-Americans understood their social relationship with the environment, even as it sanctioned policies of genocide and ecocide and promoted aggressive resource extraction. Dramas about the frontier experience aided in the construction of an imaginary west, consisting of a wide-open, empty wilderness that cried out for "use." Whether characterized by man's conquest over nature, a victory of the plow in which the pioneer is hero (e.g., *Little House on the Prairie*), or land won with the bullet (e.g., *The Lone Ranger*), a story of the hero-scout and later the heroic gunslinging cowboy, frontier stories defined American(ism) in tropes of individualism, ingenuity, and cunning. They reinforced the belief that those who exhibit these "American" characteristics have a right to the land and its riches as the means to prosperity. Both include a requisite violence issued from a sense of white Americans' "natural" entitlement to the land. In many ways, the story of pioneer settlement was the sequel to violent conflict, military occupation, and Indian removal. State-sponsored genocidal and ecocidal violence coupled with the belief in the righteousness of conquest that enabled general acceptance of Turner's concept of "free land" without debate. In effect, Turner's reasoning – his rhetorical erasure of the conquest – can only take place in the "empty space" that was a product of that state-sponsored genocidal violence. Insofar as theater – as a profession and cultural force – enlivened and propagated the stories that gave rhetorical support to this imperialist project, it is complicit in it. The long-range effects of those stories of conquest are both ecological and social. When genocidal and ecocidal projects were occurring in the world beyond the stage, *Horizon*'s cavalier treatment – even as tongue-in-cheek humor – shows how theater advanced the project of Manifest Destiny. Consequently, any attempt to use theater to undo or critique this history then must be cognizant of the ways in which theater continues to be too easily weaponized.

Like the shadow of the moon during a full eclipse racing across the land at 2,000 miles per hour, the sweep of social and ecological change that came with colonialization fell with seeming cosmic speed, initiating a catastrophe of human making that would mount into the next century.[25] The loss of so many diverse, ecologically sustainable, human communities was not only an incalculable human and cultural loss; it also signaled a loss for the land and its sustaining ecological systems. By the late nineteenth century, the impact of Euro-American culture on the land had begun to show, and the notion of the west as an endless repository of resources was already thinning. The fur trade in the Pacific Northwest and the 1849 gold rush in California had come and gone. The bison herds were gone; even the cattle industry, which filled the voided plains, saw a boom and bust by end of the century. By the 1890s, most of the high-grade gold deposits in California and Nevada had been blasted out, leaving scars on the land and debris in the rivers. Conflicts over water rights between farmers and mining companies accelerated as miners blasted mountainsides into rivers and streams, which, in turn, deposited those sediments on crops and orchards. More complex processes extracted silver, copper, iron, lead, and new gold deposits, but chemical agents used in processing the toxic metal bullion ended up in groundwater. The forests of the Great Lakes region had been overcut, and erosion was beginning to threaten water supplies of some eastern farms and towns. The timber industry was moving into California, Oregon, and Washington, where it depended on big capital and the railroads.

Ecological stresses and social upheaval were being felt across the continent. On the east coast, immigrants streamed into the county looking for work and housing. In the 1880s and 1890s, approximately 10 million new immigrants came to the United States fleeing economic hardship. Part of a burgeoning American industrial sector, these immigrants worked for low wages in inhumane conditions (see, e.g., Zinn 2005, 262–265). Factory effluents and human sewage were polluting air and water in urban areas, making cities death traps of disease and despair (Malone and Etulain 1989, 2). A spirit of rebellion was fomenting across the country, one which cannot be reasoned away Turner-style as the requisite regression to lawlessness born of the frontier experience. This rebellion was, instead, the result of a pattern of industrial capitalist development that applied the same extractive policies to workers as it did to the land itself.

Bloody Bloody Andrew Jackson reprise

The American frontier narrative reinvents itself like a bacterial infection made stronger by each successive round of antibiotics. It continues to inform American behaviors, values, policies, and institutions in our present moment. Limerick (1994) observes that "the term 'frontier' blurs the fact of conquest and throws a veil over the similarities between the story of American westward expansion and the planetary story of the expansion of European powers" (75). The frontier constitutes a collective mythos and ethos still operative on Wall Street; in Washington, D.C.; and on Broadway. U.S. history of the twentieth century has not only refused to recognize the lie at the center of the frontier ethos but continues to export the narrative across the globe. The American exceptionalism that disguised the conquest of indigenous lands also disguises the Americanizing of the world. And both projects are fueled by the labor of colonized peoples.

We cannot take responsibility for what we refuse to acknowledge. What we *can* do, however, is to use the theater to expose ongoing settler colonialism, and to call for reparations and new visions. Some settler nations have made attempts to acknowledge and repair historic wounds, albeit sometimes with considerable internal or international pressure. Both Australia and Canada have issued formal proclamations of apology to their respective aboriginal peoples; New Zealand incorporates Maori social structures and cultural concepts into its official conservation strategy. Yet the United States has resisted admitting, much less mitigating, historic and ongoing injury to indigenous communities within its borders.

Rhiana Yazzie (2014), artistic director of the New Native Theatre in Minneapolis, was optimistic when she first heard about *Bloody Bloody Andrew Jackson* (2014):

> Maybe it will acknowledge the thousands of Native Americans he killed. . . . What a wonderful opportunity and contribution to American theater to see a play responsibly take up these important issues, issues that have determined Native American inclusion and access. . . . But that's not what happens, instead [it] reinforces stereotypes and leaves me assaulted, manipulated and devastatingly used.

Yazzie argues that spinning false histories – even on stage – contributes to the perpetuation of narratives that *continue to do harm*: "Andrew Jackson was not a rock star . . . the 'Indian Removal Act' is not a farcical backdrop to some emotive, brooding celebrity. Can you imagine a show wherein Hitler was portrayed as a justified, sexy rockstar?" Indigenous people remain targets of racial violence today, and inaccurate or generalized history, coupled with racist tropes, perpetuates cycles of injury, erasure, and distortion.

History, Yazzie (2014) points out in her critique, is always the living backdrop to representation. She notes that as part of the choreography for the "Ten Little

Indians" musical number, for example, ten Indians, played by white actors, are executed in ten different ways:

> [T]he last death is a hanging. Wow. How does that land here in Minnesota? Our state holds the record for the largest mass hanging in U.S. history when 38 Dakota men were executed. . . . There has to be a better way to make a political point. The first step is to be smarter about your subject matter. Learn about the culture [or place] you're trying to make a point about. Do not stand idly by while history is whitewashed and Native culture – already imperiled by hundreds of years of misrepresentation – is further debased as a theatrical device.

To ignore the responses of those who are still the targets of settler colonialism's injury is an extreme form of white privilege. One of the central aims of ecodramaturgy, then, must be to take responsibility for a history of performance that participated in and was complicit with the project of Manifest Destiny, including state-sponsored acts of genocide, land theft, and the generations of breakage (both personal and ecological) that have since come to pass. Native theater artists have been working hard at this and yet – if *Jackson* is any signal – theater by, for, and about Euro-Americans has a lot of catching up to do. Stereotypes, even in satire, are offensive to those they depict; in the collective imagination, those stereotypes tend to reinscribe even when the artists may have sought to convey irony.

It is incumbent that theater folk, who (like I am) are descendants of European settlers, take active and vocal responsibility for the frontier narrative that has wreaked havoc on lives and land. One of the ways to do this is to better understand the facts of history in contrast to the hegemonic narratives that have framed that history. Those old stories have preserved settler entitlement for five centuries. Artists must recognize and help disarm the exploitive narratives that live and breathe around us to this day. If we create work to satirize the past, we must be sure that our meanings are landing justly and compassionately for those who have been most injured, and for those whose histories we also necessarily represent, so that we stop the cycles of violence that this story of origin has wrought and perpetuated. Daly, Cody, and Turner, along with the many other storytellers of the late nineteenth century, together shaped a national ideology that is collectively maintained and operative today. The stagecoach logo of Wells Fargo Bank, the horse and rider of Marlboro cigarettes, and the bull of Merrill Lynch perpetuate its existence, like bits of cultural DNA, by carrying the full load of the narrative's ideology and instruction about what it means to be American. Public figures also invoke the frontier narrative's symbolic power intentionally as well as inadvertently. Indeed, both Slotkin (1992) and Dunbar-Ortiz (2015) have observed the myriad ways in which the role of Indian gets transferred and applied over time to new enemies of the nation-state, be they indigenous, black, immigrant, Muslim, or female. Through its aggregate of images, ideals, values, associations, historical pasts, and collective fictions, the ideology of the frontier will continue to do violence to people and lands until it is consciously and conscientiously dismantled by those who benefit from living under cover of its tales of bravado.

Notes

1. As Brantley describes, the musical began as a concert version and formed part of the Public Theater's 2008 "Lab" season.
2. In *Mythologies,* Roland Barthes (1972) describes "the language of myth," by which ideology becomes ingrained through the cultural process of mythologizing. This mythologizing, in turn, works to naturalize ways of understanding the world within popular belief motifs, images, language, and narratives shared by people who claim a common identity. In this way, Barthes maintains, certain big ideas or widely held beliefs live on as stories or master narratives in wide swaths of the public imagination.
3. To date, no extant script of the *Wild West/The Drama of Civilization* has been found. Detailed descriptions, however, can be found in Percy MacKaye (1927), J.A. Sokalski (2007), and Frank Christianson (2012), which also contains photo illustrations from the production that toured England during the 1887–1888 season. Since the England tour directly followed the Winter 1886 production at Madison Square Garden (the subject of this chapter), see Cody's account of the show's elements in Christianson's edition for details that might otherwise be gleaned from a published script. Sokalski is also a critical resource for apprehending the visual power of MacKaye's transformation of Cody's *Wild West* show. Finally, William Brasmer (1977) recognized that the longevity and narrative structure of Cody's show should be credited to Steele MacKaye.
4. Throughout this chapter, I use the term *Indian* when referring specifically to the white social construction and *Native, indigenous,* or specific tribal names when referring to peoples and tribes. For recommendations on the preferred use of these terms, see Michael Yellow Bird (1999), 1–2.
5. For a more detailed analysis of the frontier myth, see Richard Slotkin's three-volume work (1973, 1985, 1992).
6. In 1886, Cody played in a revival of *The Prairie Waif* alongside one of his *Wild West* stars, Buck Taylor. Other plays in which he performed include such favorites of the day as *The Prairie Waif*, *May Cody; Lost and Won*, *The Knights of the Plains*; *Buffalo Bill's Best Trail*; *Buffalo Bill at Bay*; and *The Pearl of the Prairies*. While these titles are not known today, late twentieth century television westerns, such as *Little House on the Prairie*, *Bonanza*, and others, continued in their tradition. For more context around Cody's stage trajectory, see Donald Russell (1960), 17.
7. With chilling effect, Kato Buss (2011) has unpacked Cody's *Red Right Hand* in light of its historical context, Cody's celebrity, and the "tangled mess of performed identities" that complicated the interface between the reality and fiction in the theft of the Black Hills (134).
8. Mary Austin and Helen Hunt Jackson's *Century of Dishonor* (1871) remains an important example of activism of the era by Anglo-Americans in solidarity with indigenous resistance to U.S. government mistreatment, abuse, and treaty violation.
9. Anglo-Saxonism – the philosophy of white supremacy that defined progress as an evolutionary process leading to increasingly complex forms of civilization, and thus sees Anglo-Saxon nations as the pinnacle – originally emerged from England's dominance during the period of European colonization, in which England and its military mastery of the sea sought control of global commerce for the purpose of accumulating wealth into its national coffers.

    Historian Richard Hofstadter (1992) describes this thinking as a belief that the spread of English culture, language, and settlement "was the Manifest Destiny of the race.

Every land on the globe that was not already the seat of an old civilization should become English in language, tradition, and blood. . . . From the rising to the setting sun, the race would hold the sovereignty of the sea and the commercial supremacy" (177).

For a discussion of ever-shifting constructions and conceptualizations of "whiteness" from colonization of the Americas to the late twentieth century, see Judy Helfand (n.d.).

10 By 1885, the influential writings and lectures of John Fiske and Rev. Josiah Strong in the United States had fused Spencer's social application of Darwinian theory with white Christian nationalism. As Hofstadter notes, the line between a vision of a democratic Christian society and militant imperialism was thin. For a fuller sense of this rhetoric, see Hofstadter (1992), 177–178. Popular culture of the nineteenth century is also replete with rhetorical and visual references and allusion to Darwin's theories. See, for example, Jane R. Goodall (2002).

11 *Horizon* is only one of numerous dramas that ingrained the frontier as the nation's sustaining mythos during the late-blooming but explosive U.S. industrial revolution. See Frank Murdock's *Davy Crockett* (1872) for an example of another play that followed *Horizon*'s model.

12 For more on the role of these constructions, see Vine Deloria, Jr. (1988) and Richard Slotkin (1992); for more about the role of labor in western expansion, see Richard White (1993) and Howard Zinn (2005).

13 Anecdotally, my grandfather, Joe Parsons, made his career in Vaudeville of the 1920s, and I recall my mother telling stories of his greasepaint assortment containing "Indian Red" and "Chinese Yellow." Even as an undergraduate in theater in the 1970s and 1980s, I was required to purchase a Ben Nye makeup kit that came with a similarly identified phenotype options.

14 In this subplot, *Horizon* echoes James Kirke Paulding's *The Lion of the West: Retitled The Kentuckian, or, A Trip to New York; a Farce in Two Acts* (1954). First performed in 1831, *The Lion of the West* centers on an English nobleman who pursues the daughter of an eastern businessman into the then frontier of Kentucky. In Daly's story, the eastern merchant becomes a western railroad developer, and the backwoods Nimrod is replaced by the military issue Captain Alleyn.

15 The bison of North America, commonly called "buffalo," numbered in the millions prior to colonization. Buffalo, a distinct species, is found in parts of Asia and Africa (White 1993, 216–220).

16 The vast majority of railroad investments came from New York rather than from Philadelphia or Boston (White 1993, 246–258).

17 In *The Drama of Landscape* (1998), Garrett A. Sullivan, Jr., describes a notion of "absolute landscape" as a relationship between people and land that emerged with the "transition from precapitalist to capitalist conceptions of property" (8–9). Sullivan observes that when "rights in land became more absolute and parcels of land became more freely marketable commodities, it became natural to think of the land itself as property . . . [as] portions of commercial capital . . . that were exchanged in the market" (14).

18 Theater historian Rosemarie Bank observes that characterizations of Indians were born from white subjectivity and reinscribed the centrality of white culture. As such, we cannot view them as a representation of an Indian "self" but rather as part of a larger story, a master narrative that helped to justify genocide on North American soil as a necessary component to U.S. national destiny. See Bank (1993, 2011).

19 For more discussion of the semiotic gendering of landscape, a topic that I address in Chapter 2, see Carolyn Merchant's *The Death of Nature: Women, Ecology and the Scientific Revolution* (1983).

20 Hobson Quinn (1927) reinforces this identification in his description of landscape in *History of American Drama* (14).

21 On the idea of property, see also Dunbar-Ortiz (2015), 34–39, 205–208, 218–236.

22 In 1886, Steele MacKaye wrote the following to his wife: "There is a battle about to be fought between Capital and Labour that is of much importance to the right and wholesome progress of the race" (Percy MacKaye 1927, 70). In Chapter 3, I discuss in more

detail the struggle between those who sought to exploit the land for profit and those on whose labor that profit depended.
23 See, for example, Chauncy Yellow Robe's (1914) "The Menace of the Wild West Shows," 39–40. Yellow Robe does not discuss the experiences of Native performers in Cody's show, but he does suggest that the continuance and proliferation of Wild West shows are "a detriment to the progress of the American Indian" and a "curse." He also cites occasions in which Native children were killed in mock battles with white children playing cowboys and Indians in Wild West fashion.
24 This semiotic identification between Custer and William Cody was rooted in fact. In 1872, before Cody's career in show business began, he served as guide to a hunting party for the Grand Duke Alexis of Russia. A number of U.S. brass, including Generals Sheridan and Custer, accompanied him. The hunt was widely publicized, and press made much of the visual as well as the mythical doubling of the two famous Indian fighters, Custer and Cody. After the hunt, Custer began to dress in Codyesque fashion. Slotkin (1985) notes that Cody was key in the cultural mythologizing-as-martyrdom of Custer's death at Battle of Little Bighorn/Greasy Grass. Cody began to modify "his own appearance and dramatizing his adventures in a manner designed to emphasize his Custer likeness and his close relation to the Custer story" (408).
25 The speed of the moon's shadow falling over the earth in a solar eclipse varies, depending on geographic location. The 2017 eclipse, for example, moved at speeds between 1,100 and 5,000 mph, according to Weather.com.

References

Archuleta, Margaret, Brenda J. Child, K. Tsianina Lomawaima, and Heard Museum. 2000. *Away from Home: American Indian Boarding School Experiences, 1879–2000*. Phoenix, AZ and Santa Fe, NM: Heard Museum; Distributed by Museum of New Mexico Press.
Austin, Mary, and Helen Hunt Jackson. (1871) 2001. *Century of Dishonor*. Scituate, MA: Digital Scanning.
Bacon, J.M. 2019. "Settler Colonialism as Eco-social Structure and the Production of Colonial Ecological Violence." *Environmental Sociology* 5 (1): 59–69. https://doi.org/10.1080/23251042.2018.1474725
Bank, Rosemarie K. 1993. "Staging the 'Native': Making History in American Theatre Culture, 1828–1838." *Theatre Journal* 45 (4): 461–486. www.jstor.org/stable/3209016
Bank, Rosemarie K. 2011. "'Show Indians'/Showing Indians: Buffalo Bill's Wild West, the Bureau of Indian Affairs, and American Anthropology." *Journal of Dramatic Theory and Criticism* 26 (1): 149–158. https://doi.org/10.1353/dtc.2011.0016
Barthes, Roland. 1972. *Mythologies*, translated by Annette Lavers. New York: Hill and Wang.
Blackmon, Douglas A. 2009. *Slavery by Another Name: The Re-Enslavement of Black Americans From the Civil War to World War II*. New York: Anchor Books.
Blackstone, Sarah. 1986. *Buckskins, Bullets and Business: A History of Buffalo Bill's Wild West*. New York: Greenwood Press.
Blackstone, Sarah. 1988. *The Business of Being Buffalo Bill: Selected Letters of William F. Cody, 1979–1917*. New York: Praeger.
Brantley, Ben. 2009. "Old Hickory, That Emo Punk, Singing and Dancing to Fame." *New York Times*, May 17.
Brasmer, William. 1977. "The Wild West Exhibitions and the Drama of Civilization." In *Western Popular Theatre*, edited by David Mayer and Kenneth Richards, 133–156. London: Methuen.

Buss, Kato. 2011. "Performance on the Plains: Staging the Great Sioux War in Buffalo Bill's *Red Right Hand*, 1876." *Journal of Dramatic Theory and Criticism* 26 (1): 129–139. https://doi.org/10.1353/dtc.2011.0010

Cajete, Gregory. 2000. *Native Science: Natural Laws of Interdependence*. Santa Fe, NM: Clear Light.

Calloway, Colin G., ed. 1996. *Our Hearts Fell to the Ground: Plains Indian Views of How the West Was Lost*. Boston, MA and New York: Bedford/St. Martins.

Christianson, Frank. 2012. *The Wild West in England*. Papers of William F. "Buffalo Bill" Cody. Lincoln, NE: University of Nebraska Press.

Cody, William F. 1978. *The Life and Times of Hon. William F. Cody, Known as Buffalo Bill, the Famous Hunter, Scout, and Guide: An Autobiography*. Lincoln, NE: University of Nebraska Press.

Cronon, William. 1996. "The Trouble with Wilderness." In *Uncommon Ground: Rethinking the Human Place in Nature*, 69–90. New York: W.W. Norton & Company.

Daly, Augustin. 1978. *Horizon*. In *Six Great American Plays*, edited by Allan G. Halline, 335–375. New York: Modern Library.

Darwin, Charles. 1859. *On the Origin of Species by Means of Natural Selection, or the Preservation of Favoured Races in the Struggle for Life*. London: John Murray, Albemarle Street.

Deer, Sarah. 2015. *The Beginning and End of Rape: Confronting Sexual Violence in Native America*. St Paul, MN: Minnesota Women's Press.

Deloria, Philip J. 2004. *Indians in Unexpected Places*. Lawrence, KS: University of Kansas Press.

Deloria, Jr., Vine. 1988. *Custer Died for Your Sins: An Indian Manifesto*. Norman, OK: University of Oklahoma.

Dunbar-Ortiz, Roxanne. 2015. *An Indigenous Peoples' History of the United States*. Boston, MA: Beacon Press.

Felheim, Marvin. 1956. *The Theater of Augustin Daly*. Cambridge, MA: Harvard University Press.

Fenimore Cooper, James. 1985. *The Leatherstocking Tales*. New York: Viking Press.

Friedman, Michael, and Alex Timbers. 2010. *Bloody Bloody Andrew Jackson*. New York: Music Theater International.

Gast, John. 1872. *American Progress*. Painting. Autry Museum of the American West, Griffith Park, LA.

Goodall, Jane R. 2002. *Performance and Evolution in the Age of Darwin*. London: Routledge.

Gram, J.R. 2016. "Acting Out Assimilation: Playing Indian and Becoming American in the Federal Indian Boarding Schools." *American Indian Quarterly* 40 (3): 251–273. https://doi.org/10.5250/ameriniquar.40.3.0251

Hall, Roger A. 2001. *Performing the American Frontier: 1870–1906*. Cambridge: Cambridge University Press.

Helfand, Judy. n.d. "Constructing Whiteness." *Race, Racism, and the Law*. https://racism.org/articles/race/66-defining-racial-groups/white-european-american/378-white11a2

Hofstadter, Richard. (1944) 1992. *Social Darwinism in American Thought*. Boston, MA: Beacon Press.

Huhndorf, Shari M. 2001. *Going Native: Indians in the American Cultural Imagination*. Ithaca, NY: Cornell University Press.

Jennings, Julianne. 2012. "The History of 'Ten Little Indians': How Did the Genocidal Nursery Rhyme Come About?" *Indian Country Today*, October 11.

King, Thomas. 2005. *The Truth About Stores: A Native Narrative*. Minneapolis, MN: University of Minnesota Press.

Kirke Paulding, James. 1954. *The Lion of the West: Retitled The Kentuckian, or, A Trip to New York; a Farce in Two Acts*, edited by James N. Tidwell. Stanford, CA: Stanford University Press.

Levine, D.M. 2010. "Native Americans Protest 'Bloody, Bloody Andrew Jackson'." *Politico*, June 24.
Limerick, Patricia Nelson. 1994. "The Adventures of the Frontier in the Twentieth Century." In *The Frontier in American Culture*, edited by James R. Grossman, 66–102. Berkeley, CA: University of California Press.
MacKaye, Percy. 1927. *Epoch: The Life of Steele MacKaye, Genius of the Theatre*. New York: Boni and Liveright.
Malone, Michael P., and Richard W. Etulain. 1989. *The American West: A Twentieth-Century History*. Lincoln, NE: University of Nebraska Press.
Marx, Leo. 1964. "The Machine." In *The Machine in the Garden: Technology and the Pastoral Idea*, 145–226. New York: Oxford University Press.
Merchant, Carolyn. 1983. *The Death of Nature: Women, Ecology and the Scientific Revolution*. San Francisco, CA: Harper Collins.
Murdock, Frank. 1972. "Davy Crockett." In *Davy Crockett & Other Plays*, edited by Isaac Goldberg and Hubert C. Heffner. Princeton, NJ: Princeton University Press.
Nash, Roderick. 2001. *Wilderness and the American Mind*, 4th ed. New Haven, CT and London: Yale University Press.
Quinn, Arthur Hobson. 1927. *A History of American Drama*, 2 vols. New York: Harper Brother Publishers.
Review. "'Horizon,' at the Olympic Theatre." 1871. *New York Times*, March 23.
Russell, Donald. 1960. *The Lives and Legend of Buffalo Bill*. Norman, OK: University of Oklahoma Press, 1960.
Russell, Donald. 1961. *The Wild West: A History of the Wild West Shows*. Fort Worth, TX: Amon Carter Museum.
Sayre, Gordon M. 2005. *The Indian Chief as Tragic Hero: Native Resistance and the Literatures of America, from Moctezuma to Tecumseh*. Chapel Hill, NC: The University of North Carolina Press.
Slotkin, Richard. 1973. *Regeneration Through Violence: They Mythology of the American Frontier, 1600–1860*. Norman, OK: University of Oklahoma Press.
Slotkin, Richard. 1985. *The Fatal Environment: The Myth of the Frontier in the Age of Industrialization, 1800–1890*. Norman, OK: University of Oklahoma Press
Slotkin, Richard. 1992. *Gunfighter Nation: The Myth of the Frontier in the Twentieth Century*. Norman, OK: University of Oklahoma Press.
Smith, Henry Nash. 1950. "The Myth of the Garden and Turner's Frontier Hypothesis." In *Virgin Land; the American West as Symbol and Myth*, 150–160. Cambridge, MA: Harvard University Press.
Sokalski, J.A. 2007. "Civil War Cycloramic Oratory and *The Drama of Civilization*." In Pictorial *Illusionism: The Theatre of Steele MacKaye*, 131–178. Montreal and Ithaca, NY: McGill-Queen's University Press.
Stegner, Wallace. 1992. *Beyond the Hundredth Meridian: John Wesley Powell and the Second Opening of the West*. New York: Penguin Books.
Stone, John Augustus. 1941. "Metamora, or, The Last of the Wampanoags: An Indian Tragedy in 5 Acts." In *Metamora & Other Plays. America's Lost Plays*, Vol. 14, edited by Eugene Richard Page. Princeton, NJ: Princeton University Press.
Sullivan, Jr., Garrett A. 1998. *The Drama of Landscape: Land, Property, and Social Relation on the Early Modern Stage*. Stanford, CA: Stanford University Press.
Turner, Frederick Jackson. [1921] 1956. "The Significance of the Frontier in American History," in *The Turner Thesis: Concerning the Role of the Frontier in American History*, edited by George Rogers Taylor, 1–18. Revised Edition. D.C. Heath and Company, Boston
Twain, Mark, Bret Harte, Frederick Anderson, Mallette Dean, Vivien Dean, and Perry G. Davis. 1961. *Ah Sin: A Dramatic Work*. San Francisco, CA: Book Club of California.

Wadsworth Longfellow, Henry. 1855. *The Song of Hiawatha*. Boston, MA: Ticknor and Fields.

White, Richard. 1993. *It's Your Misfortune and None of My Own: A New History of the American West*. Norman, OK: University of Oklahoma Press.

White, Richard. 1994. "Frederick Jackson Turner and Buffalo Bill." In *The Frontier in American Culture*, edited by James R. Grossman, 7–66. Berkeley, CA: University of California Press.

Winslow Martin, Edward. 1873. *History of the Grange Movement: Or, the Farmer's War Against Monopolies; Being a Full and Authentic Account of the Struggles of the American Farmers Against the Extortion of the Railroad Companies; With a History of the Rise and Progress of the Order of Patrons of Husbandry, Its Object*. Philadelphia, PA: Chicago National Publishing Company.

Wister, Owen. 1945. *The Virginian*. New York: Grosset & Dunlap.

Worster, Donald. 1992. *Under Western Skies: Nature and History in the American West*. Oxford: Oxford University Press.

Yazzie, Rhiana. 2014. "New Native Theatre Protests *Bloody Bloody Andrew Jackson*." *Minneapolis Star Tribune*, June 14.

Yellow Bird, Michael. 1999. "What We Want to Be Called: Indigenous Peoples' Perspectives on Racial and Ethnic Identity Labels." *American Indian Quarterly* 23 (2): 1–2. https://doi.org/10.2307/1185964

Yellow Robe, Chauncy. 1914. "The Menace of the Wild West Shows." *The Quarterly Journal of the Society of American Indians* 2: 39–40.

Young, Harvey. 2013. *Theatre & Race*. New York: Palgrave Macmillan.

Zinn, Howard. 2005. *A People's History of the United States*. New York: Harper Perennial.

2
THE SABINE WILDERNESS AND PROGRESSIVE CONSERVATION

David Belasco's *The Girl of the Golden West* and William Vaughn Moody's *The Great Divide*

In Act II of William Vaughn Moody's (1906) play *The Great Divide*, Ghent, an outlaw-turned-entrepreneur, has hired an architect to remodel his rustic cabin into a spacious home. He has planned this remodel as a surprise for his wife, Ruth. In the following scene, Ghent reviews blueprints with the Architect, while the Contractor looks on:

ARCHITECT: I have followed your instructions to the letter. I understand that nothing is to be touched except the house.
GHENT: Not a stone, sir; not a head of cactus. Even the vines you've got to keep exactly as they are.

The Contractor cautions that, given those constraints, they will have to hold off the workmen with a shotgun. How can a man build a home without some destruction or displacement of other living things, without changing the land? The Architect is as perplexed as the Contractor is indignant, but responds diplomatically: "As you please, Mr. Ghent. The owner of the Verde mine has a right to his whims, I reckon" (59). As a metaphor for nation-building after the close of the frontier, Ghent's blueprints for remodeling his rustic cabin into the grand home he envisions allude to a larger public debate about the value, use, and management of land – particularly land west of the Mississippi. Ironically, while Ghent instructs the Architect to be careful of cactus plants and stones, his own Verde Mine would have been extracting gold from a hard-rock mine in Arizona, using processes that ruined topsoil and polluted groundwater, to say nothing of the ill effects on nearby cacti.

The Progressive era was characterized by the struggle for control of natural resources. Much of the land west of the Missouri River was still under the control of the federal government; it was the literal and figurative stage where those

DOI: 10.4324/9781003028888-3

struggles played out. In the early twentieth century, environmental and human consequences of industrial capitalism became increasingly visible as U.S. industries, fed by new ore and oil deposits, improved mechanization, and the transcontinental railroads, surged to catch up with European markets. Immigrant and working-class families struggled in unhealthy conditions in the eastern cities, while the boom-and-bust *modus operandi* depleted resources and despoiled lands across the United States. The gold rush was over as fast as it had begun, leaving dredged mountainside and piles of waste despoiling watersheds; the timber industry in northern states got its first taste of deforestation and accompanying erosion; the fur trade faced near collapse as some species had been hunted to near extinction; cattle on the plains had seen its own boom and bust as the healthy grasslands, once sustained by bison, declined. The United States struggled, then as now, with questions about the relationship between resource exploitation and the collective good; about how to balance present needs with those of future generations. Preservationists argued for the aesthetic and recreational value of land, while conservationists stressed the utilitarian value of land and calling for the protection of natural resources. The conservation movement of the Progressive era – often called first-wave environmentalism – sought to slow the grab-and-greed approach of extractive industries, raising questions about who would make decisions about land use: private landowners and businessmen like Ghent or government experts? Applying scientific principles to the management of land and resources would, conservationists argued, ensure that those resources remained in abundance for society's use in the present and the future.[1] Progressive conservation did not refute frontierism but sought rather to retool it through applied science and government management. Neither side considered the sovereignty of the indigenous peoples or the intrinsic rights of the land.

Theatrical representations of the west on stage reflected the sentiments and values that informed the emerging conservation movement. At the turn of the century, stories of making a home in the new land gained appeal alongside those that retold tales of conquest. Augustus Thomas's *Alabama* (1891) sought to heal Civil War wounds, while *In Mizzoura* (1893) and *Arizona* (1899) recounted stories of settling, and critics praised Thomas's dramas for their "regional" sensibility (see, e.g., Hobson Quinn [1927]; Montrose and Brown [1934]; and Mordden [1981]). Even as conflicts faded from newspaper headlines, the exploitation of indigenous people as subjects of curiosity continued in a variety of venues, expositions, and sideshows; meanwhile, white constructions of indigenous characters in plays continued on stage. The litany of plays and other performances that perpetuate settler constructions of indigenous people at the turn of the century included Colonel Frederick T. Cummins's *Indian Congress*, which toured from 1899 through 1903, playing first at the 1901 Pan American Exposition and later at Madison Square Garden (Russell 1961, 64–70). Geronimo and Chief Joseph were among the numerous prisoners of war to be exhibited at the Louisiana Purchase Exposition in St. Louis in 1904, and William Cody's *Wild West* continued to tour through 1906 (Christianson 2017, 1–12). Meanwhile, some dramatists looked back and seemed to acknowledge, if not condemn outright, the patterns of land theft, violation, and displacement. David Belasco's (1936) *The Rose of the Rancho*, to name one example, explores the occupation of Spanish California by white settlers.[2] In a similar manner, a number of stage plays revisited the earlier white construction of the "noble savage," such as William DeMille's *Strongheart*, which appeared in 1905, and Mary Austin's *The Arrow Maker* in 1911. Perhaps the most striking example of these so-called Indian plays was the popular *Squaw Man* by Edwin Milton Royle, which appeared in 1906 (Mordden 1981, 46–47).[3] Viewed as a kind of sequel to Daly's *Horizon* (discussed in the previous chapter), *Squaw Man* recounts a tale of an Englishman who flees into the American West after a business deal goes sour, where he discovers a new identity and falls in love with an "Indian maiden." As the play unfolds, she saves his life and later takes her own in an act of despair over the departure of their son back to England. While these plays, and others, express a kind of "settler grief" about a history of devastation and loss, they nonetheless did not represent white resistance to the project of colonization. Instead, they tended to solidify (often through the death of a central indigenous character) the nonnegotiable price of (white) civilization's advancement.

The two plays examined in this chapter, David Belasco's *Girl of the Golden West*, produced in 1905, and William Vaughn Moody's *The Great Divide*, produced the following year, were among the most popular plays of the era, reflecting the tensions between romanticization of a lost frontier past and the ongoing projects of settlement, industrial expansion, and extractive capitalism. Belasco's *Girl* is a romantic look back to the California gold rush era; it represents a kind of a nostalgic reinforcement of a frontier mythos. Moody's story of a national divide, on the other hand, serves as an answer to Belasco's drama through his posing of troubling questions about the problem of making a home atop the legacy of a violent past. Critics heralded *Divide* as the first real "American drama" – that is, a play that is neither sentimental melodrama nor a mere replica of European drama. As a diptych focused on the resource-rich lands of the western United States, *Girl* and *Divide* together reflect changing public values about that land.

Belasco's *Girl* comes to a close as the lovers travel east, ostensibly toward civilization and a new beginning. The union between Johnson and Minnie is consensual, and the direction of their journey indicates that they plan to proceed along the path of civilized norms. Moody, on the other hand, begins *The Great Divide* where Belasco's left off. Ghent and Ruth's marriage comes about through theft and violation, and throughout the play these two characters struggle to cope with their own story of origin.

The newcomer's Eden

During the Progressive era, wilderness enthusiasts increasingly found themselves at odds with proponents of the newer ideas of conservation-for-use. The two viewpoints, which nevertheless shared an unquestioned sense of Anglo-American entitlement, differed mainly in their priorities regarding the purpose and use of land. Both views emerged in sync with long-ingrained settler stories about what and whom land was for. Before exploring how theater participated in the conservation/preservation debate, this section sets the historical and analytical stage to show the way perception, behavior, and policies are saturated with underlying cultural narratives.

Progressive conservation, like the era of westward expansion that preceded it, was steeped in stories that Euro-Americans brought with them, stories that shaped their sense of relationship with the land. Foremost among those stories, the Judeo-Christian story of the expulsion from the Garden of Eden in which the male god Yahweh bestows a world replete with abundance on a man and woman. All they need for their sustenance is there in the Garden. The god commands the people not to eat from the tree of knowledge, but when tempted by a serpent the woman eats the forbidden fruit and then offers it to her mate, who also eats it. On news of this offense, the godhead banishes the man and woman from the Garden, cursing them to earn their sustenance by the sweat of their brow (Gn 3:23). The story then follows the couple's offspring into generations of conflict as the decedents of these original humans work to recover and restore the Garden they had lost. This lapsarian narrative (i.e., focused on a lapse or break in trust between the god and the people) sets human beings on a generations-long path of recovery or *reclaiming* their edenic state through human toil (Merchant 1996, 132–159). Like a layer of geologic sediment that lies below the U.S. relationship with the land, this origin story seeps up carrying images and values not only about dominion and entitlement but also about work, industry, and technology and about gender that infused both conceptual and practical aspects of early U.S. environmental policy. The edenic narrative empowered Euro-American settlers to understand themselves as doing God's work by *reclaiming* the wilderness under the dominion of Anglo-Saxon Christendom. It is no wonder, then, that one of the early pieces of legislation to authorize massive engineering projects to contain and control western rivers – to make the desert bloom – was entitled the Reclamation Act. Passed in 1902, it authorized formation of the Bureau of Reclamation (BOR) to undertake irrigation, flood control, river development, and hydroelectric dam projects that would reclaim the "useless" arid lands of the west, making them suitable for farming and human habitation. Overall, approximately 30 BOR projects got underway in western states between 1902 and 1907 (Bureau 2018).

The newcomer's story also had particular consequences for women (and implications for the female characters in the plays discussed later), who were equated with the land in this fall from grace. In "Reinventing Eden: Western Culture as a Recovery Narrative," historian Carolyn Merchant (1996) describes the totalizing mythos of the edenic recovery story, in which woman appears in three guises. As original Eve: nature is virgin, pure, and light – land that is pristine or barren but that has the potential for development. As fallen Eve: nature is disorderly and chaotic, a wilderness, wasteland, or desert requiring improvement, dark and witchlike, the victim and mouthpiece of Satan as a serpent. And finally as mother Eve: improved nature, a nurturing earth bearing fruit, a ripened ovary, maturity (137). Within this tri-fold construction, nature (and woman) exists solely for man: to be visually admired, to offer redemption, and to be of material (reproductive) use. The idea of wilderness connoted notions of femininity as "virgin" land waiting for masculine agency to conquer, tame, and make it fruitful, as well as masculinity linked to savage wilderness' capacity to inspire fear and awe in an experience of its "sublime" – indeed, almost divine – beauty. Through this lens, pastoral land carried valences of fruitful motherhood – a product of that masculine mastery – and the notion of Man's ideal and destined home.

In a similar vein Merchant writes civilization, like motherhood, was gendered female even as it was credited to and issued from male agency. John Gast's painting, *American Progress*, for example, depicts progress as a classically robed European female figure. These kinds of associations translated into geographic stereotypes as eastern cities were feminized as hubs of genteel culture and refined sensibilities in contrast to rough-edged, harsh, more masculine qualities of the western wilderness. Yet as industrial capitalism in eastern cities increasingly compromised the ecologies of land and waterways, as well as the health of people living and working in the cities, society simultaneously blamed civilization for emasculating men, citing evidence of ill health, high crime rates, indigence, and immoral behavior. The wilderness movement and the idea of national parks arose in response to this kind of anticivilization critique, positing that men who had grown soft and weak as a result of civilization might recover their manliness by returning to the wilderness to reenact the activities of settler ancestors.[4]

Understood from within the framework of the edenic story, human beings are not merely the passive recipients of nature's benevolence but rather agents whose enterprise must restore the bounty of the Garden. Human enterprise and all its technological products thus become attempts to recover from the fallen state and to reclaim the metaphorical Garden (Marx 1964, 73–144). As Merchant (1996) explains, "[t]hese meanings of nature as female and agency as male are encoded as symbols and myths into American lands as having the potential for development, but needing the male hero" (137). The idea of "Man" as the managerial agent of change is predicated on his status as set apart from the environment he seeks to change; thus, this ideology implies and perpetuates a patriarchal and paternalistic otherizing that makes the objectification of nature not only possible but also desirable. For Progressives, technology was the means by which to reclaim the Garden from the wilderness. Both *Girl of the Golden West* and *The Great Divide* valorize technology as an expression of the very directive implicit in the Eden story – namely, recovery through human endeavor. Both plays reflect the tension between preservation and conservation as they implicitly feminize land and mask the material-ecological effects of extractive industries.

The objectification and binary thinking implicit in the edenic story became part of U.S. environmental policy during the Progressive era. Since 1860, wilderness enthusiasts had argued for the aesthetic and inspirational value of lands that possessed sublime beauty. By the turn of the nineteenth century, the United

States had established a precedent with the creation of Yellowstone and Yosemite National Parks, and numerous new parks were added (Nash [1967] 2001, 141–181; and Albright and Cahn [1985]). The early twentieth-century wilderness preservation movement galvanized around the enigmatic figure of John Muir, whose spiritual leadership, scientific study, and prosaic writing became the voice for preservation to politicians, press, and public (see, e.g. Nash [1967] 2001, 122–140]; Shabecoff [1993]; and Gottlieb [1993]). President Theodore Roosevelt also used his bully pulpit to make the case for additional national parks, claiming that Yosemite and other parks "should be kept as a great national playground. In both, all wild things should be protected and the scenery kept wholly unmarred" (Nash, 168).

Meanwhile, the Antiquities Act of 1906 authorized the president to set aside lands with particular historic and/or scientific value, protecting certain natural sites from vandalism and former indigenous cultural sites from artifact theft.[5] One aim of wilderness preservation involved freezing time to preserve a "relic" of the past. Land in its "original" state constituted a representation of American history and heritage. Parks would need sustained maintenance to preserve and accentuate their picturesque and sublime aesthetics to ensure that they continued to serve their sociocultural narrative function as representations, as vessels of community identity, and as sacred texts inscribed with cultural narratives (see, e.g., Whiston Spirn [1996]). The tendency to fix the natural world in an eternally static state not only denies the dynamic ecological systems that are alive and always in flux in the land, but also the longstanding complex interactions, or relatedness, among indigenous people and their homelands. Like their reflected images in landscape paintings and theatrical production, national parks functioned as *representations*, as narrative sites that speak back to visitors reinforcing a nationalist narrative while erasing other stories. As William Cronon (1996) writes, the first national parks and many others that followed were predicated on the forced deterritorialization of indigenous inhabitants.[6] The federal government systematically removed tribes from Yellowstone, Glacier National Parks, and other scenic reservoirs of Anglo-Saxon identity.

As the effects of industrial capitalism became increasingly visible, a new discourse arose that stressed the conservation of resources to ensure continued prosperity and economic growth. Gifford Pinchot, Theodore Roosevelt's director of forestry, sought to infuse a spirit of conservation into the national psychology. In the face of resistance from Congress and industry to any government oversight of resource use, Pinchot and Roosevelt took their crusade to the American people, preaching a "gospel of efficiency" (Hays 1999). Soon, in turn, conservation became popularly understood as the scientific development of the nation's natural resources (Pinchot [1947] 1987, 84–95). Pinchot argued that

> [t]he forest and its relation to streams and inland navigation; to water power and flood control; to the soil and its erosion; to coal and oil and

other minerals; to fish and game; and many other possible uses or waste of natural resources . . . were not isolated problems.

(322)

As a result, these tasks required a single natural resource manager: namely, the federal government. By 1905, Pinchot had succeeded in arresting control of federal lands in the west from the General Land Office, placing those lands under control of the Forest Service and later instituting a scientific forest management system that encompassed everything from mineral rights to watersheds. As president, Roosevelt brought together numerous civic commissions, which included teams of scientists and engineers as well as business leaders and government officials, to study the nation's natural resource reserves. Claiming political neutrality, these supposedly objective "experts" would bring scientific efficiency to land and resource management.[7]

Throughout this period, Roosevelt and Pinchot each impressed on businesses and lawmakers the need to protect and, more important, to efficiently utilize resource reserves. They maintained that this was the only method for ensuring a continuous supply of raw materials for the industrial growth on which the nation's future prosperity depended. In *Conservation and The Gospel of Efficiency*, Samuel Hays (1999) describes Roosevelt's public conservation message as a "problem of national efficiency, the patriotic duty of insuring the safety and continuance of the Nation" (125). In short, Pinchot and the president effectively reframed conservation as a matter of national security. Neither Congress nor local communities readily embraced Roosevelt's first programs and initiatives. In 1908, for example, Congress refused to fund Roosevelt's Inland Waterways Commission and instead passed a measure *prohibiting* the president from declaring additional national forest reserves in western states (see, e.g., Hayes, 122–174). Ironically, preservationists found allies in big business, while conservation became an argument for public (mediated by government experts) control and development of natural resources.

The debate between preservationists and conservationists came to a head between 1906 and 1914 in the battle over Hetch Hetchy dam.[8] Twenty miles north of Yosemite, but within park boundaries, Hetch Hetchy Valley was filled with granite monoliths sculpted by glaciation, plunging waterfalls, and pastoral meadows. The valley had received its own cadre of tourists, naturalists, and artists, including Albert Bierstadt who painted *Sunrise in the Hetch Hetchy Valley* in 1872 (Turner 1985, 286–287). In 1906, a devastating earthquake and fire leveled more than 500 city blocks of San Francisco, bringing the city's need for water into sharp focus and renewing efforts to secure water from the Tuolumne River with a dam and hydroelectric facility in Hetch Hetchy Valley (White 1993, 213–214). John Muir (1973) warned that the project would obliterate one of America's "natural cathedrals" and that the valley was "Yosemite's twin jewel" (192). Pinchot disagreed, arguing that the reservoir of much-needed water was the valley's best use. The controversy railed for years in the press and estranged the once-strong professional and personal relationship between Muir and Pinchot and their allies. In 1914 Congress passed the Raker Act, which authorized Hetch Hetchy Dam, and putting the public's need for water and hydropower above aesthetic and recreational use – a precedent that would authorize multiple hydropower dams on the big rivers of the western states (see, e.g., Nash [2001], 161–182). Muir died the following year, while Pinchot's thinking continued to inform U.S. policy well into the Franklin Roosevelt's administration in the 1930s. The resulting bitterness on both sides produced a lasting schism in U.S. environmental thought born of binary, reductive, and wholly anthropocentric thinking – namely, that land could connote *either* the raw material of nation building *or* a scenic wonder for visual consumption and recreational use.

In 1916, this binary was institutionalized in the formation of two distinct governmental agencies: the National Forest Service, which Congress authorized in 1905 and which President Theodore Roosevelt subsequently moved from the Department of Interior to the Department of Agriculture; and the National Park Service, which President Wilson authorized in 1916 within the Department of Interior (see, e.g., "Quick" [2018]). The legacies of the utilitarian versus scenic framework concretized in governmental structure have shaped and vexed environmental policy and popular perception of land use for the remainder of the twentieth century. In an era that lacked comprehensive environmental policies, much less precedents of environmental law, cultural narratives supported a tradition that sanctified one parcel of wilderness while sanctioning the wholesale destruction of another. In light of this pivotal moment in U.S. environmental history, I examine implicit gendering of land in *Girl* and *Divide* to demonstrate how these and other dramas of the period worked to rationalize and mask violence and violation of both.

Technology facing West

The Girl of the Golden West first opened in November of 1905 (Belasco 1929, 357). The play ran for two years at the Belasco Theatre in New York, and it earned recognition as a bridge from melodrama to theatrical realism on the American stage (Mordden 1981, 42–54). In 1910, Giacomo Puccini set Belasco's story to music, and Puccini's *The Girl of the Golden West* is still widely produced today. The Klondike gold rush of 1898 had revived interest in California's days of 1849, and *The Girl of the Golden West*, like other gold rush melodramas, was peopled with "western types" including miners, outlaws, compromised women, and the newest of the stock characters, the sheriff, as it presented a romanticized view of that era of profiteering. Set during the California gold rush of 1849–1850 and centered on the character of the bandit or rogue agent and his lover, Belasco's *Girl* echoed Charles E.B. Howe's *Joaquin Murieta de Castillo, or The Celebrated California Bandit*. Set in the Polka Saloon in the idyllic mining town of Cloudy Mountain, where the saloon's proprietress, Minnie (i.e., "the Girl") provides a sense of home for a group of rowdy miners. The Girl is, in turn, all three valences of the feminine described by Merchant. In her maternal function, she is a civilizing force in a rough and wild world; she teaches the miners to read and write; she records and safeguards the miners' gold until the Wells Fargo stagecoach arrives (and profits from the share of their earnings spent in her establishment). As one reviewer notes, Minnie is "indomitably persevering and brave under difficulties, but not without a woman's feelings when difficulty is over" (Delano 1857, 66). The plot barrels forward when the Wells Fargo agent arrives to warn the miners of a notorious bandit operating under the alias of Ramerrez. Meanwhile, a mysterious stranger in their midst offers to lead the Sheriff, Rance, to the bandit's hideout. It turns out that the stranger is really the Girl's former lover, Johnson, under the alias Ramerrez. After Johnson/Ramerrez is wounded in a gunfight, Minnie hides him upstairs in her private quarters. Meanwhile, Rance, who is also in love with her, returns to the saloon looking for the bandit. Minnie denies any knowledge, but when a drop of blood falls from the ceiling, Rance knows she is lying. In the most famous scene in the play, Minnie begs to play Rance in a hand of poker, betting that if she wins, her lover will be set free. Minnie wins by cheating, pulling an ace from her stocking. The Sheriff continues to pursue Ramerrez/Johnson through subsequent scenes that include a blizzard created with all the special effects for which Belasco was well known. When the lovers ultimately triumph, the play closes as the sun rises in the east, and the couple sets out, leaving the west to return to the civilized east. This final scene – achieved through a spectacular demonstration of theatrical technology – paints a recovery narrative in which past deeds, however violent, might be redeemed in light of a new, brighter future.

Belasco's narrative is consistent with Frederick Jackson Turner's frontier thesis. The pioneer regresses to a "primitive" or "fallen" condition but then transforms as he builds his own civilization. Belasco demonstrated this intersection at several turns. In *Girl of the Golden West*, the fallen road agent, Ramerrez, is transformed by his encounter with the Girl. Redeemed through her (i.e., returned to his "true" identity as Johnson), both the outlaw and the Girl are reclaimed. In the end they leave the wilds of the west, journeying east as man and wife. In Belasco's drama, the virginal aspect of Minnie both purifies and redeems Johnson in the end; her maternal aspect heals him, just as her ferocious nature protects him. In this case, it is Johnson who comes to Minnie's cabin, Johnson who wins her, Johnson who is changed, and Johnson who, at the end of the play, takes the goodness and beauty he has reaped in the west back to the civilized east. Yet Minnie is not without some agency. First of all, she hides Johnson in her saloon. In a later scene that reviewers commented held such suspense that "you could hear a pin drop" (66), Minnie wagers nothing less than herself. Proving herself an inventive and expedient proverbial child of Turner, Minnie pulls a pair of aces from her stocking to win the last hand. She has been won by the potential of transformation of Ramerrez (into Johnson) and his potential renews her. Minnie's "eyes fill with tears" because she knows "what I might have been . . . what we – might have been? And I know it when I look at you" (348). The audience then holds out hope that this pair – the outlaw and the prostitute – might be brought back together to *reclaim* that lost future. In these ways and others, *Girl* illustrates the Progressive conservation notion that nature invites human mastery over it; it is nature's plan, so to speak, for human beings to improve on what Progressives understood as the untapped potential of the natural world. In her relationship with Johnson, she is both the force that remakes him and the virgin terrain that is remade by him, transformer, and transformed, tamer as well as tamed.

Even as Belasco's play further illustrates the Progressive notion that nature desires human intervention, *Girl* also suggests a natural world that has life and purpose of its own. After losing the poker game, the sheriff permits Johnson/Ramerrez to stay with Minnie until he heals, warning that afterward, the bandit will need to pay for his crimes. Once healed, Johnson/Ramerrez must again flee for his life, this time into the night and a raging blizzard – the visage of nature as violent, fierce, and uncontrollable. If we consider Minnie as a signifier of wilderness, then we must take into account how she pushes back against the feminized passive land, perhaps implying that the land is also a living presence with agency and can be equally fierce, violent, and cunning – the very aspect of nature's that Progressives believed could and should be managed, and thus transformed, by man. That mastery and technological prowess were nowhere better illustrated in the play than in the production's scenography and lighting.

At the time stage productions drew on the visual iconography popularized by western landscape artists, like Albert Bierstadt, Thomas Moran, and others who sought to portray an untouched, virginal wilderness. Belasco's panorama and scenic devices, in particular, were realized in the tradition of romantic scene painters for whom the stage was a kind of living canvas. The visual opulence of Belasco's stage for the production of *Girl* reflected the influence of American landscape painting of the nineteenth century, in which paintings of real places invoked biblical narratives. The panorama that served as a prologue signified both a mythic and material landscape. The spectator's eye was drawn up Cloudy Mountain where a "little world by itself" (311) was nestled in the Sierras.

In her reconstruction of *Girl*, Lise-Lone Marker (1975) stresses Belasco's attention to nature's detail, as she describes the panorama, unfurling on vertical rollers, that took viewers along a winding trail through a ponderosa forest backed by snowcapped peaks and blue sky:

> It is night and the moon hangs low over the mountain peaks. The scene is flooded with moonlight, contrasting oddly with the cavernous shadows. The path to the cabin of the Girl is especially flooded and the light pours down from the mountain above to form this. The sky is very blue and cold. The snow gleams white on the highest peaks. Here and there pines, firs, and manzanita bushes show green. All is wild, savage, ominous.

(141)

Marker's description, like Belasco's theatrical genius, constructs an idea of wilderness in the visual vocabulary of the paintings of Moran and Bierstadt and tuned to the aesthetic tastes of eastern audiences.

In Act I, Minnie's description of her surroundings echoes Muir's religious sentiment about wilderness:

> If I want to, I can ride right down into the summer at the foothills, with miles of Injun pinks just a-laffin'- an' tiger lilies as mad as blazes. There's a river there too – the Injuns call it a 'water road' – an' I can git on that an' drift and drift, an' I smell the wild syringa on the banks – M'm! And if I git tired o' that, I can turn my horse up grade an' gallop right into the winter an' the lonely pines and firs a- whisperin' an' a-sighin'. Oh, my mountains! My beautiful peaks! My Sierras! God's in the air here, sure. You can see Him layin' peaceful hands on the mountain tops.
>
> *(357)*

Like the land grant mogul in Daly's *Horizon*, Minnie marks her world with possessive pronouns (e.g., "*my* mountains"), a discursive marker signaling that aesthetic use of land was as predicated on possession. If the beauty of the land was to remain intact, preservation advocates argued, the new U.S. nation would need to possess the land.

Reviewers characterized *The Girl of the Golden West* as Belasco's farewell to the Old West (see, e.g., Marker [1975]; and Delano [1857]), yet *Girl* was also, perhaps more significantly, an homage to the new west. Critical response touted Belasco as a theatrical "wizard" because of his advancements in special effects, especially stage lighting. At a time when technology had captured the imagination of Americans, Belasco's production was emblematic of Progressives' enthusiasm for technology. For Progressives who believed that God had designed the human species to improve on creation and that virgin land must always be wedded to human enterprise, Belasco's electrically powered stagecraft aptly demonstrated American progress.

The wonders of technology, like the fierce beauty of the High Sierra, thus took on qualities of the sublime in Belasco's roller coaster of technological effects. Simultaneously demonstrating the utility of beauty and the beauty of utility, two scenes in particular stand out as striking examples of Belasco's theatrical staging technique – the blizzard of Act II and the sunrise in the Sierras that brings the curtain down. In her reconstruction of the 1905 production, Marker (1975) argues that the blizzard scene demonstrated an environment that has "virtually assumed the role of the dramatic antagonist" (154). The storm drives the lovers into one another's arms, forming a bond that will prove as strong as the force of nature that threatens them (Belasco 1929, 349). William Winter's (1981) account of the scene, among other similar reviews, likewise reveals a fascination with Belasco's engineering genius:

> a force of thirty-two trained artisans – a sort of mechanical orchestra, directed by a centrally placed conductor who was visible from the special

station of every worker. The perfectly harmonious effect of this remarkable imitation of a storm necessitated that at every performance exactly the same thing should be done on the stage at, *to the second*, exactly the prearranged instant.

(207)

For an audience keenly interested in technology's capacity to tame the natural world, Belasco's mimetic blizzard was an apt and captivating demonstration of sublime technology in the "second nature" of artifice (205).

As eastern cities transitioned from gas lighting to electricity, and hydroelectric projects in western states had begun to promise wider electrification, Belasco showcased this new technology for the theatergoing public. On stage, electric lighting became a scenic wonder of its own as in the final tableau in which the Girl and Johnson leave the Sierras; they ride east into the rising dawn made possible by electricity. Not only a practical tool but also an instrument of imagination, artistic virtuosity celebrated electricity's metaphoric and material victory over natural cycles of day and night. Belasco experimented incessantly in scenic lighting, studying methods and equipment. He worked closely with his engineers to perfect "the right, soft coloring and tone of a distinctly California sunrise" (Marker 1975, 160). The legendary achievements of the wizard of the switchboard signified the new dawn of prosperity promised by the very technology that was used to produce the illusion. Marker describes the unfolding final scene this way:

> The blue of the fading night gradually gave way to the breaking dawn. A glimmer of the first dim rays of the rising sun was seen on the foliage of the foothills. Slowly the light became stronger, illuminating everything in the clear, soft hues of early morning and casting the two figures into bold relief against the horizon. Outcasts from paradise, on their way to a new life in the east, the lovers turned to look back at the mountains in the brilliance of the sunrise.
>
> *(160)*

The trajectory of the play's final scene moves from west to east, suggesting a kind of reverse Manifest Destiny. The rising civilization born of industrialization symbolized by the dawn is the natural destiny of the nation and the uncontestable direction of progress. The costs in human and ecological destruction, as with the real impacts of mining, are occluded in the play just as they were in the Progressives' vision of the future. Belasco's contemporaries, however, understood their moment in history as a new dawn for a new nation in a new century. The west would serve as the foundation of immense natural resources for a growing industrial nation that would take its place in the world as the pinnacle of Anglo-Saxon civilization.

In *The Girl of the Golden West*, Belasco engineered the two archetypal extremes of nature – sublime beauty and vicious wilderness. His ability to duplicate the visual effects of weather and sunlight with such precision rendered nature a seeming captive on his stage and an apt metaphor for Roosevelt's engineers and efficiency experts who hoped to bring the powers of nature under the rein/ reign of applied science. With Pinchot and W.J. McGee at the helm of policy-making, Roosevelt's "conservation crusade," which began in earnest in the final months of *Girl*'s New York production, became a moral crusade. Pinchot (1990) preached that the "first duty of the human race is to control the earth it lives upon" (86). To be clear, while Pinchot's application of conservation was a particularly American project, the ideas were part of, and influenced by, an Anglo-Saxon, European tradition in which technology would produce a modern utopia, an Arcadia on earth.[9]

Pinchot himself had studied forestry techniques in Germany and then brought them home to the United States. Conservationists of the era aligned with the rising tide of rational reformers who believed efficiency was the path to equity and that scientific planning could solve problems as diverse as urban sanitation, crowding, and poverty, as well as flooding, erosion, and drought. Moreover, advocates thought that conservation would help to curb abuse and outlaw waste of natural resources by monopolistic industrialists through the harnessing of natural forces "which belong to us all," as well as through the efficient application of applied science (Pinchot 1990, 87). In *Machine Age Ideology*, John M. Jordan (1994) observes that in their effort to address a variety of problems from a single philosophical vantage point, these rational reformers were blind to the power politics and dogma of their own position (1–10). Progressive conservation was not about safeguarding the land and its creatures and resources, nor was it about providing for the future; instead, Progressive conservation was about national power and wealth, about building a competitive edge in the already global race for resources. In Pinchot's (1990) words,

> [t]he outgrowth of conservation, the inevitable result, is national efficiency. In the great commercial struggle between nations, which is eventually to determine the welfare of all, national efficiency will be the deciding factor. So from every point of view conservation is a good thing for the American people.
>
> *(88)*

In this great commercial struggle, then, mining became both a material industry and an ideological expression of the U.S. relationship to the land.

Masking mining the West

Turner's narrative of "free land" masked the erasure of indigenous presence on the land. Similarly, Belasco's picturesque yeoman miners masked the material-ecological realities of mining for miners and for the land. Belasco's endearing and quirky cast of characters of Cloudy Mountain paints the business of mining as an innocent and even noble pursuit – involving individual miners working their claims, "sending their gold profits back to women and children in the east" (347). White (1993) notes, however, that in the extractive economy of the west "there [was] no such thing as a subsistence miner or logger" (243). Understood in this way, the single yeoman-type miners idealized in *Girl* were primarily figments of eastern fancy.

In the decade that followed the feeding frenzy of 1849, the United States witnessed a mass migration west of would-be millionaires, and stories of the gold rush celebrated the heroism of American argonauts determined to wrest their fortunes from the land. Leaving families, farms, businesses, and sordid pasts, a brotherhood of men (and a few women) journeyed overland or by sea to be outfitted in San Francisco and then head for the Sierras to stake out a claim. The individual claims of these westward travelers on the land were part of a larger national project. In *Days of Gold: The California Gold Rush and the American Nation*, Malcolm Rohrbough (1997) describes that the days of 1849 were as much about building an empire as they were about building a fortune: "[T]he representatives of the Republic went west, among other things, to reclaim the new territories acquired from Mexico, from Catholicism and despotism" and "to establish the national presence in the eyes of the Indians of the plains" (283). The project of U.S. expansion has never been separated from fortune hunting, and Progressive conservationists like Gifford Pinchot had faith in this fundamental principle, even as they preached "equitable" resource management. This relationship of the land to wealth and influence was not what conservationists hoped to change; instead, they merely wanted to shift the focus from personal to national empire building. Conservation promised, according to its foremost scientist W.J. McGee, "the use of natural resources for the greatest good for the greatest number for the longest time" (Pinchot [1947] 1987, 366). Whether the conservation approach could deliver on that promise, conservationists argued, depended on how the nation would manage its resource-rich terrain.

In the context of *Girl*, Belasco's audiences understood some of the lessons of 1849, and the inequities that resulted from the gold rush were there on stage too, like a shadow. When the California gold rush had run its course, wealth had been quickly consolidated. Within two years of Sutter's discovery, one-man sites gave way to complex mining methods, requiring both machinery and large capital investments. Forty-niners like those on Cloudy Mountain were destined, more likely than not, to work for a day's wage blasting gold from mountainsides into investor's pockets. By 1851, a handful of mining companies dominated the Sierra Nevada region, and by 1860 the rush was over and only rubble remained, along with the economic and technological infrastructure of a new west with deep socioeconomic inequities that would erupt at the end of the century (see, e.g., Rohrbough [1997], chaps. 12–13; see also White [1993], Part III).

Picturesque scenic devices that romanticized extractive industry masked the material-ecological impact of mining on land and labor. In stark contrast to Belasco's theatrical treatment, mining was anything but romantic, particularly after placer mining was replaced by syndicate-run hydraulic, hard-rock, and open-pit processes. Stamp mills ran 24 hours a day, employing workers for long hours in chemically saturated conditions. (Today, twenty-first-century extractors have learned the narrative lessons of theatrical sightlines: clear-cut timber operations in the Pacific Northwest leave a row of uncut trees along highways to mask the destruction that lies behind.)

In the early days of placer mining, when stream banks were panned out, miners rerouted streams to get at the gold in the streambed itself. Partnerships formed to build dams that moved the flow of water off its original course, exposing new digs. In a conversation with Johnson, Minnie describes the "sluices" and the men who come back from a day's work covered in mud, suggesting that Minnie's miners worked as a team, if not a formal company. Diverting streams and using early hydraulic methods, miners captured water flow, forcing it through a metal nozzle into a firehose. Miners water-blasted riverbanks with high-pressure hoses, exposing fresh ore, and washing trees, topsoil, and entire riparian zones into vast muddy moonscapes (Rohrbough 1997, 202–203).[10] Belasco's stage directions tell us accordingly that "everything down R. should suggest mines" (Marker 1975, 142). Everything below a mine site risked washing away in the next avalanche of earth and debris set loose by the hoses; consequently, miners typically lived above their operations.

By 1851, hydraulic mining was more than a one-man or partnership operation; it required equipment, capital, and a hierarchical chain of command. Called "hydrolicking," this systematized mass erosion, "took mountains and washed them into rivers," exposing ever-deeper veins of ore to the industrious miner (White 1993, 233). In an early scene in *Girl*, Minnie describes the picturesque route she rides down the mountain through ponderosa forests to the flowered meadows and riverbank. Yet, if Belasco's Cloudy Mountain was a thriving "boom" town of 1849, Minnie's pleasure ride would mean picking her way through miners' blast debris. Rohrbough (1997) notes that "[h]ydraulic mining brought a violent reconfiguration of the California landscape, and it left in its wake vast quantities of waste tailings, scars that remain today" (230). The forest that Minnie describes in *Girl*, and which was depicted on the backdrop, would have been treeless and filled with smoke within a decade. Unlike Belasco's clear western sky, smoke from the smelters in mining towns often darkened the midday sky – a sign of progress.

Those who were trying to settle the land downriver from a camp like Cloudy Mountain would have had cause to be disgruntled. The fiction of Belasco's nostalgic "little world" (311), so temporally and geographically remote from New Yorkers, disguised the impact of mining on land and lives as it shored up the so-called reclamation of western lands. By the late 1860s, the gravel and debris blasted from mines in the High Sierras had clogged rivers and caused flooding that threatened farmers; it had also created shoals in the Sacramento and other rivers, making them unnavigable. Towns and cities built dikes and levees to keep out floodwaters, but levies failed on a regular basis. In California's Sacramento Valley, "examinations by army engineers revealed that by the 1890s the debris had totally buried 39,000 acres of farmland and partially damaged another 14,000 acres" (White 1993, 233). After the levee failure in the Sacramento Valley in 1881, farmers tried to sue for damages, but they could not prove which mine had buried their orchards and fields and fouled the water. According to White, "[t]he result was a long legal and political battle in California over the environmental consequences of dumping mining debris into Sierra Nevada streams. The struggle ended in a victory for the farmers" (234).[11] Conflict over the control of water would become a fixture in the political and economic life of the arid lands west of the one-hundredth meridian.

The California gold rush was perhaps the most mythologized of the mass grabs for a single resource in the history of the west, but it was only one of a stream of discoveries that fed the continuing expansion of the United States. In 1874, one of George Armstrong Custer's men discovered gold at *Paha Sapa*, the Black Hills. As the Sioux were starved and demoralized out of their resistance over the next winters, William Randolph Hearst set up a stamp mill to mine the gold of the Black Hills (see, e.g., White [1993], 258–260; and Opie [1998], 166–167). In 1898, gold was discovered in Alaska and the "ninety-eighters" soon became quite romanticized in a postfrontier American populous that had itself begun to idolize the scenic beauty of wilderness (see, e.g., Nash [2001], Chapter 9). Considered from the vantage point of its historical moment, Belasco's love story masked a characteristically American enterprise in which *taking* was conceived as heroic.

A question of marriage

In April 1906, not yet a year after Belasco's play premiered, William Vaughn Moody's *The Sabine Woman* opened at the Garrick Theatre in Chicago. New York producers recognized a possible western romance to rival Belasco (Moody 1935, 45). Retitled as *The Great Divide*, Moody's new play opened at the Princess Theatre in October 1906, fulfilling what the *New York Times* called "the hue and cry for a representative American play" ("Hue" 1906). Reviewer John Corbin of the *New York Sun* found Moody's play "subtlety veracious," "unaffectedly strong" and "unusually rich" (1906). Two years later, *The Great Divide* was still "the present high water mark in American drama" (quoted in Henry 1934, 174). Comparisons with Belasco's *Girl of the Golden West* were common and indeed the plays share much. Both reflect popular binary thinking in which nature is either demonic (the wild savage forces of nature which must be controlled, embodied in Belasco's famous blizzard) or sublime (the virgin wilderness, reflected in the sunrise over the Sierras). The American west served as the location of a recovered Eden, a wilderness to be transformed into a productive Garden through the efficient application of technology. A central female protagonist gambles with fate in both plays – in *Girl*, Minnie plays three hands of poker to save her outlaw lover; in *Divide*, Ruth gambles her body and her destiny to save her life. Yet Belasco's romanticization of the California gold rush was understood as a nostalgic backward gaze, while *The Great Divide* was concerned with the questions of how to live on the land going forward. *The Great Divide* asked questions that *Girl* had let lie. How will the nation cope with its violent story of origin? Once claimed, how will lands and resources be used for a greater good? The personal dilemmas of Moody's central characters provoke questions about how U.S. society might move from an ethos of frontierism to one of homemaking and kinship. *The Great Divide* comes close to questioning the values and assumptions of U.S. colonialism but, in the end, reaffirms an ethos rooted in the patriarchal order.

Understanding *The Great Divide* as an American tale of mythic proportions – a kind of collective dream in which the emerging American mind confronted itself – is consistent with Moody's life and beliefs about the function of the drama in society. Moody's *Letters to Harriet* (1935), for example, reveal a man who lived life as a mythic quest, dressing everyday occurrences with metaphoric meaning, envisioning his own experiences as episodes in a meta-play of cosmic dimensions that blurred reality and imagination. Along with dramatists Percy MacKaye and Josephine Preston Peabody, as well as poets Edwin Arlington Robinson, Ridgely Torrence, and others, Moody envisioned a new poetic theater. These artists cast themselves in the role of mercurial messengers of a spiritually potent theater, one that represented an alternative to the commercialism of Broadway.

Despite its change in title, Moody's play is a story of rape in the name of empire. The ancient myth of the Rape of the Sabine Women is part of the Roman story of origin, in which Romulus, one of the twin founders of Rome, sends his men to negotiate for wives with the surrounding peoples of Sabine. When those negotiations fail, Romulus directs his men to take Sabine women against their will, abducting them to serve as wives to the Romans and thus forwarding the cause of empire. The story was retold by Plutarch in the first century of this epic as part of his biographies of Roman emperors. The Sabine women became a popular mythic theme in art and sculpture of the Renaissance, with modern treatments in art and film continuing to the present.[12] During a time in which Progressive conservationists worked toward a master plan for managing western lands, Moody's version of this myth functioned as a metaphor for Euro-Americans' "marriage" to the land. A parable on a grand scale in which a woman (Ruth) is taken – figuratively, materially, and sexually – by a man (Ghent) who embodies Turner's notion of the quintessential American, *Divide* forced a confrontation with, if not a refutation of, the ethos of taking. As the nation weighed wilderness preservation against utilitarian resource conservation,

Moody's play subtly questions the viability of colonial ecological violence as the bedrock of U.S. nation building.

Set in the deserts and mining region of Arizona prior to statehood, the opening scene of Moody's *Great Divide* (1906) finds a group of young genteel easterners encountering the iconic west. Ruth, age 19; her brother, Phillip; his wife, Polly; and Winthrop, a medical student, have traveled to Arizona to visit Phillip's recent business venture – a cactus farm. Polly complains about jackrabbits and wide-open spaces, while Ruth revels in the landscape and muses about finding an "unfinished" man for a husband (16). When a boy comes with news that their foreman has been injured, Winthrop, Polly, and Phillip exit to drive the man to the nearest town while Ruth happily stays behind to enjoy the peaceful solitude and natural beauty. Stage directions tell us that she takes her hair down and her shoes off, sings, and relishes the sensual beauty of the desert night – all actions that code her as part of the feminized land itself.

Within moments, however, three intoxicated men break into the cabin. Stage directions suggest familiar gold rush era villains – each a signifier of the history of colonization in which land that was once indigenous, then part of Mexico until 1849: "The one called Shorty is a Mexican half-breed, the others are Americans [Ghent and Dutch]" (27). Her enchantment destroyed by the harsh realities of the west, Ruth douses the lantern and grabs a rifle; when the intruders light a match, she fires. When the gun fails, she grabs a knife. The stage directions describe a protracted physical struggle in which one attacker knocks the gun from Ruth's hand and another wrestles a knife away from her and begins to drag her away struggling. Ruth "breaks from his clutch and reels from him in sickness of horror." In a scene of impending gang rape, Dutch beckons Ghent to "get into the game . . . This is going to be a free-for-all, by God!" (28). Ruth pleads to the third man,

> Save me, and I will make it up to you! . . . Don't touch me! Listen! Save me from these others, and from yourself, and I will pay you – with my life. . . . With all that I am or can be.
>
> *(29)*

The scene unfolds as the men compete and gamble for Ruth, until finally, Ghent offers Shorty a strand of gold nuggets and then challenges Dutch to a dual. The scene ends as Ghent, the victor, leaves with Ruth, his Sabine woman.

In the scenes that follow, Ghent's conquest is ultimately brought to bear on the terrain of Ruth's body.[13] The male representative of the imperial equation, Ghent first legitimizes his property rights by visiting the justice of the peace in San Jacinto for a marriage license. Once he has this piece of paper – not unlike a land grant – he is free to take what is his. Later in the play, the audience learns that, after several nights of sleeping in separate tents, Ghent visits Ruth's tent and rapes her. This violation, which would not have legally constituted rape at the time as Ruth is his wife, results in Ruth's pregnancy. The play subtly questions the logic that making the land productive – symbolized in the act of impregnation – excused the violence of rapacious exploitation and even acts of genocide. As *Great Divide* unfolds, however, neither the legal legitimacy of Ghent's actions nor their productive outcome heals the wound at the center of their marriage.

In Act II, Ghent the outlaw has become Ghent the successful businessman. His Rio Verde Mine, located in a fictional mining town, hangs on the mountainside, not unlike the town of Jerome, Arizona.[14] Meanwhile, even after Ruth goes with Ghent as a matter of survival, she quickly devises a plan to earn back the gold that bought her, and thus to take back her body, her child, and her life. To gain her freedom, Ruth appropriates indigenous lifeways and cultural practices as a means to regain her autonomy and agency. She learns how to weave and make baskets from local Navaho women and then, posing as a Navaho herself, sells her work to tourists at the local hotel. While this plot element demonstrates Navaho women as highly skilled, as *present* and active participants in the local economy, it is also a troubling example of what Shari Huhndorf (2001) calls "playing native." (Indeed, even in the first scene, Ruth's flowing garments, free hair, and bare feet also perform this.) The notion that a settler woman could learn Navaho skills as easy and quickly as learning how to knit is absurd, but that Ruth's could be even more valued by tourists lends rhetorical support to the displacement of indigenous people by white settlers.

Meanwhile, Polly, Phillip, and Winthrop have searched and finally located Ruth. By the end of Act II, they pay her a visit. Ruth feigns happiness and contentment with living high up in a mining town and being married to the mine owner. In reply, Polly exclaims, "I want to live on the roof of the world and own a gold mine!" Ghent offers a tour: "We can get a better view of the plant from the ledge below," at which point Phillip demands to know how Ruth came to marry "a person of this class?" (93). Before the end of the act, Ghent discovers that Ruth has been selling her handiwork to tourists and has also saved every penny he has given her for the house: "I have paid for everything. I have kept account of it – O, to the last dreadful penny!" (110). She reveals the complex intertwining of her disgust with the "hole down there . . . belching its stream of gold" and her humiliation that "you bought me for a handful of gold, like a woman of the street! You drove me before you like an animal from the market!" (110–111). Then she throws back at him the chain of gold nuggets – "the one you bought me with . . . take it and let me go free" (111–112). Ghent asserts his will and claim of ownership: "you are mine. . . . Mine by blind chance and hell in a man's veins, if you like! Mine by almighty Nature whether you like it or not!" (114–115). Ruth takes her child and leaves with her brother to return home to Maine, setting up the play's central moral dilemma: What will be the fate of this ill-conceived marriage, begun in violence, violation, and coercion yet with a child to be raised? As Moody represents Ruth's struggle, the play raises the question of how, *or if*, settler colonialism can transition from violent conquest to homemaking.

Geographic typologies and violence

Taking their cue from the play's new title, as well as dialogue that sets eastern civil society and Christian morality in contrast to western roughness and expediency, reviewers writing in 1906 tended to frame the conflict between Ruth and Ghent as one rooted in the different sensibilities of the east, with its urban centers (as genteel, refined, morally observant, civil), and the west, with its wild open spaces (as roughhewn, expedient, lawless, uncivil). As Merchant (2003) has observed, geographies of east and west were easily gendered in the public imagination – east as feminine and west as masculine (132–159). Within the valent meanings of these stereotypes, viewers can come to understand Ruth as representative of both the untouched virgin land (that ultimately transforms Ghent) and the harbinger of civilization; Ghent, then, can be read as representative of both the brutal and unfinished wilderness and the master of industry and invention that would ultimately make the virgin land fruitful. John Corbin's review of *Great Divide* relies on this gendered metaphor for "the barrier which exists between the rigor and dry formality of old civilization and the larger and freer, if more brutal, impulses of the frontier" (Montrose and Brown 1934, 176). Hobson Quinn (1927) likewise draws on geographic generalizations with roots in Turner's frontier narrative in his consideration of the play:

> The central theme of *The Great Divide* is the contrast between Ruth Jordan, the product of generations of New England ancestry . . . and Stephan Ghent, the man in the making, the freer spirit of the west, who is the master of his own impulses.
>
> *(12)*

However, these geographic typologies – popularized through literature and drama – ultimately mask, intentionally or not, patterns of ecocide and genocide that formed the basis of U.S. western expansion. Moody's letters and poems suggest his opposition to McKinley and Roosevelt's imperialist policies in relation to the Philippines.[15] The title alone – *The Sabine Woman* – likens western expansion to Romanesque imperialism and might suggest Moody held similar views on the de facto imperial relationship between an eastern industrial capital and western lands and resources. But the neocolonial relationship between eastern urban centers and the resource-rich west, and the exploitation of those resources (including the labor that brought them to market) was the lifeblood of eastern civility on which Moody's success as a dramatist depended.

Popular constructions of land east and west of the Mississippi also served the conservation agenda of Pinchot, who argued that western lands (as unfinished wilderness) required careful management by experts. Progressive conservation was primarily concerned with safeguarding and ensuring the continuance of resources for industry – mining, cattle, timber, and hydroelectric power – all of which were examples of the eastern states' colonial relationship with western territories. Stressing the systematized use of resources for industrial growth, Progressive conservation advocated applying eastern expertise to rural western lands. Environmental historian Donald Worster (1994) notes that

> [t]he movement for organized, efficient development of resources had an unmistakably urban, professional cast, directed as it was by Washington experts determined to supply an escalating industrial demand. Meanwhile, alongside the foresters and hydrologists were appearing city professionals: engineers, physicians, sanitarians, and landscape architects.
>
> (7)

Federal conservation efforts often chaffed those in the western regions of the United States who, like Ghent, had claimed the land and profited from it and rankled those who hoped to do so in the future. Both saw Progressive conservation's push for federal oversight as top-down management by nonlocal bureaucrats and, as such, elitist. Similar sentiments would haunt governmental environmental agencies in their dealings with local communities for the remainder of the century.

These east/west typologies circulating in popular culture also worked as rhetorical strategies that simultaneously masked the neocolonial relationship between western resources and eastern capital, as it rationalized violence and violation – of a woman, in the case of the play, and of the land, in the case of ongoing extractive industries. In the first decade of the new century, big capital mining operations in Arizona, California, Nevada, Utah, and Alaska were extracting not only gold but also silver, copper, iron, lead, and coal to fuel the trains that carried equipment in and ore out (Smith 1993). In *Great Divide*, Ghent runs a mining operation that quickly makes him a rich man and allows him to save Ruth's family in Maine from bankruptcy. His largess is the product of an economic and ideological system that understands land and labor as capital. If Ghent's company were representative of historical detail, it would have resembled a war with the mountain over possession of rich veins – as dangerous to miners as it was devastating to biotic communities and watersheds. Western companies like the fictitious Rio Verde promised big returns to eastern investors and made unbridled and often reckless use of water, timber, soil, and workers (both animal and human). Equipment used in the process of ore extraction was shipped up the mountain first by packhorse and then by rail. The roads to high altitude mining camps were often littered with the bodies of horses, driven to their death and then tossed aside. Workers risked extremes of heat and cold, loss of limbs and life: "Miners died in explosions, cave-ins crushed them, fires incinerated them, machinery maimed them . . . mines crippled and killed" (White 1993, 280).

For those who did survive such perilous working conditions, exposure to quartz dust and chemicals often caused silicosis and tuberculosis. Hard-rock mining, like Ghent's operation in the play, came in two varieties, both of which required large labor forces: open pits (often called glory holes because of the number of miners who fell to their glory there) and hard-rock mines that required deep shafts and honeycombs of tunnels deep within the mountain and utilizing millions of feet of timber and resulting in vast stretches of denuded terrain. Smelting operations used chlorine, mercury, lead, and cyanide; tailings containing high concentrations of this chemical soup were returned to the environment via wastewater – an issue that became significant in environmental assessments of the Klondike gold rush of the 1890s.[16] In some mining communities that used chlorination, it was not unusual for all the plant life surrounding a mine to die off, killed by chlorine fumes and surface-water pollution. With the exception of Ruth's exclamation of her sickened feeling at the "belching" (110) of the mine, *The Great Divide* (like Belasco's *Girl*) avoids mention of these environmental impacts. Nevertheless, Ghent's success, position, and redemption traced by the arc of the play are predicated on the extractive violation of Ruth's body and the land.

A field of relations

Marriage implies/entails a relationship, and thus, for the character of Ruth, the land is more than a stockpile of resources to be efficiently utilized as the stuff of nation building. It is also more than a picturesque setting to commodify as a tourist attraction. As the night air washes over Ruth's skin, she touches the flowers much in the way she might stroke the head of a person she loved. She sings to the night and speaks directly to the flowers as if they are relations; she sings and dances within a sensible field of exchange in which the boundaries between her own body and the landscape are permeable. Ruth seems to inhabit an animated world; indeed, her ecstatic feelings for the desert bursting out in song:

> Hark, from the dark eaves the night thaw drummeth! / Now as a god, / Speak to the sod, / Cry to the sky that the miracle cometh! *She passes her hand over a great bunch of wild flowers on the table.* Be still you beauties! You'll drive me to distraction with your color and your odor. I'll take a hostage for your good behavior. *She selects a red flower, puts it in the dark mass of her hair, and looks out at the open door.* What a scandal the moon is making, out there in that great crazy world! Who but me could think of sleeping on such a night?
>
> *(25–26)*

Ruth talks *to* the land in its myriad of presences, a land composed of others with whom she resides, a community of relations of which she is part.

The dialogue referenced earlier is reminiscent of Moody's own sensibility and is inspirational for parts of the play. While on a visit to the Grand Canyon in 1904, Moody (1935) wrote the following to his wife, Harriet:

> The next morning we went down the trail to a vast plateau about four miles down, and found lodging in a camp there – a lovely spot, with white tents bosomed in clumps of willows and red-brush in glorious flowers. The full moon made such a scandal that it was impossible to sleep; I spent most of the night outside, listening to what the little brook in the willows had to tell of you, though it wasn't much compared with what the mountains and the moon spoke on the same subject.
>
> *(191)*

However, Moody's poetic acknowledgment of the myriad presences of the natural world does not press past sentiment toward a call for recognition of kinship in the play itself. Similarly, in the heated debate between preservationists and conservationists of the period, neither group moved from the objectification of nature (as an object of beauty or an object of utility) toward an admission of relationship, interdependence, or the respectful cohabitation that a "marriage" might imply.

Returning to the play, Act III takes place in the genteel family home in Maine, where Ruth is safe but not content. Her mother is worried because Ruth treats her child "as if it were a piece of machinery!" (128). Meanwhile, the doctor reports that "your daughter . . . is in a dangerous state." Unaware of the trauma that Ruth underwent earlier in the play, Polly sources Ruth's distress: "What she needs is her husband, and I have sent for him!" (129). Polly has already likened Ghent to "a volcano . . . the molten heart of the continent" (138–139), and he is modeled on Turner's quintessentially American devolution into savagery and reemergence as the visage of a successful American businessman. Now, at once predator-plunderer and parent-provider, Ghent's character embodies a paradox of settler colonialism: not only how to make a home on stolen lands but also how to sustain a relationship with someone who has been the spoils of conquest. In order to come home to the place and the person he has taken by force, Ghent must find a way to make amends and come into relation.

In Act III, Ruth's mother reveals that Ghent has been sending the family money for months and that he bought the failing Cactus Fiber company and "got a hustler to boom it" in order to return the company and its rising stock prices to Ruth's brother Phillip. His only gesture of generosity, his only act of amends – "to cancel her debt, or what she thinks her debt" – comes in the form of wealth acquisition, as if economic productivity reconciles rapaciousness (147). At least one *New York Times* reviewer found Ghent unconvincing: "From an impulsive brute, from a man whose action leaves no doubt as to the absolute meanness of his character, he is converted in a moment into one who weighs motives, who discusses futures with businesslike precision" ("Hue" 1906). Ghent's transformation from thieving marauder to responsible husband and businessman, however, is the very Turneresque transmutation from brute to suit on which the U.S. colonial project depends. Indeed, William Cody was a prime example, scalping Yellow Hand in one breath and lobbying for irrigation and investing in hotels in the next. When he and Ruth finally meet again, Ghent does not apologize nor provide – nor does she demand – any accounting of what has occurred between them. Instead, the play couches her trauma as a mere difference between eastern and western sensibilities.

At the end of Act III, Ruth reveals the layers of her trauma – not only being bought for a chain of gold nuggets but also the night in which Ghent forced himself on his newly purchased wife. In the closing scene, Ruth describes the days after her abduction:

> That night, when we rode away . . . still you spared me. . . . As you rode before me down the arroyos . . . it seems as if you were leading me out of a world of little codes and customs . . . So it was for those first days. – And then – and then – I woke, and saw you standing in my tent-door in the starlight! I knew before you spoke that we were lost.
>
> *(109)*

Ghent echoes Turner's thesis that the wilderness makes the man when he describes himself as "the man you've made of me, the life and the meaning of life you've showed me the way to!" (162). When she fires back that her own experience was one of violation, Ghent forwards an argument for American exceptionalism, saying to Ruth, "What does the past matter when we've got the future – and him?" (163). Finally, he admits to her that "[w]rong is wrong, from the moment it happened till the crack of doom . . . that seems to be the law . . . I've seen it written in mountain letters across the continent of this life" (164). Ruth, however, is unable to reconcile the violation that began her marriage with the love she sometimes felt for Ghent and thus recover the future she dearly wants.

As an allegory for the spiritual and practical predicament of Euro-American U.S. society at the dawn of the twentieth century, Moody's play provokes questions about reconciliation and reparations. What does this couple do with the violence of the past; cognizant of the rape and theft of its beginnings, how is this marriage to thrive? The play suggests, as Progressive conservation implicitly argued, that the violence of the past must be forgiven, at least set aside, in the project of building a future. Ultimately, Ruth adopts Ghent's worldview: "You have taken the good of our life and grown strong. . . . Teach me to live as you do!" (166). Rather than a transformation of Ghent's *modus operandi*, Ruth herself has a conversion: if good has resulted – defined as wealth accumulation, procreation, and participation in a set of economic relations that benefit Ruth's family, in the case of the play, and the country, more generally – a (white) man may simply take what he pleases without moral obligation. In its final moments, *Great Divide* affirms an ethos in which those who have been violated must yet bend into agreement and alignment with those who hold wealth and power.

As a national parable about American conquest and the living-with that follows, *The Great Divide* reflects an emerging environmental ethos at the turn of the century. The full arc of Moody's plot underscores the ways in which rape and rapaciousness – valorized in the myth of the Sabine women – are part of the deeply embedded mythic structures of this nation. Such mythos understands rape, theft, exploitation, and extraction as necessary aspects of nation building, with exception given to the injury and damage that may occur in the process. It is a mythos that allows for the violation of land and bodies because it conceives of those bodies and lands as objects of exploitation, resources to use in the project of nation building.

Ultimately, *Great Divide* reaffirms a Progressive-era ethos in which the "greater good" (i.e., that which forwards national interest) sanctions the former and ongoing violation of people, communities, and ecosystems. Traditionally historicized as a struggle between "private" (i.e., business) interests and "the people's" (i.e., public) interests, Progressive conservation was hardly a people's movement. President Roosevelt called for business-like efficiency in the management of the national resources (see, e.g., Pinchot [1947], 319–366). Big capital often supported conservation measures because, as Pinchot often reiterated, it made good business sense. As in the case of Hetch Hetchy, U.S. environmental policy would herald the values of the greater public good, and yet the yardstick by which that good would be measured (and was measured in *The Great Divide*) was frequently corporate profits. Ghent's goodness in the final analysis of Ruth's mother, Mrs. Jordan, had everything to do with his capacity to support them in the manner to which they had become accustomed: "He has done a great deal for us. . . . He has given us the very roof over our heads!" (150–151). The irony in the case of Moody's play is that marriage also implies kinship, but values of relationality were not, ultimately, victorious in the play.

Moody's play unveils a deep and disarming ethical question of "man's right" to the land's resources, just as it points clearly – although uncritically – to the violations on which those "rights" were founded. In the case of the play, Ghent's desire was more important than Ruth's autonomy; his taking was legitimized by the state. *The Great Divide* characterizes Americans' relationship with the land as an ambivalent marriage wrought in violation. Likewise, the play does not suggest a reversal of U.S. imperialist expansion – Moody and his characters were instruments of that conquest. The play does, however, wrestle with questions that troubled both conservationists and preservationists of the era. Progressive conservationists argued that the era of conquest was over; it was time to learn to live with the land. They did not oppose the preservation of "parks" and places of scenic beauty; they envisioned a relationship between the land and the people as utilitarian – conservation made practical sense for the continued expansion and economic gain of the nation. Meanwhile, the wilderness served a practical purpose as Ruth explains to Polly and her brother, "we may have come here to have our picture taken, but we stayed to make a living" (88). Ruth, however, is lying when she makes this comment – lying about the very foundation of that life and livelihood, and what it has cost her. Moody's play suggests that the land is not merely a collection of raw materials any more than Ruth's value can be measured out in gold nuggets.

The systematized, rational approach of Progressive conservation flew in the face of the life of chance and expediency that had characterized the frontier. Yet, greater good often translated as corporate good, as would be demonstrated through the numerous hydroelectric and other public works projects that began under Theodore Roosevelt and then Franklin Roosevelt. Federal funds were promised to conservation-wise ventures and big business benefited from federal irrigation projects, river navigation, and hydroelectric projects (Hays 1999, 133). The legacy of Pinchot's push to centralize and systematize conservation, however, fueled antagonisms between governmental agencies, small communities, and private property owners. Descendants of frontiersmen would later resent handing over control of their lands to be micromanaged by the technological elite in Washington. The legacies of Pinchot's system can be seen in the conflicts that continue between state and federal land management agencies, local communities of settler descendants, and tribal communities whose knowledge and expertise about their homelands is built on millennia of lived experience.

Notes

1 For a discussion of the managerial culture of the early U.S. conservation movement, see Samuel P. Hays's (1999) *Conservation and the Gospel of Efficiency*, as well as Donald Worster's (1994) *Nature's Economy: A History of Ecological Ideas* (1994), Parts Four and Five.
2 Belasco's themes also foreshadowed Garnet Holme's adaptation of Helen Hunt Jackson's (1925) novel *Ramona*, first published in 1884. Beginning in 1923 and continuing to the present, "The Ramona Pageant" is staged outdoors by the Ramona Festival in Helmet, California. For more analysis of those interrelated works, see Michelle Moylan (1993).
3 In 1914, a film version of *Squaw Man*, directed by Cecil B. DeMille, featured the actress known as Red Wing. Born on the Winnebago Reservation in Nebraska, Red Wing and her husband, James Young Deer, are often considered the first Native American Hollywood couple (Deloria 2004, 84–104).
4 The Boy Scouts of America, for example, was founded in 1910 to address the ill effects of urban conditions among youth by renewing and rehearsing manliness (Huhndorf 2001, 65–78; MacDonald 1993).
5 For example, in 1906 Devils Tower (located on Cheyenne River Sioux tribal homelands) was protected as a national monument, yet not until 1994 would U.S. courts rule that indigenous cultural practices had priority over rock-climbing enthusiasts ("Indian" 2018).
6 National parks established during this period included Sequoia (California, 1890), the Grand Canyon (1893), Crater Lake (Oregon, 1902), Wind Cave in the Black Hills (1903), Petrified Forest (Arizona, 1906), Mount Lassen (California, 1907), Olympic (Washington, 1909), and Zion (Utah, 1909). See "National" (2018) for more details.
7 Pinchot, for his part, was skillful in building consensus among business constituencies in the timber and grazing industries, as well within government and industry in the river-based economies of the Mississippi and Missouri River systems, where he stressed the need for "navigable rivers." For more on the movement to control rivers, see Hays (1999), 91–146; see also Roosevelt (1925), 145–156.
8 For more about the Hetch Hetchy controversy, which became a pivotal moment in the history of U.S. environmentalism, see Roderick Nash (2001), esp. Chapter 10, in which Nash provides a detailed account of the ten-year struggle that involved key legislators; businesspeople, and Muir's Sierra Club associates. Richard White (1991) similarly frames the debate around the politicization of the wilderness movement, when preservationists

cut their teeth in political strife (412–413). Political unrest in early twentieth-century California played a role in that decision. Demonstrating the federal government's stand against big utilities in the west, Hetch Hetchy set a precedent for publicly owned utilities. For more background, see Hofstedter (1955) and Nash (2001), 161–181.
9 Social theorists of the early twentieth century were not shy about advocating for modern technology's capacity to create a real-world utopia; see, for example, Wells (1905).
10 For a discussion of the effect of mining on streams, habitats and surface areas, see Duane A. Smith, *Mining America: The Industry and the Environment, 1800–1980* (1993), 6–7, 67–73. See also John Opie, *Nature's Nation: An Environmental History of the United States* (1998), 166–167, and Chapter 6 generally.
11 In 1884, a court ruled the hydraulic mines in question a "public nuisance and ordered a permanent injunction against dumping." As White notes, this case not only established farmers' rights over miners' rights; it also "established the legal precedent for shutting down a polluting industry entirely" (1993, 234). See also Smith's *Mining America* (1993), 67–73.
12 Some salient examples of artistic representations of Sabine women include *The Rape of the Sabine Women* (1583) by Flemish sculptor Gaimbolongia; *The Rape of the Sabine Women* (est. 1635–40) by Peter Paul Ruben; *The Rape of the Sabines* (1861–62) by sculptor Edgar Degas; *The Rape of the Sabine Women* (1962–63) by Pablo Picasso; and, in recent cinema, *The Rape of the Sabine Women* (2006) by filmmaker Eve Sussman. See also Plutarch (1914) for the original (albeit translated from the Latin) Roman telling of this story.

Relatedly, the Sabine women narrative was one of the many rape narratives deployed to characterize the formation of the nation. For an in-depth discussion of this important connection, see Katharina Erhard (2005), 507–534.
13 For more analysis of sexual coercion and rape as means for establishing systems of power in early America, see Block's *Rape and Sexual Power in Early America* (2006); Higgins and Silver's *Rape and Representation* (1991); and Tomaselli and Porter's *Rape* (1986).
14 Moody (1935) had traveled to Arizona and New Mexico in 1904 before writing *The Sabine Woman*, later revised as *The Great Divide*. There he visited not only Williams, Arizona, located not far from Camp Verde and the old mining town of Jerome, but also Oraibi Village, on what is now the Hopi Reservation (190–200).
15 Moody's poem "On a Soldier Fallen in the Philippines" cautions the nation to consider the cost of sending soldiers into battle and that shedding blood for imperialist causes injures the soul of the nation that sends them (1969, 29–30).
16 Alaska's Treadwell mine and the Alaska–Juneau mine processes are examined in the exhibit "Hands-On Mining Gallery" at City Museum in Juneau, Alaska, which I visited in June 1999. For more on Alaskan mining and the gold rush, see Taylor Morse (2003) and Wharton (1972). For a discussion of the contemporary impacts of gold mining, see Bland (2014).

References

Albright, Horace, and Robert Cahn. 1985. *The Birth of the National Park Service: The Founding Years*. Salt Lake City, UT: Howe.
Belasco, David. 1929. "*The Girl of the Golden West*." In *Six Plays by David Belasco*, edited by Montrose J. Moses, 310–403. Boston, MA: Little, Brown and Company.
Belasco, David. 1936. *The Rose of the Rancho: A Play in Three Acts*. New York: S. French, Ltd.
Bland, Alistair. 2014. "The Environmental Disaster That is the Gold Industry." *Smithsonian*, February 14.
Block, Sharon. 2006. *Rape and Sexual Power in Early America*. Chapel Hill, NC: University of North Carolina Press.
"The Bureau of Reclamation: A Very Brief History." 2018. *U.S. Bureau of Reclamation*.

"Chasing the Great Dramatic Will-o'-The-Wisp: The Hue and Cry for a Representative American Play – 'The Great Divide' and Other Native Work." 1906. *New York Times*, October 14.

Christianson, Frank. 2017. *The Popular Frontier: Buffalo Bill's Wild West and Transnational Mass Culture*. Norman, OK: University of Oklahoma Press.

Corbin, John. 1906. "Moody's *The Great Divide*." *New York Sun*, October 14.

Cronon, William. 1996. "The Trouble with Wilderness: or, Getting Back to the Wrong Nature." In *Uncommon Ground: Rethinking the Human Place in Nature*, edited by William Cronon, 69–90. New York: W.W. Norton & Company.

Delano, Alonzo. 1857. *A Live Woman in the Mines, Pike County Ahead!: A Local Play in Two Acts*. New York: S. French, Ltd.

Deloria, Philip J. 2004. *Indians in Unexpected Places*. Lawrence, KS: University of Kansas Press.

Erhard, Katharina. 2005. "Republicanism and Representation: Founding the Nation in Early American Women's Drama and Selected Visual Representations." *American Studies* 50 (3): 507–534.

Gottlieb, Robert. 1993. *Forcing the Spring: The Transformation of the American Environmental Movement*. Washington, DC: Island Press.

Hays, Samuel P. 1999. *Conservation and the Gospel of Efficiency*. Pittsburgh, PA: University of Pittsburgh Press.

Henry, David D. 1934. *William Vaughn Moody: A Study*. Boston, MA: Bruce Humphries, Inc.

Higgins, Lynn A., and Brenda R. Silver. 1991. *Rape and Representation*. New York: Columbia University Press.

Hobson Quinn, Arthur. 1927. *History of American Drama*, 2 vols. New York: Harper Brother Publishers.

Hofstedter, Richard. 1955. *The Age of Reform*. New York: Random House.

Howe, Charles E.B. 1858. *Joaquin Murieta de Castillo, or The Celebrated California Bandit*. City: Press.

"The Hue and Cry of a Representative American Play – 'The Great Divide' and Other Native Work." 1906. *New York Times*, October 14.

Huhndorf, Shari M. 2001. *Going Native: Indians in the American Cultural Imagination*. Ithaca, NY: Cornell University Press.

"Indian Religious Freedom at Devil's Tower National Monument." 2018. *Indian Law Resource Center*.

Jackson, H.H. 1925. *Ramona: A Story*. Boston, MA: Little, Brown and Company.

Jordan, John M. 1994. *Machine Age Ideology: Social Engineering and American Liberalism, 1911–1939*. Chapel Hill, NC: University of North Carolina Press.

Loney, Glen, ed. 1983. *Gold Rush Plays*. New York: Performing Arts Journal Publications.

MacDonald, Robert H. 1993. *Sons of Empire: The Frontier and the Boy Scout Movement, 1890–1918*. Toronto: University of Toronto Press.

Marker, Lise-Lone. 1975. *David Belasco: Naturalism in the Theatre*. Princeton, NJ: Princeton University Press.

Marx, Leo. 1964. *The Machine in the Garden: Technology and the Pastoral Idea*. New York: Oxford University Press.

Merchant, Carolyn. 1996. "Reinventing Eden: Western Culture as a Recovery Narrative." In *Uncommon Ground: Rethinking the Human Place in Nature*, edited by William Cronon, 132–159. New York: W.W. Norton & Company.

Merchant, Carolyn. 2003. *Reinventing Eden: The Fate of Nature in Western Culture*. New York: Routledge.

Montrose, Moses J., and John Mason Brown, eds. 1934. *The American Theatre as Seen by Its Critics 1752–1934*. New York: W.W. Norton & Company.
Moody, William Vaughn. 1906. *The Great Divide*. New York: Houghton Mifflin Company.
Moody, William Vaughn. 1935. *Letters to Harriet*, edited by Percy MacKaye. New York: Houghton Mifflin Company.
Moody, William Vaughn. 1969. *The Poems and Plays of William Vaughn Moody*, Vol. 1. New York: AMS.
Mordden, Ethan. 1981. *The American Theatre*. New York: Oxford University Press.
Moylan, Michelle. 1993. "Reading the Indians: The Ramona Myth in American Culture." *Prospects* 18: 153–186.
Muir, John. 1973. "Nature Versus Culture." In *American Environmentalism: The Formative Period, 1860–1915*, edited by Donald Worster, 185–190. New York: John Wiley & Sons, Inc.
Nash, Roderick. (1967) 2001. *Wilderness and the American Mind*, 4th ed. New Haven, CT: Yale University Press.
"National Park System Timeline (Annotated)." 2018. *History E-Library, National Park Service*, accessed September 30, 2019.
Opie, John. 1998. *Nature's Nation: An Environmental History of the United States*. Fort Worth, TX: Harcourt Brace & Company.
Pinchot, Gifford. (1947) 1987. *Breaking New Ground*. Washington, DC: Island Press.
Pinchot, Gifford. 1990. "The Birth of 'Conservation'." In *American Environmentalism: Readings in Conservation History*, edited by Roderick Frazier Nash, 73–79. New York: McGraw-Hill.
Plutarch. 1914. "Parallel Lives, The Life of Romulus." Vol. 1. *Loeb Classical Library*.
"A Powerful Play, Beautifully Acted." 1906. *New York Times*, October 5.
"Quick History of National Park Service." 2018. *National Park Service*.
Rohrbough, Malcolm J. 1997. *Days of Gold: The California Gold Rush and the American Nation*. Berkeley, CA: University of California Press.
Roosevelt, Theodore. 1925. "American Problems." In Vol. 18 of *The Works of Theodore Roosevelt*, 145–516. New York: Charles Scribner's Sons.
Russell, Donald. 1961. *The Wild West: A History of the Wild West Shows*. Fort Worth, TX: Amon Carter Museum.
Shabecoff, Phillip. 1993. *A Fierce Green Fire: The American Environmental Movement*. New York: Hill and Wang.
Smith, Duane A. 1993. *Mining America: The Industry and the Environment, 1800–1980*. Niwot, CO: University Press of Colorado.
Smith, Henry Nash. 1950. *Virgin Land; the American West as Symbol and Myth*. Cambridge, MA: Harvard University Press.
Taylor Morse, Kathryn. 2003. *The Nature of Gold: An Environmental History of the Klondike Gold Rush*. Seattle, WA: University of Washington Press.
Tomaselli, Sylvana, and Roy Porter. 1986. *Rape*. Oxford and New York: Blackwell.
Turner, Fredrick. 1985. *Rediscovering America: John Muir in His Time and Ours*. New York: Viking Penguin Inc.
Wells, H.G. 1905. *A Modern Utopia By H.G. Wells Unabridged 1905 Original Version*. Ann Arbor, MI: University of Michigan Library Press.
Wharton, David. 1972. *The Alaska Gold Rush*. Bloomington, IN: Indiana University Press.
Whiston Spirn, Anne. 1996. "Constructing Nature: The Legacy of Frederick Law Olmsted." In *Uncommon Ground: Rethinking the Human Place in Nature*, edited by William Cronon, 91–113. New York: W.W. Norton & Company.

White, Richard. 1991. *It's Your Misfortune and None of My Own: A New History of the American West*. Norman, OK: University of Oklahoma Press.

Winter, William. 1981. *The Life of David Belasco*, Vol. 2. New York: Moffit, Yard and Company.

Worster, Donald. 1994. *Nature's Economy: A History of Ecological Ideas*, 2nd ed. Cambridge: Cambridge University Press.

3

DYNAMOS, DUST, AND DISCONTENT

Eugene O'Neal's *Dynamo* and the Federal Theatre Project's Living Newspapers *Triple-A Plowed Under* and *Power*

Folksinger Woody Guthrie, the balladeer of workers' movements of the 1930s, was hired by Bonneville Power Administration (BPA) to promote the construction of the Grand Coulee Dam on the Columbia River in Washington State. Guthrie's popular lyrics celebrate an idea of nature – in this case, a river – quite distinct from Muir's contemplative wilderness discussed in the previous chapter:

> Uncle Sam took the challenge in the year of '33
> For the farmers and the workers and for all humanity
> Now river, you can ramble where the sun sets in the sea
> But while you're rambling, river, you can do some work for me
> Roll, Columbia, won't you roll, roll, roll.
>
> *(1991, 17)*

This is nature that possesses energy waiting to be put to work for people, conceived as a kind of co-worker in the project of nation building. Conflict between U.S. workers and industrial capitalists had increased to a boiling point since the stock market crash in 1928. As construction of what was affectionately called "the biggest thing on earth" got underway in 1933 (see, e.g., Dietrich [1995]; and White [1995]), BPA hoped that Guthrie could sell this massive hydropower development project as pro-worker. President Franklin Roosevelt promised that dam construction would provide jobs in the immediate future and that the irrigation of the high desert of eastern Washington would provide water and electricity to farmers, industries, and communities across the western states. Guthrie's songs were only one form of cultural production that celebrated the ramping up of hydroelectric development in the United States in the 1930s. Big dams and the electricity they generated were the subjects of films, photography, visual art,

DOI: 10.4324/9781003028888-4

and theatrical productions of the 1930s. Many of these works, like Guthrie's lyrics earlier, reflect changing ideas about nature, the land, and ecological systems.

During the 1920s, when technology and engineering seemed like a sure answer to social and environmental problems from poverty to soil erosion, the science of ecology found its theoretical footing. The ancient Greek word *ekos* or *oikos*, meaning "home," is the root word for both *ecology* and *economy*. The era's New Ecologists, as they called themselves, defined all relationships among living things, including human beings, in economic terms, and popularized notions of nature as a system of producers and consumers. As a subfield of economics, environmental historian Donald Worster (1994) describes how ecology in the 1930s "emphasize[d] the flow of goods and services – or of energy – in a kind of automated, robotized, pacified nature." He argues that this disciplinary cross-pollination between ecological and economic theory informed a "tendency to find in nature a utilitarian, materialist promise – a set of resources with cash value" (313). After John Muir's death in 1914, few voices challenged the policies that measured the value of a river in kilowatt-hours or irrigation gallons per minute or the land in terms of biotic capital.

Two significant ecological frameworks emerged that would shape approaches to environmental challenges for the remainder of the century, and both found their way on stage in the plays I discuss in this chapter. The first centers on the idea of nature as a matrix of economic exchanges. Worster (1994) tells the story of the development of ecological thought of this period as a building consensus of ideas around the idea of nature as a factory. British biologist Hermann Reinheimer (291–292) theorized that the Great Earth Factory ran according to the same laws of production and demand as industrial civilization (Reinheimer 1913). All organisms, Reinheimer argued, were essentially traders who earn their keep by producing food, goods, and services in a kind of natural assembly line. According to the Reinheimer framework, plants and animals form a worldwide matrix of "laboratories, factories, workshops, and industries" that contributed to the "general fund of organic wealth" on the earth (194; quoted in Worster [1994], 291). In 1926, German biologist August Thienemann introduced the terms *producer*, *consumer*, and *decomposer* (or *reducer*) (Worster 1994, 291–315). In the following year, Cambridge zoologist Charles Elton (1927) published *Animal Ecology*, which coined the idea of food chains as a hierarchical system of economic exchange among plants and the animals that consume those plants. Echoing free-market economic theory, Elton argued that the production and consumption of life-sustaining nutrients form a kind of *interspecies commerce*, in which competition improves the health of the whole community. In short, the New Ecologists envisioned nature as "a reflection of the modern corporate, industrial system," with factories that kept everything ticking along like, again, an organic machine. "Conflict could have little place in such a well-regulated system; nature did not go on strike," Worster quips, but rather "plants went on producing for the herbivores without shirking, feather-bedding, or complaining" (313). These ideas served industrial capitalism: if nature was tantamount to an assembly line of goods and services, then worker unrest could be characterized as "unnatural," a kind of aberration in an otherwise orderly system.

The second theory reconceptualized nature as a series of dynamic energy exchanges, a kind of thermodynamic river of energy. Oxford botanist Arthur Tansley (1935) theorized that Elton's food chain amounted in essence to an exchange of energy (292). A coal deposit was stored energy; a rushing river was unharnessed energy, escaping and wasted; a field of corn was caloric energy, waiting to be released as labor or transformed into another storage unit, namely, beef. Tansley's theory marked the solidification of the application of the laws of thermodynamics to ecological systems, thereby allowing the quantification of ecological outcomes (e.g., energy flow, the production of carbon, etc.). For Tansley, Worster (1994) observes, nature operated like a series of reservoirs; energy stored in plants or in the bodies of animals is released when one species consumed another. As a result, nature's energy is again stored for future use in the body of a new reservoir. The impact of these theories has been to ingrain popular and academic understandings of ecological systems as utilitarian: if nature was understood as a system of interconnected functions, it could be managed more effectively for human benefit. Employing these conceptual models, human beings could maintain a position of relative privilege as managers, master planners, and profiteers of nature's systems (297). While conceptualizations of nature rambled along in accordance with economic law, humans insisted on their subjectivity above the law, as the very engineers of their own biologic destinies.

As U.S. industrialization continued with abandon throughout the first half of the twentieth century, and as new technologies were deemed the very mode and mission of progress, mechanization was both venerated and critiqued in diverse forms of cultural production. For example, precisionist painting celebrated a new mechanized American landscape. Meanwhile, some dramatists expressed suspicion about the dehumanizing effects of technology. Philosophical and political and cultural revolutions of the 1920s, and the unfolding workers' movements of the 1930s changed the face of American theater. By that time, Belasco's romantic realism had been supplanted by the New Stagecraft of designers and the German expressionist art movement, with Edward Gordon Craig, Adolphe Appia, and Max Reinhardt among the oft-noted forces of this sea change (see, e.g., Wilson and Goldfarb [2008]; and Larson [1989]). In particular, the Little Theatre Movement of the 1910s and 1920s parted with the hegemony of Broadway, exploding with intellectual radicalism and artistic experimentation (Chansky 2004). At the same time, the work of Konstantin Stanislavski and Moscow Art Theatre, the Epic Theatre of Edwin Piscator, and European expressionism brought new tools to bear on theater practice.[1]

In the United States, these new methods strengthened the potential of theater as a means of social intervention and commentary, in large part because they provided actors the tools necessary to bring lived human experience to life and to encourage the audience to think and reflect. Taking workers and farmers as central characters, new stories represented on stages across the country explored the social conditions of marginalized groups. While social protest plays critiqued the system and exposed the injustices of capitalism, dramatists J.M. Synge and Lynn Riggs celebrated specific places, communities, and identities. New theater groups – such as the Theatre Guild and the Provincetown Players, followed later by the Group Theatre, the Mercury Theatre, and the League of Workers' Theatres – similarly galvanized around the idea that the theater could function as a crucible for social change.

The new theatrical styles that emerged often addressed ecological questions as part and parcel of stories about the conditions of life in the United States. Works such as Elmer Rice's *Street Scene* (1929), Sophie Treadwell's *Machinal* (1928), and later the Federal Theatre's *One-Third of a Nation* (1938) confronted the effects of urban-industrial habitats on the human animal, indicting a system in which many people did not have access to the ecological basics of food, clean water, and shelter. In works such as Eugene O'Neill's Pulitzer Prize-winning *Beyond the Horizon* (1920), dramatists examined the impact of capitalism on the land and the family. In that play, as in others, many playwrights represent the reality that as the land becomes a commodity, the fabric of family and community life is torn apart.

This chapter examines two plays that speak to significant environmental issues and debates of the 1930s and reflect the emerging ecological thinking that couched nature as an organic machine and natural processes as mirrors of industrial production. The first, Eugene O'Neill's *Dynamo* (1929), illustrates that the notion of nature-as-machine did not simply replace Muir's animated life force but, rather, that the two constructions merged in a distorted vision of human power and control. In many ways, *Dynamo* is a parable on the threshold between one construction of nature and another. Mapping the psychosis of the main character, Reuben, the plot shows how infatuation with one idol (i.e., "virgin" nature and animating life force) transforms into another (i.e., nature as machine, that life force harnessed by man). Like Rice's *The Adding Machine* (1929) and Treadwell's *Machinal*, O'Neill's dark portrayal of societal infatuation with electricity warned against technology's dehumanizing effects, just as it questioned the extent to which people, particularly workers, were being treated *as* machines or, worse, as one more natural resource to be mined by industrial capital. The second play, *Triple-A Plowed Under* (Arent et al. 1938),[2] examines the interrelation of land, labor, and politics in the years before, during, and just after the devastating Dust Bowl of the 1930s. *Triple-A* worked to demonstrate the systemic and intertwined nature of social, political, and ecological issues by making connections between seemingly unrelated events – such as farmers' discontent and urban unemployment or weather, war, and middlemen. As a play about these causal connections, *Triple-A* marks an important step in representing complex ecological interrelationships on stage.

Triple-A was one of several "Living Newspapers" produced by the Federal Theatre Project (FTP) during its brief life as the first federally funded theater in the United States. The FTP was a branch of the Works Progress Administration, a national jobs program of Roosevelt's New Deal. As such, the FTP's primary function was employing out-of-work theatrical artists, but under the visionary leadership of Hallie Flanagan, the FTP also sought to instill an ethos of arts as civic practice in communities across the country. One of the FTP's innovative forms was called the "living newspaper" and was designed to educate the general public on important issues of the day. These "docudrama" episodic productions did not tell a single story but rather explored an issue through a sequence of fast-paced scenes interspliced with narration and sometimes film montage. Dynamic and expansive in scope, the Living Newspapers sought to promote informed public dialogue about complex and controversial issues from public health to conservation (see, e.g., Flanagan [1940]; and Osborne [2011]). The Federal Theatre's Living Newspapers were influenced by the agitprop theater of the communist revolution taking place in Russia. Flanagan, director of the Federal Theatre, had traveled to Europe and Russia from 1926 to 1927 while on a Guggenheim Fellowship. The theater Flanagan saw and the artists she met inspired her interest in new theatrical forms and politically engaged work. Theater, she advocated, could and should be used to inspire social change, educate ordinary citizens about important issues, and participate more directly in civic life. The Living Newspapers of the FTP were at the center of the controversies over communist influence that would ultimately move Congress to stop funding the Federal Theatre in 1939.

Several Living Newspapers of the FTP focused on the intersection of natural resource management and economic policy: *Triple-A* examined the economic causes of the Dust Bowl, *Power* (1937) advocated public access to hydropower, and *Timber* (1939) was concerned with forest conservation in the Pacific Northwest.[3] These environmentally-themed Living Newspapers generally envisioned the land as an extension of human–economic relations, disseminating conceptual frameworks that informed public understanding of ecological processes and national environmental policy. For example, Elton's notion that all ecological relationships are fundamentally economic ones is palpable in *Triple-A*'s portrait of the interdependence of farmers and workers, while Tansley's framework of nature as a thermodynamic river of energy gave an implicit scientific authority to hydropower development projects celebrated in *Power*. Because ecological problems involve multiple causes and effects accumulated across time and place, they do not lend themselves to deep examination through climactic plot structures and are utterly defied by neoclassical strictures. An episodic form allows the dramatist to represent the slower and multiple effects of environmental change. Therefore, while Living Newspapers tended to valorize technological solutions, they also demonstrated how the very form of theater might effectively engage the temporal scope and material complexity of environmental issues. *Triple-A* is one example of how theater was used to respond to the complex system questions of the Dust Bowl and how it might be used to speak to the complexities of climate change in the twenty-first century.

Dynamos and other idols

The big dam projects of the New Deal celebrated in Woody Guthrie's song "Roll, Columbia, Roll" and FTP's *Power* would not have been possible without the key invention of the dynamo – the mechanism that transformed mechanical rotation into electrical current. In 1900, Henry Adams, a noted historian and descendant of two presidents, toured the Paris Exhibition and became enamored with the dynamo on display. For Adams and other Progressive era intellectuals, electricity represented a galvanizing social and political force, and the dynamo quickly became an icon for the pro-electricity movement.[4] In an essay titled "The Dynamo and the Virgin," Adams (1918) likened the modern invention and its female form to the Catholic image of the Virgin Mary (385). Adams describes the dynamo's energy as a feminine creative force with the power to magnetize masculine attention (382–389). Predicated such as it is on the Western Christianity, this rhetoric reasserts Euromerican dominion not only over a feminized landscape but also over a feminized and fetishized idea of technology as helpmate in the project of bringing nature to heel (see, e.g., "Grand Coulee" [1940]). Such visages can also be found in the numerous promotional brochures in which hydroelectric construction projects are portrayed as heroic efforts struggling against a wild and feminized "Nature" to be tamed by "Man." These brochures are but one example of the ways in which the Euro-American/Christian mythos (discussed in Chapter 2) shored up extractive capitalism. Machines and technology represented the hardworking "slaves" and compliant "helpmates" in the project of taming the forces of nature ("Grand Coulee"; see also Dietrich 1995, 249–270).[5]

With the president being an engineer himself, Hoover's Boulder Canyon Project Act became law in 1928, promising coast-to-coast electrification of the nation (see, e.g., Worster [1985], 213–261; and Dietrich [1995], 249–271). That same year, dramatist Eugene O'Neill wrote *Dynamo* (1929), a cautionary tale about society's increasing infatuation with technology. Among other themes, *Dynamo* exposes the dangers of blind faith in the marriage between capital and mechanization. As the New Ecologists increasingly understood nature as an organic machine, machines themselves were envisioned as extensions of nature. Technology merely augmented natural forces by bringing them under the guidance and control of human beings. Indeed, Richard White (1995) has observed that machines during this period, "could replace bodies. Machines overcame nature [and] they seemed but a new manifestation of natural forces" (30). Likewise, in *Rivers of Empire*, Worster (1985) notes that Marxist theory conceived of machines as a kind of "second nature" (26) – the product of human effort applied to natural resources. Popular, scientific and philosophical discourse alike viewed science and human managerial expertise as the tools to bring the environment into efficient working order, ultimately "improving" on the work that nature had begun. In *Machine-Age Ideology*, John M. Jordan (1994) observes that the "storm of invention" and consequent material progress of the early twentieth-century United States saturated popular culture with "an inorganic machine aesthetic" (3), one that ultimately fetishized and feminized inventions like the dynamo. O'Neill, in particular, seized on the dynamo as the central metaphor for his story about both the corrosive effects of the nation's puritanical values and the destructive power of an equally blind faith in technology.

The plot of *Dynamo* is reminiscent of *Romeo and Juliet* – two families, neighbors in a small east coast town, experience the fallout of their bitter differences through the lives of their children. Reuben, the play's protagonist, is the son of the religious Reverend Light, who raised his son to be suspicious of electricity. Reuben is in love with Ada, the archetypal "girl next door," whose own father (Mr. Fife) is both an atheist and an engineer at the local hydropower plant. Divided by their extreme ideologies, Reuben's and Ada's respective parents forbid their relationship. To escape his overbearing and religious parents, Reuben runs away from home and has a conversion to an extreme rationalist worldview. When he returns from his exodus, he has become cold; he abuses Ada and seeks counsel from the dynamo in the power plant, which has become for him a kind of goddess. As in Shakespeare's tale of star-crossed lovers, *Dynamo's* pair also perish; Reuben eventually takes Ada to the power plant to introduce her to his idol. But Ada openly mocks his new obsession. Meanwhile, Reuben believes that the dynamo has rejected Ada, and claiming a mandate from the dynamo herself, he shoots and kills Ada. The play concludes after Reuben sacrifices himself to the goddess he claims to worship, and he is electrocuted.

Throughout the play, Reuben endows the dynamo with a feminized capacity to restore the human spirit, and his reverie echoes the poetry of the wilderness cults. Like those who made pilgrimages to the wilderness in order to wash away the soil of civilization, in Act 3, Scene 1, Reuben finds his absolution in the machine that embodies the essence of that very wilderness:

> It is like poetry. Her song in there – Dynamo's – isn't that the greatest of all – the poem of eternal life? And listen to the water rushing over the dam! Like music! It's as if that sound was cool water washing over my body! – washing all the dirt and sin away! Like someone singing me to sleep – my mother . . . Dynamo – the Great Mother.
>
> *(132–133)*

The preceding excerpt illustrates how Reuben's Dynamo represents a fusion of Henry Adams's (1918) notion of the dynamo-virgin and Worster's (1985) analysis of the Marxist concept of "second nature." The machine, like the Virgin, becomes the sacred mediator between man and nature. In *Dynamo* in particular, the machine has replaced the godhead figure of religious idolatry:

> [Dynamo is a] great dark idol . . . like the old stones statutes of gods people prayed to . . . only it's living and they were dead . . . that part is like a head . . . with eyes that see you without seeing you . . . and below it is like

> a body ... not a man's ... round like a woman's ... as if it had breasts ...
> a great dark mother! ... that's what dynamo is!
>
> *(126)*

Here, O'Neill's portrait of a young man who idolizes the Dynamo, is infatuated with technology, and possessed by an irrational faith in rational science, serves to caution society against and over-exuberant faith in technology. Indeed, Jordan (1994) notes the religious fervor with which the ministers of technology sought to reform society according to scientific principles. Under this burgeoning mythos, the engineer supplants the explorer and pioneer as hero. On this point, Jordan observes that throughout the 1920s and 1930s,

> [c]orporate capital, organized research, and applied science combined to demonstrate how humanity could understand and control the natural environment. During a structural transfiguration comparable to the Renaissance or any other cultural earthquake, citizens hailed one shining hero of the moment: the engineer.
>
> *(3)*

This class of civic heroes included inventors, scientists, technicians, and civil and social engineers of all kinds.

Returning to *Dynamo*, early in the play Reuben has a spiritual communion with the electrically charged cosmos that transforms him into a man of science. Although he breaks with traditional religion, Reuben has subscribed to a new faith, one that gives him access to mystery. In his new "enlightened" state, he becomes a religious zealot. Reuben comes to see himself a part of a crusade; he soon desires to become one of the "savior" engineers:

> Her powerhouses are the new churches! She wants us to realize the secret dwells in her! She wants one man to love her purely and when she finds him worthy she will love him and give him the secret of truth and he will become the new savior who will bring happiness and peace to men! And I am going to be that savior!
>
> *(134)*

Machines and the engineers who build them thus become intermediaries between nature and society. For Reuben, this relationship represents a new kind of heroism. Dietrich (1995) observes the popular sense of mystery behind the means to a technological utopia that permeated popular thinking of the time:

> Electricity [had a] subtle effect on American society. An ax, a hand pump, or wood stove were things any common person could understand. Electricity was not. Edison was dubbed by the press the "Wizard of Menlo Park," implying an almost magical knowledge ordinary mortals did not

share. In 1924, *Graphic Survey* magazine noted, "Electricity is such a subtle commodity, at once so beneficent and so terrifyingly destructive, that we regard it and its high priests with a kind of awe."

(252)

As electricity signified the panacea that promised to reinvent society; it was also a harbinger of realizable utopias yet to come. A replacement for coal that burned clean with seemingly unlimited applications, electrification was, literally and figuratively, the illumination of rational thought and the foundation for a new civilization. In 1902, for example, total electrical energy generation in the United States measured 2.5 billion kilowatt-hours; by 1912, this figure had increased 500 percent to 12.5 billion; by 1922, electrical energy usage had exploded to 43 billion kilowatt-hours, which marked a 1,720 percent increase in 20 years (Hughes 1983, 4). As a mythic symbol, "electrification" became the apex of the Industrial Revolution and a miraculous development that confirmed the supremacy of applied science and, as such, came to signify the vigor of capitalism, the expression of democracy, and the victory cry of socialism all in one.

Dust and discontent in *Triple-A Plowed Under*

When Franklin D. Roosevelt took office in 1933, the United States was in the throes of economic depression, the unemployment rate had reached 25 percent, and the nation faced the very real possibility of a socialist revolution (Schlesinger 1988). As historian Arthur Schlesinger notes, the active organizing of urban workers and rural tenant farmers during the Great Depression caused membership in the American Communist Party to increase from approximately 30,000 in 1928 to approximately 300,000 by 1931. When the Communist Party backed the Farmer–Labor Party in the election of 1936, many believed a communist revolution in America was at hand (548–600).[6] Living Newspapers *One-Third of a Nation*, *Triple-A Plowed Under*, *Power*, and *Timber* examined these social problems and others, helping to popularize New Deal programs and policies, including conservation. This unique theatrical form tended to celebrate the "little guy" – worker, farmer, or citizen/consumer – as a new kind of (male) hero.[7]

When *Triple-A Plowed Under* opened in March 1936, the midwest was paralyzed by drought, dust storms, and record temperatures. Meanwhile, the nation awaited the Supreme Court's decision on the constitutionality of the 1933 Agricultural Adjustment Act (AAA), which promised to staunch the financial bleeding among U.S. farmers by using funds from a processing tax to subsidize farmers in exchange for planting only a percentage of their total acreage. The tax on the processor of grain and other farm products was unpopular among smaller-scale farmers because the acreage-reduction subsidies were calculated by taking the average of harvests from preceding years. The system benefited larger farms – those that had expanded rapidly, ignored soil exhaustion, and planted as much and as often as possible to increase profits. Smaller farmers who (either by choice or by economic circumstances) had not expanded or farmers who already practiced crop rotation as a soil conservation measure (e.g., southern sharecroppers and Mexican American farmers in the arid Southwest) were, in effect, penalized and often forced out of business (see Peña [2005], [1998]). Ultimately, the AAA (or "Triple-A" as the legislations was called) was declared unconstitutional, but the 1935 Soil Conservation Act would accomplish similar goals.

Triple-A Plowed Under participated in this public debate by mapping the history leading up to the Dust Bowl of the 1930s and then tracking what transpired in farm country across the United States. While promoting New Deal policies, *Triple-A* advocated for a deeper socialist (and, some argued, communist) platform that argued for common cause between the nation's industrial workers and its farmers. Concerned with the fundamental components of habitat to be sure (e.g., food, water, shelter), *Triple-A* reflects the ideas of the New Ecologists as it argues for a unified socialist approach to problems of both labor and land. Yet its deeper ideology is predicated on what Worster (1985) has called the U.S. "imperial and commercial" relationship with the land (97), which transcends the capitalist/socialist binary through which the era is often historicized. In *Dust Bowl* (1979), Worster similarly observes that

> [t]he American plainsmen, it must be made clear, were as intelligent as the farmers of any part of the world. They were by no means the first to overrun the limits of their environment. But the reason they did must be explained not by that vague entity "human nature," but rather by the peculiar culture that shaped their values and actions.
>
> *(96)*

While *Triple-A* indicts capitalism's profiteering, it misses the correlation between the exploitation of people and the exploitation of the earth. In *Triple-A* farmers are victims of an unpredictable nature, an act of God, rather than the cultural values and economic system to which they have subscribed – a narrative that tends to absolve farmers (and society) of complicity in what happens to the land.

Extractive capitalism and the growing socialist movement in the United States shared common ground in terms of conceptualizations and treatment of the land. In the first place, these ideological movements subscribed to a reductionist perspective that understood both nature and machines as objects (resources, tools) that exist for human use. Within the frameworks of both Marxism and U.S. expansionist capitalism, the earth's resources became the foundation of production and power, the so-called natural birthright of humans. The socialist movement in the U.S., as in the U.S.S.R., produced a climate conducive to a large-scale, unsentimental, and systematized domination of nature. Rather than understanding the material world as a highly complex system with which human beings cannot indefinitely tamper without causing significant and often unforeseen consequences, both capitalist and socialist ideologies advocated mechanized management of nature.

The countervailing views were few. Lewis Mumford, for example, advocated for a new kind of urban planning that considered air, water, and habitat quality for people and de-emphasized the panacea of technology,[8] and Robert Marshall, who continued to lobby for the preservation of wilderness ecosystems for their own sake (see, e.g., Shabecoff [2003]; Marshall [1990]), were the exceptions that proved the rule in terms of valorizing technology. Their voices were, by and large, overshadowed by dominant ideologies that shaped public opinion and policy in favor of aggressively putting the land to work. The rise in rank of the U.S. farmer within the international grain market, the politics of socialist revolution on American soil, and the general valorization of efficient, functional, and scientific approaches to economic and social problems are some of the factors affecting the failure of voices like Mumford's to gain traction in the broader national discourse.

Selling solidarity with sodbusters

In 1936, the country continued to dig itself out from the effects of the drought, while still reeling from the social and economic upheaval of the Depression. Thus, when the New York production of *Triple-A* opened at the Biltmore Theatre in New York City on March 14, 1936, the play was not only a political drama but also a political event. *Triple-A* went beyond protesting the plowing under of favored New Deal programs to serve as a bully pulpit for the Farmer–Labor Party platform. The play ran for 85 performances to full houses in New York, perhaps demonstrating just how resonant its socialist message seemed to audiences at the time (Gilder 1936, 430). On a campaign trail of its own, *Triple-A* was produced in the cities of Chicago, Cleveland, Los Angeles, and Milwaukee during the spring and summer of 1936 (Brown 1978, 10).[9]

Triple-A begins with a speed-through of the agricultural history of the early twentieth century. Through a sequence of fast-paced vignettes that include characters based on real figures of the day, *Triple-A* tells the story of how the sociopolitical and ecological predicament of the time came to be. Film montage, news footage, and projections were incorporated throughout the production, theatrical techniques that gave it a "ripped from the headlines" feel; indeed, the voice of the Loudspeaker as narrator and an omniscient person in the know was a theatrical device taken from the familiar newsreels shown in movie theaters. The cumulative effect of its documentary-style vignettes and theatrical effects argued for solidarity between the nations' farmers and workers by dramatizing common frustrations with government policies in order to build popular support for the Farmer–Labor Party.

The arid plains of the midwest came under U.S. dominion following the Louisiana Purchase of 1803; between 1890 and 1920 these lands would grow green and profitable beneath the mechanical plows of a new generation of sodbusting settlers. Following on the genocidal violence of post-Civil War militarism and economic expansion (see Chapter 1), the settler culture that brought the sodbusters to the fragile yet self-sustaining prairies was, according to Worster (1979), "ecologically among the most nonadaptive [culture] ever devised" (97). During and after World War I, the plains produced a green bonanza as European markets for American grain expanded. Encouraged by government policies, farmers took out loans on new machinery, gambled on the continued rise of grain futures, and plowed up more acres.

When the devastating drought of the 1930s arrived, along with swarms of grasshoppers, erosion, dying cattle, and clouds of blowing dust that darkened the noonday sun, the entire ordeal appeared an aberration in an otherwise ordered system. For farmers who had depended on seasonal regularity, the Dust Bowl was apocalyptic, an omen signifying the derailing of nature's machinery. However, few understood these occurrences as lessons about the interrelationship between human action and the land's natural drought cycles or as the cost of agricultural greed (Worster 1979). Indeed, the Dust Bowl was an invitation for a national ecological self-examination.

Arent's *Triple-A* (1938) opens with reference to rising grain prices during World War I as U.S. farmers plow up the prairies for planting. The first striking visual image of the play characterizes the call to plow as a call to arms, as a column of soldiers marching behind a red-lit scrim. Over the amplified sound of hundreds of marching feet, actors call out to the audience: "Your country is at war. Your country needs you. If you can't fight, farm" and "[e]very hand with a spade is a hand-grenade." Swept up in the frenzy to win the war, the farmer is tasked to "save the nation," "save democracy," "save the world," by growing "More corn! More cotton! More food, more seed, more acres! More! More! More!" As the scene builds, patriotism disguises the profit-driven incentive to expand production: "Your country needs you"; "[I]f you can't fight – farm"; "Every head of cattle can win a battle" (10). These slogans and others grow into cacophonous hysteria over the need to plant more cash crops – wheat, corn, and cotton – to win the war. Between 1910 and 1919, the midwest became a flat and expansive boomtown where industrial agriculturalists turned 80 million acres of native grassland into cash. After the war, not surprisingly, prices plummeted. As the European economy recovered and Europe was able to produce its own food once again, the price of U.S. grain dropped. In response, U.S. farmers plowed and planted *more* acres, glutting the market and contributing to the stock market crash of 1929.

Meanwhile, the U.S. landscape was ripe with other discontents. The previous 15 years had produced a groundswell of radicalism as faith in the U.S. economic system eroded. For many people, the Depression signaled the collapse of the capitalist system on which they had fixed their hopes and dreams. In *The Coming of the New Deal*, Arthur Schlesinger, Jr., (1988) writes that the New Dealers had come to Washington in 1933 charged with what Rexford Tugwell later called "the choice between an orderly revolution – a peaceful and rapid departure from past concepts – [or] a violent and disorderly overthrow of the whole capitalist structure" (22, quoting Rexford Tugwell). The AAA helped to sustain some farmers through the Depression as it answered some of the radical voices in the Farm Belt that were demanding government action. Nonetheless, the ecological disaster of the Dust Bowl in the plains, and the general drought that affected an even wider swath of the productive midwest, compounded the farmers' predicament and complicated governmental responses. During the 1930s, blowing dust fueled a growing spirit of revolution in farm communities as surely as kerosene. In some ways, it seemed as if the soil itself was staging a revolt: nature, it seemed, had gone on strike.

Triple-A argues for socioeconomic solidarity between the Farmer and the Worker who must join forces against a greedy middleman and the double-talking government. At the center of the play's argument for solidarity between farmers and workers was the question, Why are urban families going hungry while farmers are forced to watch their 'surplus' grain rot, or kill and 'plow under' pigs and other livestock in order to preserve higher prices? Meanwhile, the character of the Farmer puzzles over his own legally mandated waste: "Damn it, I bought more land and cleared all the woods on my place, and planted it to wheat, and now it's rotting in the fields" (13). As this line reveals, *Triple-A*'s Farmer is shown to have been manipulated by financial and political interests beyond his control and comprehension, rather than a willing participant in profit-driven agriculture. Workers can no longer afford the cost of bread, and so they go hungry; likewise, Farmers cannot afford to purchase manufactured goods, and so they must watch their grain rot in the fields. Thus, at a time when many U.S. farmers shared blame for the overuse of soil and a boom-and-bust commodity market, *Triple-A* worked to reframe these sodbusters as victims of poor government policy and the rural counterpart to industrial workers.

In another early scene, the Farmer and the Country Banker make reference to "the Big Boys in Washington" (13). This is the first of numerous references to big business, middlemen, merchants, and government officials that set *Triple-A*'s Farmer apart from those accountable for the system. Likewise, in Scene 10, interspliced lines of Farmers and Workers rally in common cause, foregrounding the economic interdependence between them:

WORKER: We starve and they told us you had food in your fields.
FARMER: Food is in our fields but they told us you would not pay the cost of its harvesting.
WORKER: We had no money.
FARMER: We raised eggs and milk, and you wouldn't buy them.

The scene builds into a cacophony of "Feed us!" "Pay us!" "Feed us!", faulting the missteps of a government that arguably had done little to regulate economies of boom and bust (26). Thus, *Triple-A* naturalizes, and oversimplifies, New Ecology's idea of nature as a chain of economic relations as it promoted solidarity between farmers and workers as naturally interdependent producers and consumers. In *Triple-A*, Workers, like Farmers, take on the role of plants, producing the goods consumed by all members of society – a natural circle of exchange.

As it works to build a rationale for the Farmer–Labor Party, the plot takes a turn in Scene 16, titled "Drought," and the remainder of the play maps the drought's impact on farmers through scenes in which small-scale and tenant farmers lose their land to farm consolidation as larger landowners gobble up the smaller ones. In this way, *Triple-A* functions to connect these land-justice issues to the complaints of labor in industrial centers, and finally to consumers. These widening circles of impact and common plight are shown in scenes in which protest expands to include housewives who refuse meat at high prices and organize a boycott. In yet another troubling scene, a mother drowns her child because she cannot feed him. Based on actual events, this scene in particular brought home for audiences the bankruptcy of the U.S. economic system. Like many of the Living Newspapers, *Triple-A* concludes in the uncertain present of its era. In January 1936, just three months prior to the play's premiere, the Supreme Court heard arguments on the constitutionality of the Agricultural Adjustment Act. In the final outcome, the court ruled that the power to regulate was reserved for the states. The court concluded that to levy a tax on one element of an economic chain – the processors – only to be paid back to the farmers was beyond federal authority.

Triple-A's writers historicized the farm crisis but did not examine the human behaviors that caused the Dust Bowl; moreover, they did little to recognize the ecological responsibility of the farmer or the government for the well-being of the land. The play's overarching indictment of the capitalist supply-and-demand system exempts farmers from blame by casting them as victims of the system at the same time as it characterizes the black blizzards as an act of God unrelated to human action. *Triple-A*'s messaging places blame on the middleman (e.g., the Wall Street trader in Scene 15) who rakes in profits, while workers and farmers in need of one another's products cannot afford to buy them. While this critique does call out systemic injustice against farmers and workers, *Triple-A* ultimately falls short of a more critical examination of the extractivist roots of that system. Instead, it tends to naturalize the exploitation of the land even as it condemns the exploitation of humans, sending a message that while the farmer and the worker may have obligations to one another, neither have any responsibility to the land.

At the time, mainstream U.S. agriculture was characterized by unsustainable farming practices that degraded the land on which crops were cultivated. Alternative models, such as traditional tribal cultivation or dry-land farming of Mexican American farmers in the southwest, went unrecognized. The ecological culpability of farming was undeniable, and yet in *Triple-A* the farmer is portrayed as a hero: a hardworking contributor to the prosperity and foundation of the national economy. This Farmer, the play argues, is a national icon who deserves to be fairly recompensed for his contributions. Indeed, *Triple-A* relies on the mythos of the farmer as the very symbol of the nation's entitlement to land and liberty. To critique such a figure would call into question the legitimacy of the United States of America, for if the farmer is not managing the land well, what is the justification for taking homelands away from indigenous residents, save white supremacy (Dunbar-Ortiz 2014, 170–174)? To its credit, *Triple-A* did foreground injustices of the capitalist system against small family and tenant farmers; but the play's tendency to conflate corporate and tenant farmers effectively exonerates industrial agriculturalists as a whole from responsibility for contributing to the Depression. Furthermore, the play fails to acknowledge the farming industry's ecological culpability for the Dust Bowl, a responsibility shared, albeit to different degrees, by corporate and small family farmers alike. Nevertheless, the fact that *Triple-A* explores and asserts these causal connections marks an important step toward understanding – and, most relevant here, toward finding ways to represent on stage – complex ecological interrelationships.

The cost of dust

In *Triple-A*'s Scene 13, titled "Wheat Pit," a trader's feeding frenzy drives the price of wheat higher and higher, as an image of a large clock which "[i]nstead of numerals depicts the months of the year. It has only one hand [that] revolves slowly through the playing of the scene." On the same scenic background a giant thermometer "to indicate increasing heat" does not move. Meanwhile, the rising price of wheat is cheerfully heralded by the Loudspeaker as "Fair and warmer" (30), foreshadowing the rising temperatures and the perfect storm of ecological and economic to come.[10] The scenes that follow point out what many believed to be a flaw in the AAA – the processing tax was passed on to the consumer. Thus the worker paid for the farmer's relief, while the middleman lived on the hog. Milo Reno, leader of the Farmer's Union, warns:

> I'll tell you who will suffer when the farmer and the workingman get together, as they are going to do. It's the usurer. It's the middleman who preys upon both, without performing any useful service. He's got to have his claws clipped, and he's going to.
>
> *(33)*

Milo's metaphoric language characterizes the commodity traders as predatory, like cats with claws. In Scene 14, a "shabbily dressed" Customer cannot afford even an herbivore's breakfast due to the processing tax that has been added to the price of oatmeal; in the following scene, a Man in Evening Clothes in a Park Avenue Restaurant eats from the top of the food chain. The man orders exotic and colonized prey – including "beluga caviar" and "royal squab." He tells the Woman in Evening Clothes that despite the AAA, he continues to make big profits in wheat because the processing tax gets passed on to the consumer.

The man explains how the system works: "somewhere between the harvesting of the wheat and this roll there was a processing tax . . . and it's the man who eats it who pays it." As proof of his successful white-collar hunt, he asks the Woman, "[S]o what will it be, a new car or a sable coat?" (33). Deflecting blame from farm practices toward Wall Street belies the connection between the two, in which the land was the basis of the grain commodity market. Ignoring the culpability of U.S. farmers in the causes of the Dust Bowl also disassociates them in the minds of the audience from the accompanying socioeconomic effects, including mass migration in which refugees from dust and farm consolidation displaced other workers – for example, Mexicans and Filipinos in California (see, e.g. Wald [2016]).

When *Triple-A* opened in 1936, the U.S. plains states had already experienced three years of severe drought, with no promise that the dust storms and dislocation of people would end soon. In Scene 16 of the play, titled "Drought," the narrative picks up the chorus of "fair and warmer" once again, while the temperature rises on the giant scenic thermometer. At this point in the play, history catches up with the present. This scene ends in "shrill despair" as "[t]he FARMER who is examining the soil straightens up and slowly lets a handful of dry dust sift through is fingers" (33). The Dust Bowl was "the most severe environmental catastrophe in the entire history of the white man on this continent," Worster (1979, 24) argues. As a symbol on stage, that dust pointed to not only the effects of overproduction but also the way in which land had become synonymous with settler hope, with both individual and national freedoms, with the very settler sense of entitlement to a prosperous future so often called "the American dream." It was the stuff those dreams were made of, or so it may have seemed to those who witnessed clouds of Kansas dust over New York.

The Dust Bowl was not the first settler-caused ecological disaster on U.S. soil. It is dwarfed beside the state-sponsored assault on indigenous life and lifeways during the eighteenth and nineteenth centuries (as well as before and since), which constituted the destruction of human ecology. The extractive industries of cotton and tobacco in the southern states depended on the abuse of land and the violent extraction of labor from human beings. These two – indigenous genocidal policies and the institution of slavery – constituted settler-caused environmental change on an order of magnitude unsurpassed in the history of the continent. In many ways, the Dust Bowl was fallout from these larger systems of profit making in which land and labor were formed the scaffold of wealth extraction and accumulation. The Dust Bowl stands as well among other environmental debacles on a grand scale, including the state-sponsored decimation of bison populations in the plains (carried out as a weapon of genocide), the overhunting of sea otters, the systematic eradication of wolves and coyotes from farmland, and deforestation in the eastern states, as well as the damage to watersheds from surface mining in Appalachia and hydraulic and hard-rock mining in the west, to name a few. Some of these ecological losses were understood at the time as positive signs of so-called progress; others remained local and regional losses that lacked the visible drama that captured national attention. Although its epicenter was in the Great Plains, the Dust Bowl sent shockwaves across the country. Photographs of the Farm Security Administration (FSA) documented a human and ecological disaster of immense proportion. Published in the mainstream press, these images spoke to the whole nation not only because grain and cotton commodities were central to U.S. economic prowess in the world but also because it was happening, in major part, to white people farmers who stood at the symbolic center of the American project.

Beginning with high temperatures in 1931, 1932, and 1933, by May 1934, the dust storms began and continued to worsen over the following two years. Scholarly accounts, such as Schlesinger's in *The Politics of Upheaval* (1960, 69), describe apocalyptic mushroom clouds of dust choked homes, cars, tractors, and schools in the plains states and carried dust on the wind all the way to the east coast. Kansas dust settled on the shoulders and briefcases of Wall Street tycoons, matted the hair of homemakers shopping for food in Chicago, and stung the eyes of miners on strike in Philadelphia. Oklahoma grit found its way into the mouths of the hungry and the pockets of the unemployed; Montana dust settled on eastern playgrounds, fairgrounds, city streets, and sidewalks. Miles off the Atlantic coast, Kansas's dust settled on the decks of ships bound for Europe. In the breadbasket of America, the 11 million acres of native grass that farmers had been encouraged to plow up during the Great War, 350 million tons of topsoil had up and left. With it went capital investments and cash; creditors called in mortgages and loans on farm machinery. After touring the blow area, head of Franklin D. Roosevelt's Resettlement Administration, Rexford Tugwell, called the region "a picture of complete destruction" and "one of the most serious peacetime problems in the nation's history" (Worster 1979, 12).

The worst black blizzard to strike Colorado, Oklahoma, Kansas, Nebraska, and northern Texas and New Mexico occurred 11 months to the day before *Triple-A Plowed Under* opened on April 14, 1935. On what became known as "Black Sunday," a mushrooming cloud moved silently across the Oklahoma panhandle, leaving a path of death and destruction behind. It carried away twice as much earth as had been dug up to construct the Panama Canal. For many Euro-American farmers, it seemed like an Armageddon. Eleven months later, *Triple-A*'s audience could expect the drought to continue; temperatures and dust storms did not ease up until 1939. By the time *Triple-A* was produced in Chicago and Los Angeles in the summer of 1936, papers headlined climate catastrophes for the farmer and his crops: "Blistering Weather Perils Crops. . . . No Relief in Sight" (1936) and "Weather Man Sees Little Change in Drought Conditions" (1936); *Newsweek* called the country a "vast simmering caldron" in 1936; 4,500 people died from excessive heat in 1936 alone (quoted in Worster 1979, 12).

Adding to specters of the unnatural, fields of grain, cash crops such as cotton, and animals were destroyed as part of enacting the AAA. Under reduction contracts, farmers slaughtered their own pigs by the thousands and buried them in the fields rather than bringing them to market. Many members of society were shocked, wondering why such waste could occur when so many hungry unemployed waited in bread lines. Schlesinger (1988) observes that "on the farms men reacted in strange ways. Some saw in the drought and dust the judgment of God on men who had dared plow under cotton and slaughter baby pigs. A demand arose that the AAA be abolished" (71). In a scene depicting these historical events, the play differentiates the suffering farmer from the profiteers who speculated on grain futures. Yet *Triple-A* also simultaneously portrays the drought and dust storms as acts of God rather than the result of humans' misuse of the land. In Scene 17, for example, Judeo-Christian mythology of the earth is purposed to man's benefit, as the pastor pleads with members of his congregation in prayer:

> Our great state has been burned dry. The showers of dust come in clouds so dense as to obscure the midday sun. The corn crumbles to dust at the touch of our hand, and the stalks lie dried and curling in the heat. O God, heavenly Father, who has blessed the earth that it might be fruitful and bring forth whatsoever is needful for the life of man.
>
> *(35)*

In counterpoint, voices offstage continue: "Fair and warmer, fair and warmer"; the pastor and congregation pray for "seasonable weather" so that they might "reap the fruits of our labors in the fields," while a projection of dying cattle slowly dims in on the scrim (36). The dust and dying cattle were more than an act of God; they *were* the fruits of the farmer's labor. Yet *Triple-A* largely fails to articulate human complicity with what had transpired to and on the land. Instead, the drought is represented as an aberration.

The drought of the 1930s was not an abnormal weather pattern for the arid region of the continent, and the prairie grass ecosystems had long held the land in place for the people and moving herds of bison and indigenous residents. Poor farming practices, along with a singly commercial regard for the earth, exacerbated the effects of the drought. The dust storms, Worster (1979) observes, "were mainly the result of stripping the landscape of its natural vegetation to such an extent that there was no defense against the dry winds, no sod to hold the sandy or powdery dirt" (13). In a 1933 *New Republic* interview (Bliven 1933), Milo Reno argued in the farmer's defense:

> They tell us that the farmer is just a gambler that he speculated in land at high prices and now should take his medicine like the people who got caught on Wall Street. . . . Thousands of farmers didn't speculate, or buy more land than they could handle. . . . Remember also that when the farmer speculated in land, there were other people who aided and abetted him. . . . The banks were crying for the privilege of lending him the money on easy terms.

Ironically, Reno also best expressed the singly commercial regard for the earth when he said, "God made the land. What other basis is there for its worth except its earning power?" (63). Thus, the play uses the notion of land as commodity as the ground on which farmer unrest and even violent direct action stand.

Strike!

Triple-A paints a picture in broad strokes of what the radicalization of U.S. farmers could look like, hinting at internal divisions and marked differences in circumstances and lived experience, and yet, its argument is predicated on finding common ground – a rhetorical project which often leads to erasing those very differences. When *Triple-A* opened, workers at the Firestone rubber plant in Akron sat down on the job in what became a month-long strike; while the play was under production in Chicago, the Fisher Body plant in Flint, Michigan, experienced the longest sit-down strike in history (Zinn 1995, 391). Howard Zinn notes that in 1936 there were 48 sit-down strikes; by 1937, only one year later, that number had grown to 477 (383–393). Having laid out the history and made a case for the farmer's ecological innocence and the system's guilt, the second half of *Triple-A* engages a rhetoric of revolution through a series of "Strike!" vignettes that show that the farmer is the class equivalent of the striking worker, thus linking its themes to the mounting unrest outside the theater.

Yet solidarity between farmers and workers remained a tough sell at the time. Outrage alone did not make the farmer the class equivalent of the urban factory worker. While common cause between miners in Philadelphia and autoworkers in Michigan or longshoremen in San Francisco made sense, the unification of such a heterogeneous group as American farmers did not. Unlike industrial laborers, who could make common cause as working-class people, farmers included groups as diverse as corporate tycoons and sharecroppers. Strikes and protests by farmers were riddled with discord and varied in their expressed demands. In these ways and others, *Triple-A* simplified the radicalization of farmers and their internal divisions in order to promote the Farmer–Labor Party.

When Milo Reno founded the Farmers Union in 1927, he was branded a socialist. By 1932, however, Reno was considered too accommodating by increasingly radical segments of the farm community. In Scene 4 of *Triple-A*, the character of Milo Reno bellows the farmers' demands, which include paying no taxes until farmers have "cared for our families" and paying no debts until they had received the "cost of production" (15). Writing for *The New Republic* at the time, Bruce Bliven (1933) attempted to sort the "radical" farmers from those who "were not causing any trouble" in a 1933 profile of Reno (63). Using results of a *Des Moines Register* poll, Bliven determined the farmers who were resorting to violence were a minority. "Farm unrest varies in accordance with special local circumstances," he writes. Quoting a "wise Iowan"; Bliven observes that many farmers were still "making a pretty good living" and "are not joining any protest." In addition, there were others who were "completely flat on their back" from drought, grasshoppers, and debts who "don't do any hollering" either: "It's where the farmers had something a few years ago, and have had it suddenly taken away, that the agitators find a responsive audience" (64).

Going forward in the play, *Triple-A* differentiates the activist farmers from the sellouts in scenes that depict strikes and direct action. Scene 5 concludes as Reno signs an agreement with a merchant while voices off-stage shout, "Strike! Strike!" In Scene 6, actors agitate from the theater auditorium with cries of "Forget Reno!" and "We've got to solve our problems . . . or there will be those who solve them with bayonets!" The scene builds to a pitch of protest as actors seated among the audience join in: "We ain't scared!" and "We'll fight if we have to!" (18). Scene 7, "Milk Strike," followed directly with "the ripping and smashing of boxes" off-stage and cries of "Dump the milk!" and "Turn over the truck. Push!" (19–20).[11] The mayhem depicted on stage arguably intended to rally midwest populists to the side of socialism by demonstrating common cause between urban consumers and rural farmers. But the U.S. farming community was not a single socioeconomic class and included a spectrum of classes from the corporate farm boss to the sharecropper, from the radical Minnesota dairyman to the wheat king supporters of Huey Long.

Moving away from the dairy country of the north, the next scene shifts to the cotton and tobacco-ripe south, gathering up place-specific problems into one generic farmer revolt. When *Triple-A* opened in New York, newspapers – particularly the liberal and radical press – were full of stories of violence, protest, and the inadequacy of the AAA program for black sharecroppers. In the 1930s, sharecropping in the southern states perpetuated the conditions of slavery as white landowners mined both black labor and the land through a system of sharecropping equivalent to indentured servitude and peonage. In *Slavery by Another Name*, Douglas Blackmon (2008) writes that "[r]oughly half of all African Americans – or 4.8 million – lived in the Black Belt region of the south in 1930, the great majority of whom were almost certainly trapped in some form of coerced labor" (375). Cotton and tobacco were profitable only if labor was cheap, and black sharecroppers, who were essential to the cash-crop economies that had depended on slavery, lived under a system of abuse and de facto serfdom largely ignored by the U.S. justice system:

> [B]lack men and their wives and children were compelled to remain at work for year upon year to retire so-called debts for their seed, tools, food, clothing and mules that could never be extinguished, regardless of how much cotton they grew in any year.
>
> *(376)*[12]

Financial bondage was enforced through threats of violence and terrorism. By 1930, there were nearly 2 million tenant farmers in the southern states. By 1935, the year before *Triple-A* opened, almost half of white farmers and three-quarters of black farmers were sharecroppers.[13]

In Oklahoma and Kansas, the areas of the country hit hardest by the droughts of the 1930s, nearly three-quarters of all farmers were sharecroppers, effectively upholding a racist system that reflected the legacies of the land booms of the previous generation. When the global grain market plunged after World War I and landowners were forced to cut costs, these families, in turn, were forced off the land they farmed. The cumulative effects of the drought compounded by acreage reduction contracts made conditions worse. Under the AAA, tenant farmers were required to receive part of the subsidy in proportion to the land they farmed. However, in practice, landowners did not pay tenant farmers, and as loans on machinery were called in or could not be paid, tenant farmers were forced off their land.

Scenes 19 and 20 take up the double bind of the black sharecropper, whose mule is aptly named "Guv'ment," as the loudspeaker proclaims, "Three hundred and seventy-five thousand sharecroppers lose their places in acreage reduction" (38–39). In April 1936, *The New Republic* published an editorial titled "Starvation in Arkansas," calling attention to the "immediate, day-to-day crisis [facing] the evicted sharecropper in Arkansas." The article goes on to say that

> they will be facing starvation during the coming year. Relief has been withheld from them and officials . . . have thus far said they are powerless to force local authorities . . . to extend relief . . . tenants charge that they have been wrongfully evicted from their original plantations because they joined the Southern Tenant Farmers Union.
>
> *(209)*

The American Communist Party worked to unionize the southern black sharecroppers. This *Triple-A* scene recounts the compounded risks for black sharecroppers threatened with violence by "night riders" for joining the union (Zinn 1995, 386–387; and Grubbs 1971, 30–61).

In Scene 21, "Meat Strike," *Triple-A* works hard to broaden the base of support and the popular demand for the solutions that the AAA enacted, weaving interconnection between demographic groups of producers and consumers. A group of women are caught up in the revolt, as the loudspeaker announces, "Housewives rebel!" As the scenes progress, revolution builds on stage, reaching wider circles of Consumers and Workers who join the Farmers in protest and in violent direct action. Women carrying banners demanding a 20 percent reduction in meat prices prevent a man from buying meat, calling out, "Don't let him pass!" "Strikebreaker!" and "Get the package!" (41). The Housewives rip apart the man's package like a "furious mob intent upon tearing him to pieces," and the scene ends with a crowd stopping a meat truck, and then "Soak the meat in kerosene!" (41–42). Unlike the workers and consumers *Triple-A* depicts, wealthy farmers had no vested interest in protesting against the AAA, which tended to benefit them. The "Strike!" scenes work to erase such difference and solidify the identification between the two. From farmer to worker, north to south, countryside to factory, city sidewalk to family kitchen – the revolution appeared in *Triple-A* to be an inevitable force of nature.

Visages of the *un*natural

Triple-A employs the idea of what is and is not "natural" as leverage to make its case against the capitalist system. A variety of images, scenes, and tropes unfold toward the logical conclusion that the current capitalist system is "unnatural" and the pro-socialist Farmer–Labor Party is the "natural" alternative. Among the unnatural phenomenon are farmers setting fire to their own harvests, fresh milk running white in the muddy ditches of country roads, the extraordinarily high temperatures of the drought, cattle suffocated in drifts of sand, fertile land turned to dust, and men and women pouring kerosene on fresh meat to set it on fire. The play calls up unnatural images, including the slaughter of baby pigs and the plowing under of newly planted grain, which occurred as part of the implementation of the policies under the AAA. Secretary of Agriculture Henry Wallace later said that the policies of the AAA went "against the soundest instincts of human nature" (Schlesinger 1988, 62–63).

In Scene 22, titled "Mrs. Sherwood," *Triple-A*'s critique of the system crescendos in the most *un*natural sights in the play. Based on an actual case, in this scene a woman carries her dead child into a police station where she confesses that she drowned him because she could not feed him. This is a case of nature run amuck, this plot device suggests to audiences. An officer leads the woman away, crying out, "I was hungry, hungry, hungry!" while voices chant in chorus, "Guilty, guilty, guilty!" (43). The unnatural action of Mrs. Sherwood, the play argues, indicts a system in which the economic abundance of some meant privation and starvation for others.

Triple-A's final scenes include an appearance of the character Earl Browder, leader of the Communist Party of the United States as the Supreme Court Justices posed like a dozen grim reapers in silhouettes against a huge projection of the U.S. Constitution. At the time, the closing scene functioned as a rallying cry for the Farmer–Labor Party, one of the strongest third-party movements in U.S. history. Farmers fill the stage to demand that the benefits provided under a new law – the Soil Conservation Act of 1935 – must be equal to those under the AAA. Meanwhile, Workers join them, demanding government respond to their needs for jobs, food, and a fair standard of living. The character of Secretary of Agriculture Wallace brings the two groups together in a visual representation of the unity of purpose on which New Deal policy success depended. But it remains an unsteady truce; the play concludes as Farmers and Workers, joined by a Chorus of Women who continue to press their government with cries of "We need help!" "We need food!" and "Jobs! . . . Jobs!" The play ends with a note to end with "[n]ews flashes of events that have occurred that day" to conjoin the theatrical revolution with the world outside the theater in a dynamic press for change (55–56).

Today, *Triple-A Plowed Under* stands as an important precursor to contemporary ecodrama both for what it accomplished and for what it obscured. The Living Newspaper form opened up an epic story that details the history and social context of a crisis and explained, at least in part, the interweaving of economic and social forces with impacts on lives and land. This play connects the dots between extractive capitalism and a web of social and economic relations that must be also understood as ecological relations. As *Triple-A* suggests, food – its production, distribution, and consumption – is at the very heart of ecological relations, not only for humans but for all species. Indeed, food is the very site of our rootedness in, and dependence on, the land. To examine the economic systems that impact these relations of sustenance and call out the way we are all, ultimately, related was a new theatrical accomplishment. *Triple-A* avoids a climactic plot structure in which a singular protagonist struggles. Instead, in its episodic form we can observe a dramaturgy in which the protagonist is really a *community*, a pluralist electorate. It moves away from a central hubris or mistake on the part of a protagonist, toward an exploration of multiple and intersecting causes of any given circumstance. *Triple-A*, like the other Living Newspapers of the FTP, demonstrates a daring use of the theater as a forum for civic debate, activist engagement, and even direct action that would inspire radical theater of the 1960s.

Triple-A exonerates farmers who were agro-industrial capitalists in order to align farmer and worker in a common cause. The overthrow of the capitalist system in favor of a socialist solution, signified by the play's resounding Finale in favor of the Farmer–Labor Party, did not alter the ideology that both perspectives shared: an extractivist reduction of the land to nothing but inert resources for human use. Yet, while *Triple-A* seems to overlook human complicity in the environmental catastrophe of the Dust Bowl, it does redirect public analysis from individual stories of loss to *the system* that impacts humans and land. This move toward systems thinking is ecological if only in the way it emphasizes interdependence and demonstrates that humans are part of, as well as agents in, a larger web of relations. In this way, *Triple-A* provides a useful model for theatrical explorations of the complex socioecological/socioeconomic webs of today's most pressing environmental problems.

The case for *Power*

When O'Neill's *Dynamo* appeared in 1929, Congress had recently passed the Boulder Canyon Project Act (1928), which authorized the largest-to-date hydroelectric dam in the United States (see, e.g., Worster [1985]. Aimed at flood control and improving commercial navigation on the Colorado River, Boulder Dam (later renamed Hoover Dam) would provide agricultural irrigation to California's

Imperial Valley, and electric power for the growing communities and industries throughout the southwest. Construction began in 1931, as the nation was reeling from the effects of the stock market crash of 1929, and completed in 1936, during the New Deal. When Roosevelt took office in 1933, he not only pressed for the completion of the gravity dam on the Colorado but also initiated several similar massive public works projects, including the Tennessee Valley Authority (TVA) on the Tennessee River[14] and the Grand Coulee Dam on the Columbia River. Big dams on the big rivers of the west represented a solution to a raft of social and ecological problems: flood control and irrigation for farmers in the southern states and arid west, electric power for industry, and jobs. As an example of regional power development, TVA was aimed at social and ecological damage caused by decades of post-Civil War social and environmental neglect and abuse. Electricity from TVA, its director, David Lilienthal (1953), promised, would "relieve human drudgery"; conquer poverty, disease, famine, and floods; and yield the material fruits of a new age (3).[15] Like the Hoover Dam, the Grand Coulee Dam, begun in 1933 by BPA, promised to open up and increase settlement, farming, and industrial development in the arid western states. The caution implicit in O'Neill's *Dynamo* faded in the face of the cry for jobs and rhetoric stressing economic growth and the social benefits of electricity.

In light of growing totalitarian aggression in Europe, the dams themselves became performances of state power. Touted as "the greatest things on earth" (Harden 1996), the New Deal's concrete gravity dams also served as metaphors for rising U.S. international power. In *A River Lost*, Blaine Harden notes that the big dams were "an eye popping metaphor for Manifest Destiny . . . [a] vaccine for the Great Depression and a club to whip Hitler" (79). Indeed, the word *superpower*, used in the heat of the Cold War and arms race of the late twentieth century, originated as part of the discourse of hydropower development. The words of Woody Guthrie that open this chapter promote the Grand Coulee Dam, the "biggest thing yet built by human hands," and construct the river itself as a co-worker in the project of building a new world order. Governments charged their scientists and engineers to identify, map, and develop coal, oil shale, petroleum and natural gas deposits, and hydropower. Rivers were sources of "white coal" that appeared free for the harvesting (Schwartz 1993, 202).

In 1937, as construction on the Colorado, Columbia, and Tennessee Rivers proceeded, and public debate over the first federal projects to challenge private electric companies ran high, the FTP's creative team produced another Living Newspaper titled, aptly, *Power* (Arent 1938). Fast-paced, episodic, and chock-full of current events and public figures, *Power* gave rhetorical support to the New Deal's argument that access to electricity should be a right for all, not just a luxury for those who could afford it. Productions of *Power* across the country raised awareness, educated the public about the benefits of electricity, and advocated in favor of the administration's electrification and hydropower projects.[16] *Power* celebrated the era of big dams as it promoted the idea of publicly owned utilities, and championed the necessity of electricity in every American home.

When *Power* opened at the 43rd Street Theatre in New York on February 23, 1937, the Supreme Court again weighed the constitutionality of TVA. Like *Triple-A*, *Power* worked to create solidarity between urban and rural Americans, building common cause, for example, between the City Dweller who was paying too much and the Farmer who could not get electricity at all because the private company would not pay to build the poles and wires. *Power* worked to fire up U.S. citizens by depicting (like the striking housewives in *Triple-A*) as radicalized consumers who understood and exercised their civic power. In Scene 15, for example, the Wife, Nora, complains further, asserting that electricity is a right, like clean air, water, and shelter, and that every U.S. citizen has a right to a certain standard of living. She presses him to outrage and action:

FARMER: Nora, if they don't want to string lights out to my farm, I can't make 'em.
WIFE: Who said you can't? Who said you can't go up there and raise holy blazes until they give 'em to you! Tell 'em you're an American citizen! Tell 'em you're sick and tired of lookin' at fans and heaters and vacuums and dishwashin' machines in catalogues, that you'd like to *use* 'em for a change!

(63)

The oft-quoted 1925 quip of President Calvin Coolidge, "the business of America is business," meets a compelling rebuttal in *Power*. The government is accountable to the well-being of its citizens and has an obligation, a public trust, to ensure that well-being – not just the well-being of business. Electric companies had challenged the government's role in the production and supply of electricity, asking the Supreme Court to decide whether utilities should be owned by municipalities (i.e., the public) or only by private businesses. The final scenes in *Power* show a gathering momentum of enthusiasm across the nation for municipal power, as well as the pushback that communities encountered from private utility companies. *Power*, like *Triple-A Plowed Under*, ends where civic action begins. In a rousing rhetorical commitment to inspire civic engagement, the final scene gathers the Farmer, the Businessman, the Consumer, and the City Dweller in solidarity to demand their collective right to electricity in the face of corporate interests.

The argument for public utilities, like the notion of public lands developed by Gifford Pinchot, began to set a precedent for nature as a commons. Yet, the socialist-leaning viewpoint expressed in both *Triple-A* and *Power* was predicated on a more fundamental ideology, one that conceived of the natural world as inert objects and systems ready to be shaped and managed by and for the benefit of human beings. Ironically, as Carolyn Merchant (1994) has observed, the rhetorical gendering of nature worked to further render the natural world as submissive and passive. In this 1926 dam promotional brochure, for example, a feminized nature has provided everything but the brains:

> Nature must have designed this spot to cradle a lake. As she made the sheltered valleys and wide alluvial plains, and invited man to make them

fruitful by his labors, so she hollowed out these great spaces in the granite hills and waited patiently for the men of wide vision and vast resources to wall up the narrow outlets with their titanic buttresses of steel and concrete.

The view that nature finds its fulfillment in service to humans, that human culture represent the very apex of geologic intent, undergirds both capitalist and socialist views of nature.

(quoted in Jackson 1995, 6)

In 1942, Rufus Woods (1944), editor of the *Wenatchee Daily World*, told a graduating high school class that the dam would mark their era's place in the history, not of the century, but of the millennium:

So there it stands, a monument to the idea of power and the power of an idea; a monument . . . to the magic spirit of willing men which accomplishes more than the might of money or the marvels of machinery . . . and you, class of 1942, could you come back here a thousand years hence, or could your spirit hover around this place ten thousand years hence, you would hear the sojourners talking as they behold this 'slab of concrete' and you would hear them say, 'here, in 1942, indeed there once lived a great people!'

(6)

Of course, nothing is further from the hope or ambition of people who are currently working to reverse the ecological collapse of the Columbia River – one of the most endangered rivers in the United States.[17] Rufus's hyperbole is a testament to hubris born of human-centric thinking, absent ecological ethics or reflection, and evidences the purely commodity-based idea of the natural world that saturated mainstream thinking in the United States at mid-century.

Few people in 1937 questioned an ideology that commodified nature as a stockpile of natural resources and, analogously, rivers as undeveloped sources of power. *Power*'s argument for public ownership is grounded in human entitlement, right, and ownership of the entirety of the natural world. Instead, the debate centered on whether people's right to govern and guide the streams, rivers, and waterfalls should belong to all citizens jointly, operated by the government as a kind of commons, or owned and run for profit by those with the capital to develop water resources. As it promoted Lilienthal's revolution, through which engineers made electric current from running water and turned rivers into organic machines, *Power* also asserted the shared public ownership of resources: a small but critical step in the direction of the idea of a land commons.

Perhaps ironically, the government became its own best customer as hydropower from the Columbia, Colorado, Tennessee, and other rivers were put in the service of the war effort. When the U.S. Boulder Canyon Act passed in 1928, construction of Hoover Dam by the Six Companies began as a federally endorsed corporate project. By five years later, the TVA was perceived as Americanesque socialism, and the Grand Coulee/Bonneville project was promoted as the TVA of the Northwest. But these projects were not as different as the growing anti-socialist rhetoric implied. Each brought big business together with government agencies in an alliance to master a wild river (or river system); each constituted an ecological face-lift in their regions.[18] All three projects promoted an economic boom to follow construction, yet each ultimately paid its costs, not with money from new fully electric family farms but through the manufacture of aluminum and magnesium and, at Hanford on the Columbia River, plutonium for military use.

The 1930s thus found the nation at an ecological crossroads. The drought on the plains and the erosion of soil and human life, together with the Missouri River flood, urban poverty, rural migration, massive hydroelectric development, and the corporatization of American farms, all posed the same two questions: What is society's obligation to the land? And what is our contract with one another? The Roosevelt era forwarded the cause of environmentalism in part through a growing recognition that human beings have a responsibility to protect the land from exploitation. But even as notions of private property – so central to frontierism – were under examination, what constituted exploitation was still unfolding. Nevertheless, landmark legislation provided the groundwork for later U.S. environmental policy. The Indian Reorganization Act of 1934 took a small step toward acknowledging the injustices and mending the injuries of the Dawes Allotment Act of 1887. The Indian Reorganization Act returned some lands to tribes (albeit a small fraction of the lands lost under Dawes) and provided for self-governance for tribes (Dunbar-Ortiz 2014, 171–173). The Taylor Grazing Act of 1934 and the Soil Conservation Act of 1936 put protections in place for soil. Meanwhile, the Civilian Conservation Corps (CCC) and Soil Conservation Service put millions of jobless Americans to work mitigating some of the effects of erosion and floods (Maher 2008). Under the direction of the National Forest Service and the National Park Service, the CCC built trails, roads, bridges, and visitor facilities toward the end of affording a greater number of people access to wilderness areas of the United States. Yet the most significant shift in thinking advanced by New Dealers – a shift implicit even in *Power*'s zealous support of electrification through municipal ownership – was the notion of the land as a commons (Shabecoff 2003, 84).

In the late 1930s, Aldo Leopold (1981) began to ask questions that opened new and deeply troubling philosophical and legal inquiries into the human relationship with the land in the context of democracy. Critical of values that tended to quantify land use as either resources to harvest or scenic beauty for visual consumption, Leopold began to articulate questions beyond the sphere of economic quantification. Developing what he coined a "land ethic," he maintained that society's relationship with the natural world was not only economic but also moral. Each person, and indeed society as a whole, operates within a kind of contractual relationship with the land, one that rises out of human interdependence with all land and its inhabitants. Understood in this way, it becomes incumbent on human beings, Leopold argued, to manage their affairs accordingly, to live responsibly within what was a living relationship with the land. According to Leopold, the land is not an inanimate collection of utilitarian objects but more so a set of relationships in which we must function, an ecological community in which people are merely members.

Leopold's (1981) critique, however, did not attest to or account for the underlying colonialism from which ecological violence and degradation arise, as Kyle Powys Whyte (2012) argues. Certainly, if society's relationship with the land carried moral obligation, that obligation extended to those from whom the land was taken. The idea that ecological interdependence carried with it certain responsibilities was not new. Indeed, indigenous peoples already lived by long-term land ethics. Nonindigenous environmentalists have linked Leopold's land ethic to indigenous traditional ecological knowledge, but indigenous authors have resisted this conflation of Native land ethics, not only because Leopold's ideas brush over the foundational assault on the land that was and is settler colonialism, but also because Leopold's notion of ecological stewardship does not necessarily acknowledge kinship – that the land and all its plants, animals, and presence are *relatives* (Powys White). There is profound irony in the idea of a land commons that does not acknowledge or make reparations for the historical fact that the land was stolen, taken in the violence of military force and coercion. Again, however, the problem lay in the authority of the U.S. government to dictate and control indigenous lives, lands, and futures.

The Federal Theatre's *Triple-A Plowed Under* and *Power* addressed the question of our collective obligation to one another head-on, arguing for economic policies that would benefit "the greatest number" of Americans. *Triple-A Plowed Under* and *Power* argued for a radical transformation of the capitalist system. Nonetheless, both plays eschewed the more fundamental ecological questions about society's relationship to the land by uncritically embracing economic commodification of the land for human use. That ideology draws its rationale not only from the disciplinary links between the fields of economics and ecology but also from the Judeo-Christian story of humanity's birthright to creation, as well as the veneration of the frontier as the American story of origin. Nonetheless, while the message of both plays continued to construct the land as the material basis of production rather than a living system or set of relations, the form of the Living Newspapers did challenge audiences to think ecologically – that is, to see the conditions of their lives from a systems point of view as a complex web of dynamic relations in which human action matters.[19]

Notes

1 In many narratives of U.S. theater, the methods of Konstantin Stanislavsky and the Moscow Art Theater are described as having forcefully changed the way actors, directors, and dramatists thought about character and story. Yet the influence of Stanislavsky and

the Moscow Art Theatre, Bruce McConachie (2000) has argued, may be overrated (47–68). For more on this topic, see Marianne Conroy (1993), 239.
2 All subsequent quotations from the *Triple-A Plowed Under* are taken from this edition.
3 U.S. Congress closed the FTP before *Timber* was produced. In 1991, however, the Seattle Public Theatre staged a revised version of this Living Newspaper (Cless 1996).
4 According to Marc Bloch (1959), electricity's "conquest of the earth" was an achievement that embodies "greater possibilities of shaping our immediate future than all the political events combined" (55).
5 In *Northwest Passage*, Dietrich (1995) also observes that early dynamos were nicknamed "long-legged Mary Anns" and that this fetishizing and feminizing for the dynamo was perhaps inspired by its shape – rather like a woman in bloomers with her legs in the air (249).
6 The threat of Soviet communist influence in U.S. civic society and government in the 1930s was real. In particular, several government officials working for the AAA in the 1930s were part of a covert Soviet operation called the Ware Group, who believed that the AAA was a first step toward Soviet-style agricultural cooperatives in the United States. (Schlesinger 1960, 180–207; Latham 1966; and Klehr, Haynes, and Firsov 1995). During the New Deal years, attorney Alger Hiss worked in the Justice Department and later helped defend the AAA. In 1948, Hiss was convicted of espionage for being a Soviet spy. The case helped launch the anticommunist McCarthy era (Hartshorn 2013).
7 *Triple-A* and *Power* are docudramas that echo Farm Security Administration documentary films made at the time, including Pare Lorentz's *The Plow that Broke the Plains* (1936) and *The River* (1938). For more analysis of the linkage between such films and the docudrama, see Hirsch, White, and Waltzer (1993, 239–294).
8 See Mumford's *Technics and Civilization* (1934) for his earliest published scholarship on this topic. Mumford later expanded his critique of society's overwrought faith in technology in his two-volume work *The Myth of the Machine* (1967) and *The Pentagon of Power* (1970).
9 Production dates of Chicago (July 9, 1936), Cleveland (June 2, 1936), Milwaukee (May 16, 1936) and Los Angeles (August 1, 1936) can be found in *The Federal Theatre Project* (1986).
10 Frank Norris memorialized the corruption of the Chicago commodities market in his epic novel *The Pit: A Story of Chicago* (first published in 1903).
11 Ann Folino White (2015) argues that the AAA flew in the face of morality in the United States because it sanctioned the destruction of foodstuffs – the plowing under of grain and wasteful slaughter of animals while the unemployed went hungry.
12 For more analysis and discussion of the enslavement of black bodies in U.S. history, see Devon Douglas-Bowers's (2013) "Debt Slavery: The Forgotten History of Sharecropping"; see also Blackmon (2008), esp. "Harvest of an Unfinished War," 117–278.
13 The Oklahoma Historical Society website provides a useful summary of the history of tenant farming, both before and after statehood; see also Worster (1979), esp. Parts 2 and 3, 65–138.
14 A number of recent historical accounts of TVA reexamine the project's organizational, environmental, and political aspects. See, for example, North Callahan's *TVA* (1980); see also Richard A. Colignon's *Power Plays* (1997). For a reflection of popular enthusiasm for the TVA project, see C. Herman Pritchett, *The Tennessee Valley Authority* (1943).
15 Lilienthal's work represents a kind of treatise on the significance of the TVA. First published in 1943, it was republished again in 1953 at the height of the Cold War. The additional chapters that appear at the end of the 1953 edition discuss the export of a TVA technology and its organizational model as a weapon against the spread of communism.
16 Barry Witham's (2003) case study of a Seattle *Power* production details the way in which the FTP's production was positioned to make the local case for public utilities.
17 While Woods was likely employing hyperbole here, few expected that in only a few decades many dams built during the 1930s (and after) would be found structurally and environmentally unsound, and tribal communities, sport fisheries, environmentalists,

government agencies, and, in many cases, farmers would unite in deconstruction efforts. For more context and history, see the Klamath River Renewal Corporation website.
18 Indeed, support for public hydropower in the United States crossed traditional political boundaries in the sense that it promised Prosperity for all – for workers *and* for businesses. Conservatives like Rufus Wood, utopian progressives like Lewis Mumford and A.E. Morgan, Wobblie organizers like Woody Guthrie, and big capital like Hoover Dam's Six Companies all found their way into agreement, for a time, on federally-funded power development (Worster 1985).
19 The Living Newspaper form has influenced many dramatists whose works have participated in social justice movements of the 1960s and 1970s, including Luis Valdez. See, for example, Jorge Huerta's "Introduction" (1992). See also Jan Cohen-Cruz's "Antecedents" (2005). As a way in which to stage public debate, the Living Newspapers have also influenced contemporary theater artists. See, for example, Annabel Soutar and the Porte Parole Theatre of Montreal's *Seeds* (2013), which documents the case against genetically modified canola.

References

Adams, Henry. 1918. "The Dynamo and the Virgin." In *The Education of Henry Adams*, 380–390. New York: Houghton Mifflin Company.
Arent, Arthur, Arnold Sundgaard Collection. 1938. *Federal Theatre Plays. 1. Triple-A Plowed Under, by the Staff of the Living Newspaper. 2. Power by Arthur Arent. 3. Spirochete: A History, by Arnold Sundgaard*. New York: Random House.
Arent, Arthur, William DuBois, and Rouben Mamoulian Collection. 1938. *Federal Theatre Plays. 1. Prologue to Glory, by E.P. Conkle. 2. One-third of a Nation, Edited by Arthur Arent. 3. Haiti, by William DuBois*. New York: Random House.
Blackmon, Douglas. 2008. *Slavery by Another Name: The Re-Enslavement of Black Americans from the Civil War to World War II*. New York: Anchor Books.
"Blistering Weather Perils Crops . . . No Relief in Sight." 1936. *Chicago Tribune*, July 8.
Bliven, Bruce. 1933. "Milo Reno and His Farmers." *The New Republic*, November 29.
Bloch, Marc. 1959. *The Historian's Craft*. New York: Knopf.
Brown, Lorraine, and John O'Connor, eds. 1978. *Free Adult and Uncensored*. Washington, DC: New Republic Book.
Callahan, North. 1980. *TVA: Bridge Over Troubled Waters*. New York: A.S. Barnes.
Chansky, Dorothy. 2004. *Composing Ourselves: The Little Theatre Movement and the American Audience*. Carbondale, IL: Southern Illinois University Press.
Cless, Downing. 1996. "Ecotheatre: The Grassroots Are Greener." *TDR* 40 (2): 79–102. https://doi.org/10.2307/1146531. www.jstor-org.libproxy.uoregon.edu/stable/1146531
Cohen-Cruz, Jan. 2005. "Antecedents." In *Local Acts: Community-Based Performance in the United States*, 17–34. New Brunswick, NJ: Rutgers University Press.
Colignon, Richard A. 1997. *Power Plays: Critical Events in the Institutionalization of the Tennessee Valley Authority*. New York: State University of New York Press.
Conrad, David E. nd. "Tenant Farming and Sharecropping." *Oklahoma Historical Society*.
Conroy, Marianne. 1993. "Acting Out: Method Acting, the National Culture, and the Middlebrow Disposition in Cold War America." *Criticism* 35 (2): 239–263.
Dietrich, William. 1995. *Northwest Passage: The Great Columbia Machine*. Seattle, WA: University of Washington Press.
Douglas-Bowers, Devon. 2013. "Debt Slavery: The Forgotten History of Sharecropping." *Global Research*, November 3.
Dunbar-Ortiz, Roxanne. 2014. *An Indigenous History of the United States*. Boston, MA: Beacon Press.

Elton, Charles S. 1927. *Animal Ecology*. New York: The MacMillan Company.
Flannagan, Hallie. 1940. *Arena: The History of the Federal Theatre*. New York: Duell, Sloan and Pearce.
George Mason University Libraries. 1986. *The Federal Theatre Project: A Catalog-calendar of Productions. Bibliographies and Indexes in the Performing Arts; No. 3*. New York: Greenwood Press.
Gilder, Rosamond. 1936. "The Federal Theatre, A Record." *Theatre Arts Monthly* 20: 430–438.
"The Grand Coulee: The 8th Wonder of the World." 1940. *University of Washington Library Special Collections*. Washington: Coulee Dam Souvenirs Company.
Grubbs, Donald H. 1971. *Cry from the Cotton: The Southern Tenant Farmers' Union and the New Deal*. Chapel Hill, NC: University of North Carolina Press.
Guthrie, Woody. 1991. "Roll Columbia Roll." In *Roll On Columbia: The Columbia River Collection*, 16–17. New York: Woody Guthrie Publications.
Harden, Blaine. 1996. *A River Lost: The Life and Death of the Columbia*. New York: W.W. Norton & Co.
Hartshorn, Lewis. 2013. *Alger Hiss, Whittaker Chambers and the Case That Ignited McCarthyism*. Jefferson, NC: McFarland.
Hirsch, Robert, Ken White, and Garie Waltzer. 1993. "Biographies of Selected Photographers From the 20th Century." In *The Focal Encyclopedia of Photography*, edited by Leslie D. Stroebel and Richard D. Zakia, 239–294. Boston, MA: Focal Press.
Huerta, Jorge, ed. 1992. "Introduction." In *Zoot Suit & Other Plays*, authored by Luis Valdez, 7–20. Houston, TX: Arte Público Press.
Hughes, Thomas. 1983. *Networks of Power: Electrification in Western Society, 1880–1930*. Baltimore, MD: John Hopkins University Press.
Jackson, Donald C. 1995. *Building the Ultimate Dam*. Lawrence, KS: University of Kansas Press.
Jordan, John M. 1994. *Machine-Age Ideology: Social Engineering and American Liberalism 1911–1939*. Chapel Hill, NC: University of North Carolina Press.
Klehr, Harvey, John Ear Haynes, and Fridrikh Igorevich Firsov. 1995. *The Secret World of American Communism*. New Haven, CT: Yale University Press.
Larson, Orville K. 1989. *Scene Design in the American Theatre from 1915 to 1960*. Fayetteville, NC: University of Arkansas Press.
Latham, Earl. 1966. *The Communist Controversy in Washington: From the New Deal to McCarthy*. Cambridge, MA: Harvard University Press.
Leopold, Aldo. 1981. *A Sand County Almanac*. Oxford, UK: Oxford University Press.
Lilienthal, David. 1953. *Democracy on the March*. New York: Harper & Brothers.
Maher, Neil M. 2008. *Nature's New Deal: The Civilian Conservation Corps and the Roots of the American Environmental Movement*. Oxford and New York: Oxford University Press.
Marshall, Robert. 1990. "Wilderness." In *American Environmentalism: Readings in Conservation History*, edited by Roderick Nash, 166–170, 3rd ed. New York: McGraw-Hill.
McConachie, Bruce. 2000. "Method Acting and the Cold War." *Theatre Survey* 41 (1): 47–68. https://doi.org/10.1017/S0040557400004385
Merchant, Carolyn. 1994. "Reinventing Eden: Western Culture as a Recovery Narrative." In *Uncommon Ground: Rethinking the Human Place in Nature*, edited by William Cronon, 132–159. New York and London: W.W. Norton & Company.
Mumford, Lewis. 1934. *Technics and Civilization*. New York: Harcourt Brace.
Mumford, Lewis. 1967. *The Myth of the Machine: Vol. I, Technics and Human Development*. New York: Harcourt Brace Jovanovich.

Mumford, Lewis. 1970. *The Pentagon of Power: Vol. II, Technics and Human Development*. New York: Harcourt Brace Jovanovich.
Norris, Frank. 1994. *The Pit: A Story of Chicago*. New York: Penguin Books.
O'Neill, Eugene. 1920. *Beyond the Horizon: A Play in Three Acts*. New York: Boni and Liveright.
O'Neill, Eugene. 1929. *Dynamo*. New York: Horace Liveright.
Osborne, Elizabeth A. 2011. *Staging the People: Community and Identity in the Federal Theatre Project*. New York: Palgrave Macmillan.
Peña, Devon Gerardo. 2005. *Mexican Americans and the Environment: Tierra y Vida*. Tucson, AZ: University of Arizona Press.
Peña, Devon Gerardo, ed. 1998. *Chicano Culture, Ecology, Politics: Subversive Kin. Society, Environment, and Place*. Tucson, AZ: University of Arizona Press.
Powys Whyte, Kyle. 2012. "Indigenous North American Ethics and Aldo Leopold's Land Ethic: A Critical View of Comparison and Collaboration." *SSRN Electronic Journal*, January. http://doi.org/10.2139/ssrn2022038
Pritchett, C. Herman. 1943. *The Tennessee Valley Authority: A Study in Public Administration*. Chapel Hill, NC: University of North Carolina Press.
Reinheimer, Hermann. 1913. *Evolution by Cooperation: A Study in Bio-Economics*. London: Kegen Paul Trench, Trubner, and Co.
Rice, Elmer. 1929. *Street Scene: A Play in Three Acts*. New York: Boni.
Rice, Elmer. 1929. *The Adding Machine; A Play in Seven Scenes*. New York, Los Angeles and London: S. French Ltd.
Schlesinger, Jr., Arthur M. 1960. *The Politics of Upheaval: The Age of Roosevelt*. New York and Boston, MA: Houghton Mifflin Company.
Schlesinger, Jr., Arthur M. 1988. *The Coming of the New Deal*. Boston, MA: Houghton Mifflin Co.
Schwartz, Jordan A. 1993. *The New Dealers: Power Politics in the Age of Roosevelt*. New York: Alfred A. Knopf.
Shabecoff, Philip. 2003. *A Fierce Green Fire: The American Environmental Movement*. Washington, DC: Island Press.
Soutar, Annabel, and the Porte Parole Theatre of Montreal. 2013. *Seeds*. Vancouver: Talon Books.
Staff of the Living Newspaper. 1938. "*Triple-A Plowed Under.*" In *Federal Theatre Plays*, 1–57. New York: Random House.
"Starvation in Arkansas." Editorial. 1936. *The New Republic*, April 1.
Tansley, Arthur G. 1935. "The Use and Abuse of Vegetational Concepts and Terms." In *The Future of Nature: Documents of Global Change*, 220–232. New Haven and London: Yale University Press. http://doi.org/102307/1930070
Treadwell, Sophie. 1928. *Machinal*. London, Royal National Theatre: Nick Hern Books.
Wald, Sarah D. 2016. *The Nature of California: Race, Citizenship, and Farming since the Dust Bowl*. Seattle, WA: University of Washington Press.
"Weather Man Sees Little Change in Drought Conditions." 1936. *Los Angeles Times*, August 2.
White, Ann Folino. 2015. *Plowed Under: Food Policy Protests and Performance in New Deal America*. Bloomington, IN: Indiana University Press.
White, Richard. 1995. *The Organic Machine*. New York: Hill and Wang.
Wilson, Edwin, and Alvin Goldfarb. 2008. *Living Theatre: History of the Theatre*. Boston, MA: McGraw-Hill.
Witham, Barry. 2003. *The Federal Theatre Project: A Case Study*. Cambridge and New York: Cambridge University Press.

Woods, Rufus. 1944. *The Twenty-Three Years' Battle for Grand Coulee Dam.* Wenatchee, WA: University of Washington Library Special Collections.

Worster, Donald. 1979. *Dust Bowl: The Southern Plains in the 1930s.* Oxford: Oxford University Press.

Worster, Donald. 1985. *Rivers of Empire: Water, Aridity and the Growth of the American West.* New York: Oxford University Press.

Worster, Donald. 1994. *Nature's Economy: A History of Ecological Ideas*, 2nd ed. Cambridge: Cambridge University Press.

Zinn, Howard. 1995. *A People's History of the United States*, Revised ed. New York: Harper Perennial.

4

WE KNOW WE BELONG TO THE LAND

Rogers and Hammerstein's *Oklahoma!* and Arthur Miller's *Death of a Salesman*

In 1931, the Theatre Guild of New York produced *Green Grow the Lilacs* by Cherokee playwright Lynn Riggs (1931).[1] In the midst of the Depression and Dust Bowl, producer Theresa Helburn (1960) predicted that Riggs's story with its "cowboy songs" and "rollicking liveliness" would appeal to audiences enduring hard times. But New York's *Sun* reviewed *Green Grow the Lilacs* as "hollow" and "pretentiously literary" (Lockridge 1931). Helburn was not deterred. A decade later she returned to Riggs's play as the vehicle for a universal sense of Americana that might appeal during World War II. The Guild engaged the composer/lyricist team of Richard Rodgers and Oscar Hammerstein to develop a musical based on *Lilacs*. It would be a new kind of musical, Helburn hoped, "not musical comedy in the old familiar sense, but a form in which the dramatic action, music and dancing would be aides and adjuncts of the plot" (Helburn 282). The Theatre Guild's *Oklahoma!* (Rogers and Hammerstein 1942) opened on March 31, 1943, and played at New York's St. James Theatre through the spring of 1948. Touring productions in the United States, Europe, and the South Pacific bolstered morale and gave U.S. soldiers a foot-stomping chorus of accolades for the American cause and a taste of home – whether they hailed from Oklahoma, Oregon, or Maine. Initially called Helburn's Folly, *Oklahoma!* became one of the most popular musicals of its time and a pivotal work in the history of theater in the United States.[2] On the occasion of its 1953 anniversary production in Washington D.C., Oscar Hammerstein mused that *Oklahoma!* had been continuously performed for an entire decade – that somewhere in the world audiences were still listening to Curly singing "Oh, What a Beautiful Morning" ten years after its Broadway debut. The 1955 film version brought the play to an even wider audience (Nadel 1969, 182), and *Oklahoma!* remains one of the most frequently produced, yet under-interrogated, American musicals to date (see, e.g., Carter [2007]; Evans [1990]; Kirle [2003]; Knapp [2005]; and May [2000]).

DOI: 10.4324/9781003028888-5

When *Oklahoma!* opened, the United States was in the midst of a war with Germany, and many Americans feared a Nazi invasion. Rogers and Hammerstein provided a timely and sought-after mobilizing cry: "We know we belong to the land/And the land we belong to is grand" (132). The mutating but indefatigable mythos of the frontier translated into the success of *Oklahoma!* Making the state of Oklahoma a signifier of the nation as a whole, the musical appropriated the sense of place at the heart of Riggs's play. A signifier not of some*where* but of an idealized American past, *Oklahoma!* became the yardstick by which musical theater would henceforth be measured. Roy Waldau (1972) writes that "[e]very serious play with integrated score for the next quarter century reflected qualities of *Oklahoma!*" Unlike Riggs's folk songs and ballads, which looked to the past to "recapture . . . a kind of nostalgic glow," Rodgers's music moved forward with a palpable eagerness, Waldau argues, and Hammerstein's lyrics evoked hope for the future (387). This musical ostensibly about the past was actually about the future. Likewise, Agnes de Mille's choreography was praised as "native" and expressive of "the energy and vastness of the American landscape" (Evans 1990, 146).

The artists leveraged already-ingrained mythic images and ideas at a time when those myths served particularly well as Americans negotiated an uncertain future. Theater of the 1930s had been characterized by left-leaning politics, but as the nation headed pell-mell into war, public appetites for social critique decreased in favor of nationalist morale-boosting stories, and theater critics exerted a pressure on artists to deliver popular themes. The *Christian Science Monitor* noted that the Theatre Guild's 1935 production of *Parade*, the tale of a little guy (characteristically gendered male) who fights the establishment, was "a bit to the left." But according to the *Monitor*, the Guild had "righted itself" with *Oklahoma!* (quoted in Waldau, 385). Wielding political power with ecological implications, *Oklahoma!* retooled and recentered beliefs about what and who is American. Indeed, the pervasiveness of the frontier mythos and the political climate of 1943–55 had as much to do with the musical's success as the artistic genius of Rodgers, Hammerstein, Mamoulian, and choreographer Agnes de Mille.

Riggs's *Lilacs* – a play concerned with the complex sense of Cherokee identity and place – is an ironic basis for the jingoistic U.S. nationalism of *Oklahoma!* In *Stoking the Fire: Nationhood in Cherokee Writing 1907–1970*, Kirby Brown (2018) argues that Riggs was a Cherokee dramatist of international acclaim whose artistic work is infused with *Cherokee* nationalism. In Scene 6 of *Lilacs*, Aunt Eller defends the territory law in opposition to U.S. law, saying, "[W]hut's the United States? It's just a furrin country to me!" (103). In *Oklahoma!*, however, what is clearly Indian territory in *Lilacs*, becomes simply "the territory" (85–87), a rhetorical erasure that masks the presence, contributions, and painful history of the many tribes and communities that called the land that became Oklahoma home. The story and dialogue of *Oklahoma!* was taken almost verbatim from

Riggs's *Lilacs*. In contrast to Riggs's centering of place and diverse communities, however, the musical offers up a whitewashed Oklahoma history that underscores the white supremacy implicit in U.S. nationalism. As Cherokee scholar Jace Weaver (1977) observes,

> [w]hile turn-of-the-century Indian Territory of the play's setting was racially mixed, with Natives, African Americans, and Amer-Europeans all interacting, *Oklahoma!* contains no Blacks and no Indian. The landscape has been thoroughly ethnically cleansed until it becomes the vacant landscape of the myth of dominance. Once it is emptied out, Amer-Europeans are free to occupy it without molestation or challenge.
>
> *(99)*

While *Oklahoma!* poses as a story about belonging (to the land and to one another), it carefully manages who belongs, excluding those who do not fold in. *Oklahoma!* aligned with and subtly propagated an exclusionary white America. Its production coincided with the Jim Crow South and the renewed taking of indigenous lands during the 1940s and 1950s, when state and federal termination acts dismantled tribal sovereignty and seized tribal lands by effectively ending legal recognition for many tribal governments.

Although *Oklahoma!*'s central story takes place in the year prior to Oklahoma statehood in 1907, the history that forms its backstory is soaked in cycles of land takings from Native American tribes. The 1830 Indian Removal Act that forced indigenous peoples from many regions into "the Indian territory" that included the region that would become Oklahoma. Mounting pressure from white settler encampments resulted in the 1890 Oklahoma Organic Act that authorized a land run, and set the stage for the dissolution of the Cherokee Nation in 1906 and Oklahoma's statehood in 1907. *Oklahoma!* spins this moment of deterritorialization into a story and symbol of white right to land, and the supremacy of white values regarding the land's management and use, reinscribing the values of frontierism and the colonial structures it bolsters as an enduring ethos in mainstream postwar U.S. culture.

Oklahoma! represents both an erasure of indigenous presence and a patriotic backlash against the social and environmental policies of the New Deal. In the milieu of patriotism and renewed prosperity after the war, *Oklahoma!* obfuscated the discourse that arose out of the Dust Bowl, for example, in the Farm Security Administration (FSA) photographs, in Steinbeck's ([1939] 1993) *The Grapes of Wrath*, and in the vast array of books, articles, and government reports that documented the human and ecological toll of treating the land as a commodity. Ignoring the lessons of the Dust Bowl, and indeed the more nuanced history of the geographic region itself, *Oklahoma!* supported a kind of musically-induced national amnesia, even as it helped to usher in a new era of optimism. In its nostalgic rewind, *Oklahoma!* linked a conjured past to an imagined future in which Oklahoma became a stand-in for the United States; its rousing assertion "you're okay Oklahoma, Oklahoma O.K.!" signaled not only national recovery from the depression but rising U.S. dominance in the world (132).

By breathing fresh life into the Anglo story of frontier settlement, *Oklahoma!* propagated an ideology about the land infused with nationalist ecological supremacy, at once disguising, sanctioning, and supporting the ongoing project of colonialization in postwar U.S. society. The years of its continuous production from 1943 to 1953 characterized a new era of land takings and violations of tribal sovereignties. The Indian Relocation Act of 1956 gave federal credence to the numerous termination acts already passed by many western U.S. states (see, e.g., Dunbar-Ortiz [2016]). Termination during the war years was often driven by the demand for resources by industry and the military; after the war, it was motivated both by increased industrial and agricultural development in the west and expanding westward settlement as the postwar baby boom began. Through its assertions of settler entitlement, *Oklahoma!* gave rhetorical support to termination and the resurgence of land taking in the 1940s and 1950s. The musical also envisioned traditional gender roles and white supremacy as intertwined aspects of the progress on which industrial development of western lands depended. In the analysis of *Oklahoma!* that follows, I am interested in how gender, class, race, and national origin are bound up with mainstream "American" ideas about nature, belonging, and the land – a mythologizing that has both social justice and environmental consequences.

After World War II Americans confronted new worries, including nuclear annihilation, as a result of a new superpower world order. The Nazis and fascists had been defeated, but a new Soviet enemy, and renewed fears of communism, unsettled many in the United States. Beneath the surface of optimism and the burst of consumerism of the postwar years, residues from hard times seeped to the surface. As the era became known as "the age of anxiety," advertisers capitalized on the intangible insecurities of the middle class.[3] Sources of foreboding varied from the political and economic to the personal and existential. Six months after *Oklahoma!* closed at the St. James Theatre, and only a block away, that existential despair was in plain sight in Arthur Miller's *Death of a Salesman* (1971), which opened at the Morosco Theatre on February 10, 1949.[4] These two tales about the American dream reveal an intertextual resonance and cognitive dissonance between the values *Oklahoma!* celebrates and their lived consequences in *Salesman*. Miller's play, which won the Pulitzer Prize in 1949, reveals the toll those values exact on the women and men whose labor is used and used up in building national wealth. The sense of placelessness and homelessness made plain in Salesman is ironically rooted in the frontier tale *Oklahoma!* retells. *Oklahoma!*'s mythos is implicated in Miller's play as it uncovers and mourns the subtle way that expansionist values ultimately betray home, family, and future. Willy Loman, who bought into the *modus operandi* of entitlement, taking, and lying, so basic to the frontier ethos, finds that he too is expendable.

In what follows I argue that *Oklahoma!* forwarded Turnerian ideas of settling the land (with Oklahoma as a symbol of the nation-state), settling in (to being a farmer), settling dissent (by silencing, or expelling, complainers), setting down (as in the heterosexual unions at the end of the story), settling gender (by moving women back to the domestic sphere), and settling differences (between the cowboys and the farmers), in order to settle up (by defining social relations as fundamentally economic). Then, in turn, I examine *Salesman*'s poignant rebuttal. In the arc of history represented by the juxtaposition of the two plays, the values celebrated in *Oklahoma!* became the effective cause of tragedy in *Salesman*.

If soil could speak

John Steinbeck memorialized the faces of displaced Oklahoma tenant farmers in *The Grapes of Wrath*. The novel earned the Pulitzer Prize in 1939 and the Nobel Prize in 1940. By 1940, the state of Oklahoma had become a signifier of the failed American dream. Conceived in the shadow of Steinbeck's Joads, *Oklahoma!* the musical sought to transform the name of the state into a symbol of all that was right with America. And yet *Oklahoma!* distorts the very past it claims to celebrate. The history of Oklahoma as a territory and later a state is intertwined with the history of removal and broken promises. The multiple identities that Riggs wrote about in *Lilacs* also reflect the rich and productive intersection of Native and settler communities, but in *Oklahoma!* this complex history is missing.

Before its Broadway opening in 1943, no one could agree on a title for the Guild's new musical: its out-of-town tryout (customary for New York productions) occurred under the working title "Away We Go" (Helburn 1960). In particular, librettist Oscar Hammerstein wanted to call the new musical *Oklahoma*. Concerned that "people might confuse it with 'Oakies,'" – those refugees from dust that were the subjects of *The Grapes of Wrath* – Helburn protested (289). The word *Oklahoma* was marked with a litany of the failures of the American promise. During the 1930s, the photographs taken by the Farm Security Administration of the human suffering of Oklahoma tenant farmers had saturated print media and put Oklahoma at the center of national conscience (Waldau 1972, 363). Then, Helburn recalls, "someone suggested putting an exclamation point after Oklahoma" (289). Thus, like a balm that treats the symptom while masking the causes of disease, *Oklahoma!* punctuated the positive by directing national attention away from the greatest ecological disaster of recent memory toward a nostalgic moment in U.S. history before the dust, economic woes, and the war began.

Oklahoma! used the multiple layers of the state's history to underscore the equation between land and liberty so fundamental to frontier ideology. As Howard F. Stein and Robert F. Hill have observed (1993),

> [c]ultural images are never merely images. They carry the emotional and cognitive freight of identity, of obligation, of destiny. The issue of how

Oklahomans think of themselves, of how others think of Oklahoma, and of the interplay between these is neither idle nor academic. It affects who and what Oklahomans are and become.

(230)

In 1943, the state of Oklahoma already held an iconic place in national memory, and in many ways, the musical subverted and painted over that history with a sanitized image of land-as-liberty. From the Oklahoma Boomers, who gathered at the turn of the century on the borders of what was legally Indian territory to demand it be opened to white settlement, to the oil boom of the 1920s on Osage lands in which white speculators cheated and murdered Osage people to take control of oil deposits, to the mass exodus of "Oakies" during the Dust Bowl years as displaced farmers migrated west in search of jobs, Oklahoma long had served as a stage where contested notions of American identity vied for dominance (Manes 1982, 101). Pioneer families like Laurey's, who forced open the Indian Territory with their "land hunger" and exercised what they described as a God-given right to break sod, plant wheat, and grow profits, became metaphors of the free-enterprise system. As Sheila Manes has observed, however, there are other versions of *Oklahoma!*'s history that have gone unsung (93–132).

The stories the Oklahoma land tells, from the Trail of Tears to the Dust Bowl, were results of U.S. American expansionism and imperialism. The Indian Removal Act of 1830 enacted the forced removal and migration of Native women, men, children, and elders to the Indian territory designated by the federal government. Five tribes – Cherokee, Choctaw, Chickasaw, Creek, and Seminole – were forcibly removed from areas that had been carved into the southern states after the Louisiana Purchase. Those who survived the journey to the Oklahoma territory found themselves in an ecologically unfamiliar land, primarily arid rather than humid, with flat horizons of prairie grass as far as the eye could see. Yet they found a way to inhabit the land and devised a sophisticated representative government (Debo 1998).[5] Following the Civil War, however, settlers pressured the U.S. government to open Oklahoma lands to homesteaders. From 1879 to 1889, groups of "boomers" (championed by railroad companies) lobbied for entrance to what they understood as the last "free land" in the continental United States. One energetic leader, David L. Payne, announced in 1880 his intent to violate the Intercourse Act of 1832, which forbade white settlement of Indian lands and establish his "Oklahoma Colony" in defiance of the law. The boomer vision, which runs as the base note through *Oklahoma!*, was the notion that "property" was "made legitimate only by its use" (Globe 1980, 10). In *Progressive Oklahoma*, Danny Globe describes Payne's viewpoint that

> the scattered bands of primitive Indians could never develop the land for commercial use. Therefore, continued Indian occupancy erected an artificial barrier to expanding trade and, indeed, civilization. The government

thus had the over-riding moral responsibility of fostering private ownership of land and social development through the dispersal of public property to the awaiting settlers.

(6)

Payne used the same racialized, culturally hegemonic, white supremacist argument in order to wrest the Black Hills from the Sioux in 1874. *Oklahoma!* painted a familiar picture of a racialized populist movement in common cause with industrial and commercial opportunism. While the musical does not explicitly identify Laurey's parents as Payne County colonists, dialogue does indicate that they were part of the Oklahoma Land Run of 1889. The women and farmers in *Oklahoma!* represent the populist agrarian mandate of the boomers, while the peddler Ali Hakim and the musical number "Kansas City" embody the persistent presence of commerce and the railroads.

In *The Unsettling of America*, Wendell Berry (1977) characterizes farming in the United States as an ongoing process of dislocation and dispossession (3–16). Contrary to Jefferson's dream of a nation of yeomen farmers, Berry argues, U.S. agribusiness has produced generation after generation of exiles. By 1889, the Oklahoma territory was considered the last hope for farmers (Globe 1980, 6–10). But this idealized farm life represented by Curly and Laurey in the musical would be short-lived. In 1890, landless tenant farmers in Oklahoma composed less than 0.7 percent of territory farmers. But by 1915, 50 percent of Oklahoma farmers were tenants, and 80 percent of those who did own land carried heavy mortgages (Sellars 1998, 79–90). At the time of the Oklahoma land run, a correspondent from the New York *Tribune* observed a "feeding frenzy" as the cavalry trumpeter sounded the "dinner call" and "the supreme moment had arrived at last. Away dashed the horsemen in mad gallop, lashing their horses as if life depended upon reaching the top of the hill yonder" ("General" 1889). This vision implied "expressed physical action – building towns, extending trade, marketing produce, turning the wheels of commerce" (Globe, 10). Like the frontier, "property" in this context did not denote a place but rather implied a *process of taking* informed by white supremacist hegemony and trumping all else from preexisting law, treaties, home, health, lives, and ecologies.

Ironically, the land in Oklahoma was only marginally suited to profit-driven agriculture. Gary Thompson observes that "erratic rainfall made it chancy for the region to compete well with the more reliable farmlands of the Corn Belt states" (Thompson 1993, 11). Just as Belasco's *Girl of the Golden West* hid the extractive industries in the Sierra Nevada and Rocky Mountains behind the picturesque, *Oklahoma!* reinforced an agrarian illusion of the small family farm, masking industrial farming's increasing dependence on cash crops for a global market. *Oklahoma!* seems to advocate the Jeffersonian/Turnerian conviction that human "improvements" could reclaim the arid windswept plain as a Garden of Eden. Lyrics suggest a reciprocity relationship in which "the farmer and the cowman should be friends." Each would provide for one another with the land's abundance: "Gonna give you barley, / Carrots and pertaters – / pasture for your cattle – / Spinach and termayters!" (131). Agricultural practice on the plains, however, was always envisioned as an extractive (not relational) enterprise. As early as the Boomer days, the agricultural crops in Oklahoma were primarily cash crops – cotton, wheat, some corn (see, e.g., Thompson, 3–28). In short, the illusion of the family farm masked the big business.

During the Progressive era, when farming became an industry rather than a way of life, it dislodged and uprooted human beings. Sellars (1998) describes that

> as the twentieth century dawned, more and more farmers, rather than rising to farm ownership, found themselves pushed down into the ranks of the tenant or even the migratory agricultural laborer. The causes of this decline included the exhaustion of the public domain, an increase in both legitimate and fraudulent land speculation, and a growing trend toward corporate ownership of agriculture.
>
> (79)

The experience of many farmers in Oklahoma (and elsewhere) marks what Berry (1977) identifies as a cycle perpetuated by particular economic and ecological values: "Generation after generation, those who intended to remain and prosper where they were have been dispossessed and driven out, or subverted and exploited where they were, by those who were carrying out some version of the search for El Dorado" (4). Ecological ignorance combined with greed gave rise to economic inequity and environmental catastrophes of the Dust Bowl and Depression (see Worster 1979, 44–63). But even as the lyrics recognized the power of the wind that would lift topsoil away, *Oklahoma!* resolutely overrode the ecological lessons of the 1930s in its vision of a place "where the wind comes sweepin' down the plain, / And the wavin' wheat / Can sure smell sweet / When the wind comes right behind the rain" (132). Positioned in 1906 at the top of what historians would call "the great plow up" that lead to the conditions of the Dust Bowl, *Oklahoma!*'s Laurey and Curly (or Will and Ado Annie) might be looking forward to packing up their children and possessions in a tin lizzie and leaving the land in only one generation's time.

Agrarian unrest in Oklahoma, which began in the 1880s and continued through the 1930s, was fueled by the influence that large-scale farmers and grain dealers had with the railroads. Globe (1980) observes that the "territories' agricultural products were shipped at short peak intervals [and] the railroads often failed to provide enough freight cars." Consequently, Globe continues, "it was not uncommon to see great piles of wheat and corn dumped on the ground for want of freight cars." The large landowner – often absentee – had an advantage over smaller and tenant farmers. Inequities, coupled with the drought of 1890, fueled the Populist Party in Oklahoma and elsewhere in the Great Plains states. According to Globe, "[i]f Boomers had dreamed of a market-oriented society . . . Oklahoma's farmers actually lived one, and they were its victims" (158). While tribal politics dominated the region until the 1890s, the Populist (or People's) Party gained strength in the decades leading up to statehood (see, e.g., Globe, 158–168). The Farmers' Alliance and Farmers Union organized union clearinghouses and cooperative grain storage; together, they lobbied for federal regulation and public ownership of the railroads.

By the time Oklahoma entered the Union, these radical beginnings had been overruled by business-minded landowners and grain dealers, and commercially-oriented farmers gained control of the Farmers Union (Globe 1980, 163). Although the Oklahoma Progressive Party railed against "trusts" – large corporations that artificially manipulated prices – they were nevertheless committed to "cooperation" among farmers, bankers, and merchants in order to promote "efficiency" and economic growth (163–166). While the state's Progressives sought to regulate large profit-hungry monopolies, they also reaffirmed faith in the marketplace and the free-enterprise system. In *Oklahoma!*, the Arab immigrant peddler, Ali Hakim, is essential in this Progressive economic picture.[6] Described as Persian, the depiction accentuates his otherness; unsurprisingly, he is the target of a ruse that forms one of the subplots to the main narrative.

Oklahoma!'s Hakim typifies stereotypes of characters in U.S. drama at mid-century as untrustworthy in business and sexually promiscuous. Hakim is tricked into marrying, ultimately abandoning his nomadic life and polyamorous lifestyle. In reality, Persian peddlers were among the many hardworking immigrants who came to the United States near the end of the twentieth century.[7] Consistent with the musical's assimilationist agenda, Hakim must play by the rules. In one scene, he is held at gunpoint and must agree to wed Gertie Cummings, thus symbolically marrying into the set of economic relations that make up the U.S. free-enterprise system. Beyond Hakim, a larger vision of free enterprise tempered by the recognition of mutual economic benefit between varied business sectors resounds in *Oklahoma!*: "And when this territory is a state, / And jines the union jist like all the others, / The farmer and the cowman and the merchant / Must all behave theirsel's and act like brothers" (89). In U.S. history, however, most Progressive reforms did not benefit the small farmer, the landless tenant, the peddler, the sharecropper, or the day laborers in the wheat, corn, cotton, and oil fields.

Who belongs to the land?

Oklahoma! is focused on the shift from the nomadic life of the cowboy to the settled life of the farmer – the same trajectory of identity formation articulated by Frederick Jackson Turner. In *Oklahoma!*'s opening scene, the audience meets Laurey, an Oklahoma settler's daughter, and her Aunt Eller, who manage their farm with the help of Jud, a hired hand. In love with Laurey, Curly is a buoyant cowboy who works for rancher Ike Skidmore. Together, the senior Aunt Eller and Skidmore represent the principal (read Anglo) populations of the Oklahoma territory at the time – the boomers who wanted to farm and "improve" the land

and the cattlemen who often leased rangeland from plains tribes in the 1870s and 1880s. Meanwhile, Jud, a hired hand, is also infatuated with Laurey, and the rivalry between the suitors is the center of the plot. The distinction between the two men is, on the surface, one of temperament; yet the protagonist/antagonist polarity disguises a deeper politics infused with racialized economic privilege, including entitlement to land.

Historically, the development of barbed wire precipitated rancorous and sometimes violent animosity between farmers and ranchers in the west. In Riggs's original text, stage directions indicate that a discarded coil of barbed wire entangles Aunt Eller on her way to the box social (85). Curly mourns the passing of the cattle era, telling Laurey, "[n]ow the cattle business'll soon be over with. The ranches are breakin' up fast. They're puttin' in barbed w'ar, and plowin' up the sod fer wheat and corn" (107). Much of the humor, likewise, revolves around competition between these two groups: "One man likes to push a plow, / the other likes to chase a cow, / . . . / The cowman ropes a cow with ease, / The farmer steals her butter and cheese" (87). Much the spectacle of *Oklahoma!* features separate choruses of cowboys and farmers (Evans 1990, 145–148). Agnes de Mille's choreography accentuated the differences through two contrasting movement vocabularies for the dancers and interspersed dance numbers with "free-for-all" fight sequences. The construction of difference between the two socioeconomic groups is critical to the play's later argument that "the cowman and the farmer must be friends" (89) – in other words, partners in the commerce of the new state.

The characters of Curly and Jud – the two forces vying for Laurey's love (and with it the land her family settled) – exemplify how the play's final celebration of Oklahoma as "a brand-new state" carried implicit messages about who belongs to this new civic amalgamation. In a pivotal scene between these two male rivals (discussed in detail later in this chapter), Curly twirls and tosses his lasso, looping it over a beam, as if to tie a noose, as he continues to taunt Jud: "You could hang yourself on that, Jud" (63). Curly, the archetypal U.S. cowboy, cajoles Jud into fantasizing his own funeral: "You never know how many people like you till you're daid" (64). Curly sings until Jud joins in, participating fully in his own imagined demise. "Pore Jud is daid, / pore Jud Fry is daid!" (67). Like the defense of *Bloody Bloody Andrew Jackson*'s (Friedman and Timbers 2010) musical number "Ten Little Indians" (discussed in Chapter 1), the humor of this scene pales when we remember the history of U.S. state-sponsored lynching of blacks (as well as Mexicans and indigenous peoples) by Anglo whites, employed as a method of white supremacist control. What does Rodgers and Hammerstein's *Oklahoma!* reveal about the intertwined ideas regarding racialized identities, society, belonging, and the environment during and following World War II? Who is Curly, and what did he represent in 1941 when the Theatre Guild first produced *Oklahoma!*? Who does Curly become in the course of the musical, and what has he come to represent as the musical continued to gain currency through the war years and into the 1950s? In a similar vein, who is Jud? What did he represent in those years and afterward, and why does the musical imply (through the preceding enactment) that he should go hang?

The role of Curly, almost always cast (and read) as a Euro-American/white, is the personification of frontier Americanness outlined by Fredrick Jackson Turner in 1893 (1956, 2–4).[8] Jud meanwhile is a "hired hand," who lives on Laurie's farm and performs odd jobs. He is not a landowner; he is a laborer. In the 1930s, when Riggs wrote the play from which the musical was adapted, Oklahoma was a site of rising radical dissent, where the International Workers of the World (IWW) had a strong following. While *Green Grow the Lilacs* (1931) does not specify Jeeter/Jud's identity beyond his status as a hired hand, the stage directions and dialogue in Riggs's play, as well as Hammerstein's musical adaptation, suggest a man of mixed ancestry.[9] For example, Riggs describes Jud/Jeeter as "earth colored" and "dark" (1931, 40); Curly describes Jud as "a bullet-colored man" (1942, 21). The soil in much of Oklahoma is reddish in color, while bullet casing is typically copper or brass – a descriptor that also associates Jud with violence.[10] Later, in the Smokehouse scene, Curly baits Jud to talk about and nearly confess to setting a nearby farmhouse on fire, marking his as the "kind of a man" (70) who would resort to violence to get what he wants.

Violence was part of the body politic and came in many forms and for many reasons in the region. Prior to statehood, violent tactics were part and parcel of the graft that defrauded tribal landowners. Regional populist movements, like the Boomers and, later, the Sooners, often led to violent conflict over land and grazing rights. During the Great Depression and the Dust Bowl, violent direct action was used as a form of resistance and dissent by the urban unemployed and rural sharecroppers. Such violence was often only surpassed by retaliation from industry and law enforcement. Given the dissolution of the Cherokee nation in 1906, but also in the context of mid-twentieth-century racial tensions in the U.S. generally, I argue that the conflict between Jud and Curly signified both racial and political tensions in which the role of "savage" was applied to any and all who resisted the hegemonic order.

Reading cowboy

When *Oklahoma!* opened in 1943, director Rouben Mamoulian called it "aboriginally American" (Evans 1990, 143), possibly an implicit (if problematic) nod to Riggs, but also implying the white settler descendants that had been Americanized through Turner's process of regression and transformation (see Chapter 1). This "true-blooded" American was also understood in contrast to the enemies that U.S. soldiers were facing on the battlefields, seas, and in the air. In a London performance on April 30, 1945, the day Hitler took his own life, the "[a]pplause for the play was so enthusiastic that a stage pistol had to be fired to regain composure of the audience" (Nadel 1969, 182) as they cheered the U.S. cowboy who had become symbolic of the soldiers who stood between Britain and the Germans (see Waldau 1972, 385–386). For a decade, touring companies in Europe, Australia, and South Africa continued to perform the show that had become the face of U.S. largess after the war. A *London Times* reviewer (April 23, 1945) claimed that *Oklahoma!* "offered a romanticized view of the heartland of the United States from whence had come the battalions that had saved the British world" (quoted in Thompson 1993, 3). The legitimate basis for this gratitude notwithstanding, *Oklahoma!* also helped to identify the cowboy and the American frontier narrative as synonymous with U.S. geopolitical power.

Meanwhile, in the United States, Curly is not only the character with whom audiences of the time identified and also wanted to become. Quoting Roland Barthes's *Mythologies* (1972, 124), Richard Slotkin (1994) observes that "myth uses the past as an 'idealized example,' in which 'a heroic achievement in the past is linked to another in the future of which the reader is the potential hero'" (18–19). Rodgers, for his part, understood that in the "midst of a devastating war" people needed not only a past full of heroes but also one that allowed them to envision themselves as heroes: "People said to themselves, in effect, 'If this is what our country looked and sounded like at the turn of the century, perhaps once the war is over we can again return to this kind of buoyant, optimistic life'" (quoted in Evans 1990, 228). As the fulfillment of Turner's "American character," Curly possesses an unstoppable optimism about the American project, singing even in the face of an uncertain future, "Oh, what a beautiful mornin', / Oh what a beautiful day. / I got a beautiful feelin' / Ev'rythin's goin' my way" (Slotkin, 4).

Curly is also a romanticized personification of what Worster has called the "unmistakably capitalist institution" of the American cattle industry (1992, 37–40) – an industry that White (1991) notes resulted in "ecological and economic catastrophe" (225). As with the California gold rush, the late nineteenth-century cattle bonanza enjoyed a short boom period followed by a rapid bust, the unbridled uptake of a specific resource (in one case, gold ore and, in the other case, prairie grass). Worster (1992) notes that

> [l]ivestock became a form of capital in this innovative system, capital that was made to earn a profit and increase itself many times over, without limit. But the animals were only one part of the capital – a mere mechanism for processing the more essential capital.
>
> *(40)*

Unlike the sustainable biological relationship between tribal populations, bison, grass, and climate, the cowboy's home on the range depended on the notion of the range as a laissez-faire commons at the service of an international commodity market that, in turn, depended on the railroads for transport of cattle to the stockyards of Chicago and Kansas City.[11] When the refrigerator ship was invented in 1879, the American cattle industry became global.[12] The grass that fed the cattle was free (but quickly depleted); cowboys were hired to mind herds of hoofed profits until calves had matured to maximum yield.

Yet even when Curly must give up his occupation, his "cowboy" identity remains intact by virtue of its elastic application to wider American life. In *Oklahoma!*'s Act II, Aunt Eller auctions off the box lunches. In this scene, the cowboys and farmers bid on the baskets as if they are bidding for livestock. In order to maintain his bet against Jud for Laurey's basket, Curly must sell his saddle and his horse, symbolically divesting his cowboy identity to embrace his new identify as a farmer. Fashioned after Owen Wister's 1902 novel, *The Virginian* (1945), Curly is the enterprising young man who is ready to throw his weight with the new state: "country a changin', got to change with it" (123). In Wister's best-selling novel, a demigod of manliness leaves his home and family in Virginia for the wide-open spaces of Wyoming. There, Jefferson Davis, "the Virginian," meets with adventure, romance, and success. Drawn by the freedom and independence of life in the west; but, in the end, the Virginian becomes a civilizing force. Similarly, in *Oklahoma!* Curly relinquishes his cowboy life, settling down to prosper in agribusiness: "I got to learn to be a farmer . . . buy up mowin' machines, cut down the prairies! Shoe yer horses, drag them plows under the sod" (123). Like Wister's *Virginian*, Curly transforms along a Turnerian trajectory. Grown in the soil; a free, independent spirit, Curly moves in the "natural" direction of taking a wife, and settling down to become part of the project of nation building. Like the Virginian, Curly bridges a violent past and a prosperous future.

By the mid-twentieth century, however, the cowboy represented the central heroic figure of American popular culture, metaphorically conflated with the U.S. service personnel who were in harm's way fighting for the land and the freedoms that Curly and Laurey represented. That identification with servicemen plays out in the smokehouse scene discussed earlier, when Curly proves his superior marksmanship:

> I wish you'd let me show you sumpin. . . . They's a knot-hole over there about as big as a dime. See it a-winkin? I jist want to see if I c'n hit it. (*Unhurriedly, with cat-like tension, he turns and fires at the wall high up.*) Bullet right through the knot-hole, 'thout tetchin, slick as a whistle, didn't I?
>
> *(71)*

Like Buffalo Bill Cody and other gunfighter heroes who settled into pastoral success, Curly has roots in a lineage of bloody conquests. Curly embodies traces of Custer himself, who often figured in iconography as the boy-hero with curly auburn hair.[13] In Riggs's (1931) *Lilacs*, Old Man Peck (who becomes Skidmore in Hammerstein's version) sings 12 verses commemorating a "blue-eyed boy" who died in "brave Custer's charge" (97). The ballad disappears in Hammerstein's version, but Custer's martyrdom is nonetheless invoked as the price paid for the freedom and opportunity that Curly and Laurey will enjoy – a message that was not lost on the servicemen and women who saw special discounted matinees before shipping out.

Cowboy and frontier heroes became increasingly alluring in the 1950s. Mushrooming colonies of suburban homes and a plethora of convenience appliances and automobiles became metonyms for the promise of the land espoused in *Oklahoma!*.[14] The postwar infatuation with nineteenth-century western heroes maintained an aura of Turnerian innocence around U.S. technological advancement, economic dominance, and military posturing, as the expansion of U.S. economic interests around the world also accelerated the development of the land west of the Mississippi River. By the time the cowboy came to be identified with Americanism in the United States, the life he symbolized had long disappeared. A victim of its own misuse of wildlands and animals, the cattle bonanza of the late nineteenth century buried the cowboy life almost as fast as it had produced it (White 1991).

By 1890, the open range had succumbed to ecological collapse. The tyranny of barbed wire was over, and the big drives were a faint memory (though later romanticized in television programs of the 1950s and 1960s such as *Rawhide* and *Bonanza*). The playing up of individualistic cowboys and farmers in *Oklahoma!* disguised a highly mechanized postwar agribusiness that renders crops of corn, wheat, cotton, beef, pork, and poultry – interchangeable commodities in a global marketplace. The musical's message of cooperation ("the farmer and the cowboy must be friends") is less about community and more about business agreements – economic relations predicated on the reification of a variety of "others," including human beings, animals, and the land. Taking its cue from assembly-line mass production, the cattle industry did not pass away as Curly feared, but instead transformed as the cowboy became the farmer's customer. The industrialization of agriculture and the rise of the factory farm went hand in glove: old antagonisms yielded to new profits as ranchers purchased grain from farmers and cattle were processed in feedlots.

Why did the cowboy – a mere wage-earning laborer of the range whose lifestyle lasted only a decade – become a signifier of American identity around the world? In *West Fever*, historian Brian Dippie (1998) comments on Frederic Remington's sculpture *The Bronco Buster* (1895): "Devoid of space, Remmington's horse and rider are free to occupy any space the imagination desires" (41). With the wild horse replacing the land as a signifier of wilderness, we can lift the image of the cowboy in his context on the western prairies and apply it freely to any and all circumstances in which the values of the frontier will forward U.S. interests. The cowboy could, as it were, ride out of his original frame and into another, taking his expansive horizon with him. Thus transmuting the values of frontierism into conditions of character rather than place, in the twentieth century, the cowboy became an ethos in itself: a behavior, a way of doing business. His horse became a shifting signifier standing in for just about anything he is called upon to master: mountains and rivers, political foes or market rivals, nuclear fission and bioengineering, outer space or cyberspace. The ideology that constructs the land not as a place to inhabit but rather as a place to ride through ultimately has an ecological impact.

Race, radicalism, and the face of Jud

Oklahoma!'s success and its rousing call to certain American values are particularly significant against the contemporaneous backdrop of anticommunist fervor of the 1940s and 1950s. Nineteenth-century frontier narratives typically feature two types of antagonists – the outlaw and the Indian. In certain cases – the valorizing of Jesse James and the Dalton Gang, for example – the Anglo outlaw became the protagonist himself, a kind of Robin Hood of the range. In the twentieth century, the "savage" radical or "red" enemy of the state became a third generic antagonist. Slotkin (1994) argues that the binary of cowboys and Indians became a transferable metaphor applicable whenever a perceived threat to the hegemony occurred; he argues that the role of savage transferred to other marginalized groups who were always constructed as the antithesis to the cowboy/patriot (Part VII). Thus, the role of the Indian in the cultural performance of the frontier myth could be filled by Mexicans, blacks, Chinese, radicals, "foreigners" of all types, and landless tenant farmers – in short, any group that poses a perceived threat to the hegemonic U.S. project. In 1943, Jud signified all the forces that threatened national security, whether or not those enemies were located on home or foreign soil. The layers of meaning in the character of Jud deserve excavation here, for within them are fault lines and contrasting sedimentations that illuminate not only Jud's role as the story's villain, but also the ways in which the United States has vilified imperialism elsewhere in the world while ignoring it on home soil.

During World War II, Hitler's aggression had presented the United States with a visage of evil and a clear and present danger to the land and liberty that Laurey and Curly represented. Jud served (and has continued to serve) as a dark blank slate onto which U.S. audiences could project the face of that (and other) enemies. Variously marked as earth, wild beast, savage, Indian, black, Mexican – and as Wobblie, socialist, or communist – Jud becomes a shifting signifier for anything and anyone that must be expunged or controlled. Jud is characterized as animalistic and associated in dialogue and stage directions with various varmints that Curly must subdue. His living space is a place "where meat was once kept" (61), and Laurey tells Aunt Eller that Jud prowls her windows, stalking her like "sumpin' back in the bresh som'eres" (24). Curly likens him to a snake and to other animals who live in dark holes in the ground, and at the end of the scene in which Jud is killed in the chivaree, "the howl of a coyote drifts in . . . desperate and forlorn" (124). Indeed, the coyote and wolf were often characterized as "enemies" of midwest farmers and ranchers; in the 1930s, the U.S. government executed a comprehensive extermination policy (see, e.g., Worster [1994], 258–290). These and other animal tropes mark Jud as one to be feared and subdued, particularly in relation to his sexual desire. The French postcards, pinup, and copies of the *Police Gazette* that Jud keeps link him to the "savages" who threatened to run off with white women in D.W. Griffith's *Birth of a Nation* ([1915] 2002).

After the Civil War, racist land practices continued through sharecropping and tenant farming. Racist ideologies are predicated on an identification of marginalized groups – people of color and/or radicals – with the "forces of nature" that must be controlled or eradicated. In addition, racism's aim is extractive – like the land, bodies are either part of what must be eradicated (in order to gain access to the land and its resources) or labor mined in the process of that extraction, as in the plantation south, where slave labor was used to mine the land, growing a cash crop for international markets. Maintained by violence and opportunism, racist practices are derived from an ideology that constructs the whole of the natural world as "other" and, therefore, exploitable.

Yet Oklahoma was also one of the regions in the west that African Americans formed communities in which they flourished in the period following the Civil War. Charismatic leaders like Edward Preston McCabe, the first to hold elective office outside the Reconstruction south, had hope for a black-dominated state (see Globe 1980, 115–144). These ambitions and wealth accumulation prompted white racist acts of terrorism against black communities. In 1921, for example, the same righteous Boomer spirit that motivated Payne to penetrate and seize lands in Indian Territory, spurred Anglo whites to invade the black business district of Tulsa, Oklahoma, after a black man was falsely accused of rape. The whites burned the area known as Black Wall Street to the ground, killing 300 people (see, e.g., Ellsworth [1982]). The potential resonance of this history for audiences in 1942 and beyond complicates the nationalist resonance of *Oklahoma!* with white supremacist overtones.

Indeed, Jud's earth/bullet-colored otherness suggests a racialized difference that is compounded when economic class is considered. Even when he is played by white actors (as has mostly been the case), his identity shifts fluidly.[15] As a landless hired hand, Jud is also like the rank-and-file Wobblies of Oklahoma in the early twentieth century – "a red," the shorthand for socialists, communists, fellow travelers, and radicals of leftist persuasion. The real-world counterparts of *Oklahoma!*'s Jud, like those of Steinbeck's Joads, fell under abuses inherent in a system bent on turning land to profit as quickly as possible. In the decades between the period in which Riggs set his play (1900) and the historical moment of its first production in 1931, discontent rose up from the soil in the west and south. Protesters climbed up out of the mines, walked out of the forests and oil fields, and shut down mills, canneries, and shipyards.

A closer look at a few particular scenes demonstrates that Jud may be not only a mixed-race character, but also a radical. Whether Wobblie or communist, Jud evokes a connection to "reds" through language and images associated with Oklahoma's radical history. Between 1890 and 1906, Sellars observes, Oklahoma had come "to be seen as a hot bed of radicalism . . . old style agrarians, populists of both radical and conservative stripes, socialists, and union labor . . . shared a sophisticated understanding of the new economic order" (1998, 6–7). Profit-driven agriculture drove small and tenant farmers off the land and increased inequities, which fueled radical discontent among landless farmers and workers like Jud. By the early twentieth century, the IWW organized a stronghold in Oklahoma among farm- and oil-field workers. The Oklahoma Socialist Party, made up primarily of tenant farmers, grew to be the strongest in the country (Manes 1982, 110–124).

In Scene 2, the Smokehouse scene, Curly's stage business and dialogue invoke potent images in light of both white dominance and the violent suppression of dissent. Curly spies Jud's rope and, taking it down, remarks:

> That's a good-lookin' rope you got there. (*He begins to spin it.*) Spins nice. (*He tosses one end of the rope over the rafter and pulls down on both ends, tentatively.*) 'S a good strong hook you got there. You could hang yerself on that, Jud.

Through his playful actions with the rope, Curly not only suggests that Jud hang himself; Curly goes through the motions of lynching Jud, even instructing him on how to knot the rope and kick the chair out from under him:

> It ud be as easy as fallin' off a log! Fact is, you could stand on a log – er a cheer if you'd rather – right about here – see? And put this here around yer neck. Tie that good up there first, of course. Then all you'd have to do would be to fall off the log – er the cheer, whichever you'd ruther fall of. In five minutes, or less, with good luck, you'd be daid as a doornail.
>
> (63)

The scene's implicit rehearsal of lynching – horrific enough to be sure when Jud is read as a man of color – also alludes to the silencing of diverse political voices through terrorist policing and vigilante enforcement.

A patriotic fervor during World War I inspired vigilante groups, composed of business leaders and police, to drive IWW organizers out of the state. Wobblie opposition to conscription branded them as anti-American and an enemy to be driven back using the tactics of war. The Green Corn Rebellion of 1917 fed antiradical feelings in Oklahoma when angry tenant farmers clashed with police and local "councils of defense." In *Oil, Wheat and Wobblies*, Sellars (1998) explains that

> by the summer of 1917 rumors of a radical uprising had spread throughout Oklahoma. . . . Events came to a head on August 2 when the Seminole County Sheriff and his deputy were ambushed near the Little River. Within hours raiding parties had cut telegraph and telephone lines, burned railroad bridges, and allegedly dynamited oil pipelines near Healdton.
>
> (77)

The IWW later denied the violence, but the revolt nonetheless fueled antiradical sentiment across the state. Sellars notes that "[s]ome editors openly called for lynch mobs" to deal with the rebels. In November of the same year, a group of

Wobblies and other radicals were escorted by police from Tulsa City Hall to the jail when they were intercepted by the Knights of Liberty, a group that included "local businessmen, oil company executives, and police officials" in black robes and hoods. Sellars notes that

> what followed was a hostile charivari, a vicious ritual in which the vigilantes beat, whipped, and tarred and feathered their captives, symbolically reducing them to the level of animals to be driven from the community of humans. "It was a party, a real American party," one newspaper called the incident, soon known as the Tulsa Outrage.
>
> *(78)*

The vigilante attack inspired a chain reaction of raids on union locals that eventually resulted in a "series of conspiracy trials intended to destroy the radical labor union" in Oklahoma (3). Events in Oklahoma fueled antiradicalism in the plains states and mountain west. In 1917, for example, IWW leader Frank Little was lynched in Butte, Montana. In *Oklahoma!*, the kangaroo court that acquits Curly and, by implication, convicts Jud at the end of the play found real-world counterparts in the vigilante justice system that used terrorism to suppress radicals in Oklahoma and blacks in the south.

In *Oklahoma!*, there are three classes of people who work the land: wage laborers, hired both by tenant and land-owning farmers; cattlemen or cowboys like Curly, whose particular use of the land had already succumbed to ecological collapse; and the family farmer, represented by Laurey, Aunt Eller, and the other farmers. "Miners," who work in the oil fields, and merchants, like Ali Hakim, complete the community portrait. Two other groups – radicals and Native Americans – remain unrepresented and silent; however, their absence provides a glaring clue as to what the play masks. Wobblie organizers appealed to those who worked in the oil, wheat, corn, and cotton fields – workers like Jud. But while Wobblie demographics were diverse, so were the lives of the Oklahomans they sought to organize. Elsewhere in the play, Jud's identity seems to slip between that of a wage-earning laborer and a tenant farmer. Sellars (1998) notes that failure to recognize tenant farmers as possible union members contributed to the ultimate failure of the IWW:

> The union's insistence on recruiting only wageworkers meant it failed to see the tenants as the agricultural proletariat they were quickly becoming. . . . [M]any tenants and small farmers took wage-paying jobs in industry and suffered the same hardships as the hoboes who composed much of the IWW. Both wageworkers and tenant farmers were at the mercy of modern business practices that prevented them from gaining even the smallest margin of economic self-sufficiency. . . . For a populous, both in Oklahoma and nationwide, in the midst of a wartime patriotic frenzy and . . . a steady stream of news stories about the "lawless" IWW, the finer

distinctions among the Socialists, the Wobblies and the WCU [Working Class Union] were meaningless.

(92)

In short, all radicals were lumped together as a threat.

Acts of violence were also committed by Wobblies as direct action against a system that used men's bodies like it used the land – as one more resource to be mined and exploited. Within the musical, allusions to the violent tactics of Wobblies and other radicals – such as Jud's story about the fire on the Barlett farm – mark the character of Jud as the "kind of man" to whom the IWW and Socialist Party appealed (69). Conversely, Curly's mischievous instructions on lynching and the men's pre-wedding-night chivaree, in which Jud loses his life, were constructed as harmless pranks. Similarly, vigilantes justified violence against Wobblies and other radicals, in part, because many radicals advocated direct action to redress injustices. In the Smokehouse scene, Curly also probes Jud about his past, linking him to the kind of violence typically, and often erroneously, assigned to Wobblies: "Whur did you work at before you come here? Up at Quapaw, wasn't it?" Curly asks. Jud's reply includes bitter complaints about how he was treated at his last job: "Lousy they was to me. Both of 'em. Always makin' out they was better. Treatin' me like dirt" (68). Curly asks Jud if he believes in getting even:

JUD: 'Member that f'ar on the Barlett farm over by Sweetwater?
CURLY: Shore do. 'Bout five years ago. Turrible accident. Burned up the father and mother and daughter.
JUD: That warn't no accident. A feller told me – the ha'ard hand was stuck on the Barlett girl, and he found her in the hayloft with another feller.

In the language of imagery in the play – in which the land is overtly feminized – the violence carries a political meaning, suggesting that the landless "h'ard hand" plotted an attack on another man's property: "It tuck him week to get all the kerosene – buying it at different times," Jud tells Curly, implicating himself as the arsonist. "[H]e was a kind of a – a kind of a murderer, too. Wasn't he?" Curly responds, completing an argument that justifies Jud's death, and in an analogous way, justifies state-sponsored and vigilante violence against radicals (69).

Oklahoma! reinscribes the idea of rugged individualism and masculinity in opposition to all that Jud signifies: all the complaining "snakes" of society, those who blame others – a common criticism leveled at Wobblies, unionists, and other radicals. At the end of the scene, Curly returns to animal tropes to condemn Jud and defines American manhood in relation to the land: "In this country, they's two things you c'n do if you're a man. Live outdoors is one. Live in a hole is the other" (69). Within this polarity, Jud is the creature who lives in a hole "a-crawlin' and festerin'," while Curly asserts an identity linked to the land itself – what became known in the 1950s as "the great outdoors." Under the Turneresque ideology that puts a premium on individualism, men like Jud are the victims of their own negative attitudes, what Curly calls their own "pizen." "Somebody comin' close to his hole! Somebody gonna step on him!" Curly chides Jud (71). Within this discourse, someone like Jud cannot be the victim of a *system* under which both worker and land are exploited; instead Jud's "kind" is perpetually positioned as a complaining malcontent.[16]

Sanitizing the region's radical history, *Oklahoma!* constructs a homogeneous America, one in which "[t]erritory folks should stick together, / Territory folks should all be pals" (87). But Jud is excluded from this circle, and his death in the chivaree on Laurey and Curly's wedding night is an inevitable outcome of the amalgam of patriotism, frontier opportunism, and rugged individualism celebrated in *Oklahoma!* As the scene unfolds, the men bang pots, toss dolls in the air, and shout obscene innuendoes at the couple. Suddenly, Jud appears and begins to threaten Laurey. Curly defends her, and then a fight ensues, during which "Jud pulls out a knife and goes for Curly. Curly grabs his arm and succeeds in throwing him. Jud falls on his knife, groans and lies still" (138). Jud's death not only marks a victory over the forces he represents but also poses a question about the extent to which Curly is responsible or should be held accountable.

The differences between the handling of Curly's "acquittal" in Riggs' original text and in Hammerstein's musical adaptation point to the increased momentum of U.S. anticommunist sentiment in the decade between the 1931 production of *Green Grow the Lilacs* and *Oklahoma!* in 1943. In Riggs' *Lilacs*, Curly is arrested and spends three days in jail. Unable to bear being away from Laurey any longer, he compromises his chances of having the charges dismissed and breaks out of jail, becoming for a time an "outlaw." In a discussion of the letter of the law versus the spirit of the law in Riggs's script, Aunt Eller notes that "if a law's a good law – it can stand a little breakin'" (159). The end of the play witnesses Curly agree to go before the magistrate after just one night of bliss with his new bride, and we are left wondering *whether or not* he will be acquitted. In *Oklahoma!*, however, Jud's death does not merit true investigation, and Curly does not go to jail. Rather, the acquittal occurs on stage in real time. The character of Cord Elam, the community's "Fed'ral Marshal" protests that "you have to do it in court," but Aunt Eller insists, "[L]e's do it here and say we did it in court." In an intertextual reference to Riggs's earlier script, Aunt Eller says, "Well, le's not *break* the law. Le's just *bend* it a little" (143). The character of Andrew Carnes, who acts as judge, coaches Curly in what to say: "You had to defend yourself, didn't you? . . . Never mind the furthermores – the plea is self-defense" (144). In another possible reference to Riggs's earlier text, Laurey demands that the ambiguity at the end of *Lilacs* be put to rest. She pleads, "Well then, *say* it!" (145). At this moment, the entire community shouts, "Not guilty!" Thus, Jud's death becomes a kind of community purification, dismissing not only law but also accountability itself in the resounding unified cry of "Not guilty!" The blithe dismissal of Jud's death as accidental epitomizes American exceptionalism, for though it was an act of self-defense, it implies that deaths – perhaps those caused by state-sponsored acts of genocide or the advancement of U.S. economic interests – are also accidental and dismissible.

Manufacturing desire

Throughout the history of the United States, musical comedy has been a potent force for maintaining and propagating American values, its policing function made more insidious by its lighthearted tack. Against the backdrop of Oklahoma's history, the musical *Oklahoma!* helped to codify who and what constitute American while at the same time subverting a more complex image of Oklahoma and the United States. As it parlayed a heroic past into a vision of future prosperity, *Oklahoma!* held out the promise of a piece of the land and the good life to every returning serviceman; in this way, the musical served to shore up the exploitation of land and people on which that prosperity depended. Like the king and queen of a springtime fertility rite in which the fecundity of the land is signified in heterosexual coupling, the union of Curly and Laurey doubles for the "marriage" of the Oklahoma territory to the Union. In the final scene, in which the townspeople celebrate Laurey and Curly future, elders Aunt Eller and Ike Skidmore sing: "They couldn't pick a better time to start in life! / It ain't too early and it ain't too late. / Startin' as a farmer with a brand new wife – / Soon be livin' in a brand-new state!" (131).

Metaphorically, GIs returning from World War II envisioned a new beginning also as many migrated from the heartland to the cities or the suburbs in search of the so-called American dream. The metaphorical identification between a frontier past and a horizon of new opportunity, along with the identification between the cowboy and the serviceman – who had been riding the range and was ready to settle down – was a potent social force that translated into goods sold and houses built and occupied. Developers who promoted postwar "tract" housing made use of the deeply ingrained values associated with having a plot of land to call one's own – a legacy of Turner's thesis.

As a parable that drew a metaphoric connection between the boomer pioneers of Oklahoma and industrial boom in the western states during and after World War II, *Oklahoma!* also gave symbolic support to that industrial development, with its requisite ramped up resource extraction. The war effort transformed the west from a region dependent on and dominated by eastern capital. As a result, areas from Arizona to Colorado to Washington boasted their own urban industrial centers and vast labor markets. But this development came at ecological and human cost. In "When Green Was Pink: Environmental Dissent in Cold War America," Jeffery Charles Ellis (1995) observes that in the 1950s, "[m]any of the representations of the nation's major institutions and industries conflated environmentalist concerns with a Communist-inspired attempt to undermine the American way of life" (9). As individuals, unions, or citizen groups complained about industrial effluents, worker safety, or environmental degradation, their complaints were often vilified, conflated with anti-Americanism, and ultimately silenced by business and government alike.[17]

Unlike Steinbeck's *Grapes of Wrath*, the economic climate into which *Oklahoma!* played was not one parched by the failure of the system but one quenched by a singular national focus and a steady flow of government contracts and subsidies to industry and farmers. One of the basic tenets of the American free-enterprise system according to its architects, Ellis (1995) argues, is that "for the system to run smoothly, government, business, labor, science and the American consumer had to work together" (22). In the spirit of postwar consumerism, Act II of *Oklahoma!* emphasizes the importance of the merchant via the character of the peddler, Ali Hakim. He too must take his place in the economic schemata: "The farmer and the cowman and the *merchant*/Must all behave theirsel's and act like brothers" (89). But in order for the system to work, the farmers, cowmen, merchants, and industrialists needed *consumers* to complete the loop. Robert Gottlieb notes that a member of President Eisenhower's Council of Economic Advisors remarked that the ultimate goal of the economy is "to produce more consumer goods. This is the goal. This is the object of everything we are working at; to produce *things* for consumers" (Schlesinger 1963, 83; quoted in Gottlieb 1993, 342). Their strategy proposed to shift a war-based economy driven by consumers.

Oklahoma! in the postwar era helped to transform the ordinary citizen into this all-important consumer, beginning with Will Parker's return from Kansas City. After having an experience not unlike thousands of Americans who flocked to the New York World's Fair in 1939 to see "the world of tomorrow," Will sings in amazement:

> Up to then I didn't have an idy
> Of whut the modern world was comin' to!
> . . .
> Ev'rythin's like a dream in Kansas City.
> It's better than a magic lantern show!
>   (18)

Automobiles, telephones, skyscrapers, plumbing, and radiators are among the turn-of-the-century wonders that Will describes. Curly is enthusiastic about the way American life is "changin' right and left!"; he believes that people must "keep up 'th the way things is goin' in this here crazy country!" (123). When Germany invaded Poland in 1939, and then Japan brought the United States fully into the war in 1941, the wonders of the 1939 World's Fair were put on hold. Manufacturers retooled to meet wartime needs and rationing called on Americans to limit consumption. After the war, however, Will's news that "[e]v'rythin's up to date in Kansas City . . . with ev'ry kind of comfort ev'ry house is all complete" must have fanned the coals of desire for the things Americans had gone without but knew were possible (17). Ironically, the images and values of the frontier – a place of hardship and making do – served the new consumer age well. A willingness to adapt and to try new things, an appreciation of ingenuity and invention, a lust for new horizons – all aspects of the frontier myth – helped to produce a society of eager consumers that would safeguard the national economy. In short, after the war, purchasing became patriotic.

As manufacturers produced the goods, advertisers manufactured the desire for the goods through radio, television, and print. It was the age of the advertising man. Gottlieb (1993) observes that "advertising's role was . . . to generate new values associated with the drive to establish markets for new products" (75–76). In *Oklahoma!*'s Act I, and with the arrival of the peddler Ali Hakim, Laurey is filled to bursting with desire that conflates things with experiences:

> Want things I've heard of and never had before – a rubbler-t'ard buggy, a cut-glass sugar bowl. Want things I cain't tell you about – not only things to look at and hold in yer hands. Things to happen to you. Things so nice, if they ever did happen to you, yer heart ud quit beatin'.
>
> *(34)*

Laurey's lines demonstrate the very strategy of U.S. advertisers who sought to sell products *as* experiences.

Nothing in the musical communicated the conflation of consumption and desire as well as "The Surrey with the Fringe on Top." Hammerstein's lyrics are prescient of the images and narratives that would be used to sell automobiles as image of success:

> The wheels are yeller, the upholstery's brown,
> The dashboard's genuine leather,
> With isinglass curtains y'can roll right down
> In case there's a change in the weather –
> Two bright side-lights, winkin' and blinkin'
> Ain't no finer rig . . .
> Fer that shiny little surrey with the fringe on the top!
>
> (9)

And like the automobile, the surrey is not merely a means of transportation; rather, it is an event in and of itself: "All the world'll fly in a flurry / When I take you out in the surrey." Automobiles gave those who could afford them more access to the land – as the General Motors jingle "See the USA in your Chevrolet" suggested.[18] In Laurey's words, "[t]he river will ripple out a whispered song. . . . In that shiny little surrey with the fringe on the top!" (10). The car and the highways built during the Eisenhower administration provided mimetic adventures for vacationing families who retraced the routes of explorers and went through the motions of pioneer life in campgrounds built by the Civilian Conservation Corps. But the cost of Americans' four-wheeled entitlement was high, and the automobile figured at the center of many of the environmental ills of the 1960s and 1970s. Its path of environmental destruction included the bifurcation of public lands by new highways, as well as air and water pollution. Oil shortages in the mid-1970s precipitated new oil drilling in Alaska, resulting in dozens of oil spills along coastlines.

The new frontier

Creators and critics heralded *Oklahoma!*'s authenticity, calling it an "indigenous American musical" (Evans 1990, 2). Richard Rodgers himself claimed that the musical was an attempt to find "cultural truths in the idea of homeland" and that it "gave citizens an appreciation of the hardy stock from which they'd sprung" (quoted in Evans, 228), *Oklahoma!* distilled so-called Americanism and set it to music, just as Theresa Helburn had hoped. The mythic figures of the cowboy and his "girl" invoked the endless horizon of opportunity symbolized by the American frontier gave new life to an old myth for a technological postwar United States. Slotkin (1994) writes that "[s]uch metaphors are not merely ornamental. They invoke a tradition of discourse that has historical roots and referents, and carries with it a heavy and persistent ideological charge" (18). As the "language of historical memory," Slotkin writes, myths are never merely stories from the past, transmitted and transmuted for nostalgic pleasure alone, but rather carry a load of instruction and ideology (19). *Oklahoma!* celebrated not only the end of the World War but also victory over the Depression, the drought, and the land itself. *Oklahoma!* reinscribes and valorizes Turner's frontier qualities that, along with a Codyesque bravado, became emblematic of an American spirit that won the war. American ingenuity was seen as an eminently powerful thing, as U.S. dollars helped rebuild Europe and Japan. The American cowboy spirit had built the bomb and now those cowboys sat in the saddle of planetary destiny.

The sociocultural messaging conveyed by *Oklahoma!* helped to generate a postwar mandate for increased development of western lands and for the export of U.S. technologies around the world. In the United States, *Oklahoma!* argued for the ongoing reclamation of western lands as it subverts the causes of the Dust Bowl and the images of mass exodus of tenant farmers and migrant workers from the plains states. In effect, the musical preserves the illusion of the "family farm" even as agribusiness in the Great Plains became increasingly corporatized and mechanized for the production of cash crops as cotton, corn, wheat, and beef. *Oklahoma!* also subverted Oklahoma's radical past, presenting instead a landscape in which radicals either conform to U.S. frontieresque ideals of private property and bully entrepreneurship – or literally die. *Oklahoma!* reinscribed a feminine gendered landscape against which U.S. notions of male dominance would continue to be defined, as it centered heterosexual marriage as the metaphor for an American relationship with the land (Fearnow 1997).[19]

Oklahoma! is a testament to the power of stories to nourish hope, but also to shape the direction of cultural priorities. For early to mid-twentieth-century citizens, who were recovering from hard times and involved in a war with formidable enemies, *Oklahoma!* functioned like a religious icon or a picture of a family member or ancestor; the play provided a distilled image of all that was worth fighting and living for. For all of these reasons, *Oklahoma!* can be seen as propaganda for a hegemonic order determined to preserve economic growth at any cost, return women to the domestic sphere, and construct an enduring notion of Americanism that could be used to marginalize, or even expunge, anyone who criticized the system. In its selective mythologizing of history, *Oklahoma!* erases many other histories and with them deceptively absolves the entire nation of responsibility for crimes. Whether Helburn and the Theatre Guild intended this outcome or not, *Oklahoma!* is a powerful example of how stories shape and participate in the body politic and precisely why artists must be mindful of that power.

"The woods are burning, boys!"

On February 10, 1949, approximately six months after the New York production of *Oklahoma!* closed (but while touring productions continued around the nation and the world), Arthur Miller's (1971) *Death of a Salesman* opened at the Moscow Theatre. Directed by Elia Kazan, Miller's story of an aging traveling salesman and father served as an indictment of the American dream and the unsustainable system on which it depends. As with *Oklahoma!*, its reception is the stuff of legends, and its themes resonate across time. Few American dramas have received as many revivals as Miller's portrait of an ordinary man and the personal cost of an ethos that defines success as economic gain at the expense of home, health, and kinship.

Miller's work exposed the rupture between the mythic origins of the "American dream" and the deeply personal consequences of the ideologies (social and environmental) that *Oklahoma!* espoused. Like *Oklahoma!*, Miller's *Salesman* may be familiar to most readers; thus, rather than a detailed plot summary, certain elements about Willy Loman, an aging traveling salesman, his wife, Linda, and their two adult sons, Biff and Happy, are key to remember. Throughout the play, Willy confuses past and present, converses with memories, and becomes increasingly depressed and ultimately unable to continue the long drives to his sales territory in Boston. Willy's delusions increase over the course of the play as he moves in and out of reality. He conjures his older brother, Ben, who was successful in ways Willy was not; he alternately makes excuses for his failure to measure up, disguising his insecurities behind bravado. He blames his father, belittles his sons, and is cruel to his wife. Through a litany of self-aggrandizement, Willy projects his overblown dreams onto his sons. Finally, he plans and carries out his suicide.

Willy believed in the dream that Curly signifies, yet the promise of *Oklahoma!* is ultimately shattered in *Salesman*. The section that follows tracks Willy's poisonous dreams and argues that their toxicity informs and infects a U.S. ethos with both human and ecological consequences. *Oklahoma!*'s chorus of cowboys and farmers sing "everything's up to date in Kansas City!" (17–18), and later

Curly heralds a new era, saying, "[C]ountry's changin', gotta change with it!" (123). But Willy feels trapped by that change and the ethos that puts a premium on expediency and profit. "I don't want change!" he cries out as the world he thought he knew crumbles (10). Miller reveals the seeping terror of Willy's gradual awareness that he, too, is a victim of unsustainable, extractive, and exploitive planned obsolescence. Fueled by an ideology that condones and fosters waste in the name of efficiency and understands newness as a status symbol, Willy Loman's tragedy is the product of an economic system that throws away a man like a piece of rubbish.

After *Death of a Salesman* won the Pulitzer Prize for Drama in 1949, Columbia Pictures produced a film version in 1951. At the time, however, the nation was on the verge of an "anti-communist rage that threatened to reach hysterical proportions," Miller recalls (2016, 252) recalls. "In less than two years *Death of a Salesman* had gone from being a masterpiece to being a heresy, and a fraudulent one at that" (255). At the time, the House Un-American Activities Committee and Senator Joseph McCarthy's hearings sought to root out and villainize members of the U.S. Communist Party and its sympathizers. Due to growing anticommunist sentiment, many critics were mindful to absolve the play of social commentary. According to a February 21 (1949) issue of *Newsweek*, for example, Miller was "writing not as a critic of an economic system but as an observer with a profound compassion for plain people" and that his "vivid emotion-shattering, and deeply moving play . . . *eschews social significance* to explore the hopes and frustrations, the sins and loyalties and brief pleasures of a very ordinary middle-class family" ("Magnificent Death" 1949, 78, my emphasis). Given that only ten years earlier those families had lost savings, jobs, homes, dreams, and lives in the Depression and Dust Bowl, the reviewer's claim that an ordinary middle-class family is not an institution of social significance is particularly ironic. When Miller refused to sign an anticommunist loyalty statement presented by Columbia Pictures, the film version of *Salesman* premiered with an accompanying educational short that touted the joys and fulfillment of being a salesman in the United States that directly undercut the prescient themes of the play.

Following the McCarthy era, however, *Death of a Salesman* has been recognized as an examination of the American dream from a systems-based perspective. Still, critics have often interpreted Willy and his saga as the story of a man overwhelmed by the technological age and a man who cannot adapt to change. Willy's incapacity, they argue, marks his weakness and dementia. Barclay W. Bates (1983) discusses, for example, Willy as "the archetypal cherisher of the pastoral world" who "dreams of fulfillment in the countryside" and longs for "release from the complexity of urban life and obligation" (60). This type of reading is infused with frontieresque values that equate progress with the relentless demonstration of survival of the fittest. Miller (1987) writes that the play "involved the attempts of his sons and his wife and Willy himself to understand what is killing him" (182). When we consider the complexities of what is happening to Willy Loman in light of the paradox of the frontier ethos as it plays out in the experience of one ordinary man, Willy's lapses are much more than nostalgic weakness. Indeed, they become harbingers of ecological collapse sited in the personal but signaling what Miller (2016) called the "unbroken tissue that [is] man and society, a single unit rather than two" (182). Certainly the environment is included in this unbroken tissue of interdependence.

The material-ecological fabric of Willy's life is unraveling from the start of the play and was captured in Joe Mielziner's scenography. Two alternating scrims, one painted with the brick walls of the apartment buildings that have grown up around Willy's Brooklyn home, and another reminiscent of Willy's landscape of longing, invoked a dreamlike pastoral of arching elms (see, e.g., Mielziner [1965]). These two primary scene paintings faded in and out, signaling the lapses in Willy's temporal continuity and setting up a visual dialogue between past and present, between the landscape of Willy's sensory memories and the present in which he cannot find himself. "Remember those two beautiful elm trees out there?" Willy reminisces. "When Biff and I hung the swing between them?" Lost in reverie, he adds: "More and more I think of those days, Linda. This time of year it was lilac and wisteria. And then the peonies would come out, and the daffodils. What fragrance in this room!" His memories of a vital and fecund time are more than pastoral longings; they are outcries in the face of something that is killing him. Willy complains that there is not enough light in his yard, nor enough nutrients in the soil to grow the garden he remembers. Willy is exiled in a landscape that provides little of the sustenance that he remembers: "They boxed us in here . . . there's not a breath of fresh air in the neighborhood. The grass don't grow anymore, you can't raise a carrot in the backyard" (11). Both materially and metaphorically, Willy cannot survive the disappointments and losses that have resulted from his own faith and participation in an ethos of extraction. Willy is lost, and the dissolution of his sense of place is personally devastating. Later he pleads with his sons: "The woods are burning, boys, you understand? There's a big blaze going on all around" (100). He dares them (and the audience) to bear witness to his personal catastrophe born of his inability to reconcile his values and beliefs (of masculinity, opportunism, frontierism) with his lived experience (of loss, failure, and alienation).

Willy's longings and delusions are not merely metaphors, nor are they exaggerations. Rather they signal the era's urban expansion in which more lands were gobbled up by developers. Dependence on synthetic and often toxic chemicals accompanied development, and air and water pollution seemed, at first, a small price for the prosperity of the postwar years. By the late 1940s, the automobile – the instrument that made Willy's livelihood possible – had polluted the air in metropolitan areas to such a degree that scientists began to call attention to the health risks. Leaded gasoline, the staple of the era of high-octane vehicles, with their celebrated "pickup," produced leaded emissions. "Black Monday" in 1943 was a foreshadowing, Robert Gottlieb (1993) writes, of the postwar phenomenon of smog:

> Similarly, the subsequent air pollution episode in Donora, Pennsylvania, in 1948 and the deadly 1952 London fog, in which air contaminants were thought to have contributed to several thousand deaths, also symbolized the potential reach of these new hazards, linked to the evolving postwar urban and industrial order.
>
> *(77)*[20]

The reciprocity between human and land that Willy claims to have once known is broken.

Willy's family is imperiled by values handed down from father to son, from brother to brother. The Loman men subscribe to a definition of masculinity and meaning that corresponds to Frederick Jackson Turner's description of American character traits:

> coarseness and strength combined with acuteness and inquisitiveness; that practical, inventive turn of mind, quick to find expedients; that masterful grasp of material things, lacking in the artistic but powerful to effect great ends; that restless, nervous energy; that dominant individualism, working for good and or evil, and withal that buoyancy and exuberance which comes with freedom.
>
> *(1956, 37)*

Willy and his brother, Ben, learned the making and measure of a man from their "adventurous" and "wild hearted" father who walked "away down some open road" (41–42). The dialogue is riddled with tropes of violent conquest. Ben claims he struck it rich "principally diamond mines," telling his nephews Biff and Hap, "Why boys, when I was seventeen I walked into the jungle and when I was twenty one, I walked out. . . . And by God I was rich" (46). Like the brash Captain Alleyn in Daly's 1871 *Horizon*, who believes "a soldier ought to fight" (1978, 347), Willy's figures of speech suggest the applied methodological violence required of frontier heroes. He brags to his sons that he "knocked 'em cold in Providence" and "slaughtered 'em in Boston" (27). He tells Ben that he has tried to imbue his sons with the same manly spirit "to walk into a jungle!" (46) where "the sky's the limit" (79). Uncle Ben warns his nephews, "Never fight fair with a stranger, boy. You'll never get out of the jungle that way" (43). In the geo-temporal confusion of the play, the jungle is not only Africa (diamond mines) and Alaska (gold and timber) but also the jungle of Wall Street or any other place where success is measured in profit. Yet Willy is confused when he himself becomes a victim of extractive expediencies when Howard fires him. He pleads, famously saying, "[Y]ou can't eat the orange and throw out the peel – a man is not a piece of fruit!" (75). Willy is both the victim and the carrier of an infectious myth that has wreaked havoc across the small sphere of his life.

Yet, as I discuss throughout this book, personal dreams are made from the stuff of collective myths. In Willy's case, his dreams are a complicated weave of contradictory elements. Willy is unable to transmute the endless frontier into rooted habitation. Like Moody's *Great Divide* (Chapter 2), *Salesman* raises a critical question about how or whether extractive capitalism and its innate violence can ever be reconciled with the values of home and kinship. Willy Loman's experience resonates with all who have been removed from their homelands by the force of empire; with the Oakies and Exodusters, whose roots in the soil were shaken loose by farm consolidation; and with migrant workers who followed the road in seasonal cycles, compelled by the next job and the next, never home. Unlike these displaced and dislocated ones, Willy did not leave his home; rather, his home left him. Attributing Willy Loman's tragedy to his personal failings – his delusions of grandeur, his moral expediency, his miscalculated parenting – is to ignore the ethos of frontierism that informs his sense of self.

Dreams that kill us

"A terrible thing is happening . . . attention must be paid" (50), Linda Loman cries out in frustration at her sons' disregard for Willy's suffering. Her plea asks for attention to be paid to the lived experience of people who are impacted by an exploitative system in which human life – in this case, that of a middle-class white man – is regarded as expendable once the value of his labor has been extracted. In *The Unsettling of America*, Wendell Berry (1977) has observed that amid the waves of migration that characterized American expansion, a counter and compelling force shows itself as "the tendency to stay put, to say, 'No farther. This is the place.'" Berry continues by noting that "every advance of the frontier left behind families and communities who intended to remain and prosper" (4). The human impetus to inhabit (like Willy's) rises out of our material-ecological interdependence with the natural world and is consequently at odds with the master narrative of the frontier. Thus, in *Death of a Salesman*, an internalized frontier ethos tells Willy that he should travel, sell more, be more successful than he is. Yet Willy's sensorially experienced life fills him with longing that flies in the face of expansionism. Willy's memories of the environment as it once was are expressions of an ecological self (see, e.g., Roszak [2002]; Orr [1995]; and Thomashow [1995]).

As architects of the yeoman dream, both Thomas Jefferson and Frederick Jackson Turner believed that habitation followed naturally in the wake of conquest. But the values that propel conquest are not the values that sustain home and relations. Americans' confusion on this point has been systemic and institutionalized over many decades and centuries. As a process, the frontier had always been about expansion and possession for the purposes of power and profit. The constantly expanding market, Berry ([1966] 1977) argues, was worth more than gold, deriving value from the very dynamic of expansion itself (5). By its ever-expanding borders, the frontier has engendered a consumptive and exploitative relationship with the land. On the other hand, habitation and what Berry has identified as a relationship to the land as "homeland" *can be* sustainable. The values of habitation arise from a lived awareness of the permeability between human culture and the natural world. As Berry explains, the idea of home/homeland implies and requires relationships "based upon old usage and association, upon inherited memory, tradition, [and] veneration" (4), all of which are missing in Willy's world.

In the ecology of Willy Loman, as in the ecology of the nation, the process of habitation is at odds with the mechanism of the frontierism. Willy Loman and his family are caught between these two forces that, Berry ([1966] 1977) reminds us, "describe a division not only between persons, but also within persons." Berry identifies this predicament as a division between the tendencies of "exploitation and nurture," saying that "[t]he standard of the exploiter is efficiency; the standard of the nurturer is care" (5). Willy and his family are caught between the force of the frontier that demands "ever expanding markets" and his longing to "take hold," growing in and with a place. Willy himself is caught between his biologic and perhaps spiritual longing to come home and a frontier ideology that compels his uprootedness. As his delusions worsen, he passes in and out of various pasts. Irretrievably lost in a landscape of perpetual change, he cannot place himself in time or space. Willy confides to Ben, "I feel kind of temporary about myself" (45). Willy longs for the smell of grass, a look at the moon, the feel of the wind, the scent of home, love, and family; he longs to see his offspring thrive, to have knowledge of his own roots, and to eat cheese that has not been "Americanized." Willy's life is part of the wreckage piled by the undertow of progress.[21] But the ethos of the frontier – conquest, expansion of markets, increased wealth extraction – mandates that in order to go home, one must first leave home.

Within the play, Willy's life has become the site of a terrible paradox. In private conversations, he must negotiate between the forces of exploitation and the values of habitation. It is a conflict in which the two are ingrown and grafted one to the other so that life is lived as a contradiction in terms. The dialogue is rife with the paradoxes, the insoluble riddles left in the wake of the frontier: Biff is a "lazy bum" and "a hard worker" (10); the Chevy is "the greatest car ever made" (27); and "that goddam Chevrolet, they ought to prohibit the manufacture of that car!" (30), and theft is understood as "initiative" (23). In Willy's world, Alaska is Africa and Brooklyn is the frontier. Willy defends his choice to stay put by identifying his neighborhood with the frontier Ben signifies. "It's Brooklyn, I know, but we hunt too . . . there's snakes and rabbits . . . Why Biff can fell any one of these trees in no time!" But as Willy attempts to live according to the dictates of the frontier, he violates the obligations of habitation. The two value systems cannot be reconciled. According to Turner's thesis of "free land" available for conquest and settlement, timber, gold, diamonds – or whatever commodity the land held – were awaiting extraction. This entitlement is predicated on ever-expanding territories where, as Uncle Ben claims, "there's a fortune to be made." To prove his self-reliance to Ben, Willy commands his sons: "Boys! Go right over to where they're building the apartment house and get some sand,"

and later, "You shoulda seen the lumber they brought home last week. At least a dozen six-by-tens worth all kinds of money." Here, in the frontieresque escapades of the Loman boys, the doctrine of discovery[22] that sanctioned the taking of lands and resources from indigenous tribes is exposed as theft plain and simple. The improvements Willy has made on his house are achievements he can lay his hands on, and yet his home is made from stolen goods. Willy's neighbor Charley warns him that "if they steal any more from that building the watchman'll put the cops on them!" (44). Nevertheless, Willy and Ben understand their hunting and harvesting as the requisite behavior of real men. On the frontier, as in business, ethics are a luxury, jettisoned in the face of necessity or opportunity. While the values inherent in the commodification of land and labor have been the foundation of Willy Loman's life's work and child rearing, those same values ultimately destroy his livelihood and his family.

In one of his many moments of delusion, Willy calls out to the apparition of his brother, Ben: "How did you do it?" Willy demands. The memory consumes him, and the lure of the profit-rich frontier is visible. According to Ben, there are profits to be made in timber in Alaska. "God, timberland! Me and my boys in those grand outdoors!" Alaska now is the "new continent" where a man can "fight for a fortune" (78). Ben warns that "[t]here's a new continent at your doorstep, William. You could walk out rich. Rich!" Willy places his bets on the patterns of habitation he has established. "We'll do it here, Ben! You hear me? We're gonna do it here!" (80) As Berry ([1966] 1977) has observed, many who pressed into the frontier were also looking for a home. Berry warns, however, that the mechanism of the frontier promises that one day, those individuals and communities too will be driven from their homeland by a new generation of conquerors: "Time after time, in place after place, these conquerors have fragmented and demolished traditional communities, the beginnings of domestic cultures. They have always said that what they destroyed was outdated, provincial," much like Willy Loman (4). Ironically, because the frontier is self-perpetuating, those who choose to stay and inhabit the land, who say with Willy Loman "we'll do it here," will be overrun in some future "land rush" as wave after wave of frontiersmen chase dreams of riches. This pattern represents "an intention that was *organized* here almost from the start" of the white man's presence on the land. Ben's mythical expeditions to Alaska and Africa are this self-same mandate for "constantly expanding markets" (5).

Willy passes a definition of success – as a victory over one's foes – to his sons. In the Loman family, the battlefield is reenacted on the football field, where Biff is Adonis, and in the marketplace, where a man must become number one. In defense of the mythos around which he has organized his life, Willy must continually lie about his sons' achievements, claiming Biff is "working on a very big deal" and is "doing very big things in the west. . . . Very big" (84). Willy encourages his son to make it up as they go. In preparation for a job interview, Willy tells Biff, "Tell them you were in business in the west. Not farm work" (61). Lies, exaggerations, erasure of the past – all consistent with the ethos of the frontier – demand a constant state of impermanence while simultaneously holding out the fantasy of permanence. Home is the mirage in the desert. Within this system, Berry ([1966] 1977) observes that the

> only escape from . . . victimization has been to "succeed" – that is, to "make it" into the class of exploiters, and then to remain so specialized and so "mobile" as to be unconscious of the effects of one's life or livelihood.
>
> (5)

The fruits of conquest (bodies, boomtowns, hydraulic dams, and lakeside property) are called into question when Happy observes that his boss buys estate after estate but cannot seem to take up residence – to abide – in any of them: "[H]e lived there about two months and sold it, and now he's building another one. He can't enjoy it once it's finished" (17). Linda asks the pointed question: "Why must everybody conquer the world?" (78). Meanwhile, later in the play, Charley provides the only answer that makes sense within a system in which profits are paramount: "The only thing you got in this world is what you can sell" (90). The cost is total: "You end up worth more dead than alive" Willy observes, fixing his mind on the "diamond, shining in the dark" (91) and thus plans his suicide.

Like Willy, Biff is both the instrument and the victim of the ideology that perpetuates his exile because the frontier is an unstoppable process of expansion that denies or reshapes the past according to expedient ends. The hope "that there's still a possibility of better things" to come perpetuates a condition of impermanence and exile (Miller 1971, 47). Biff is harnessed to the same ideology that kills his father, a belief that the livable life is out beyond the horizon, where the long-term consequences of frontierism have not yet become visible. Biff tells his mother, "I just can't take hold, take hold of some kind of life" (48). He dreams of raising horses while moving from one odd job to another as part of the development projects in the west. Biff encourages his brother to "come out west with me" where "we could buy a ranch. Raise cattle, use our muscles" (170). For Biff, the illusion of freedom is still wed to his father's requirement for success: "we'd be known all over the counties!" (17). The possibility of a new beginning encoded in the frontier calls him, as it would beckon to young families after the war who moved west to find in suburbia what Willy had lost in New York. Mainstream U.S. culture of the 1950s would hold onto its pastoral dream while simultaneously poisoning, paving over, or otherwise compromising the land on which that dream is fixed. Suburbia was conceived as a place that was not the city but not the country, as a kind of midlands made possible by the nation's increasing dependence on the automobile. During the Eisenhower years, new highways became the arteries of national commerce. Richard Slotkin (1994) notes that the mythos protects itself from accountability through a "one-directional" relationship with history (41). The same frontieresque ethos that killed Willy's elm trees, that blotted out the sky, and that boxed him in would soon proliferate in Los Angeles, Phoenix, and cities and suburbs across the country.

Willy Loman's story is, however, a kind of accounting – a localized, personalized consequence of the frontier ethos. In his final hours, Willy is possessed by the necessity to plant: "I'd better hurry. . . . I've got to get some seeds, right away. Nothing's planted. I don't have a thing in the ground" (115). And so Willy Loman measures what hope is left him in a final effort to plant, to take root, to render sustenance from a home his own values have helped destroy. It is a desperate return to the land, the one link to permanence, the biologic interconnection between the body and the land. Willy turns to the small plot of arable land in a kind of grotesque prayer of supplication to the household gods. Pathetically, he goes through the motions according to the rules written on the back of the packages distributed by seed companies, while talking to a conjured Ben about the insurance money that will come to his family if his suicide is successful. In a last and hopeless attempt to wrest from the soil the kind of sustenance that the mirage of the frontier promised and to leave that promise for his family, Willy paces off rows: "Carrots . . . quarter inch apart. Rows . . . one-foot rows. (*He measures it off.*) One foot. (*He puts down a package and measures off.*) Beets. (*He puts down another package and measures again.*) Lettuce." As if trying to squeeze hope from the Turnerian promise that has been implicit in his life, Willy reasons his decision with an imaginary Ben that "a man has got to add up to something . . . it's a guaranteed twenty-thousand-dollar proposition" (119). Willy's earlier plea to Howard echoes here as well: "There were promises made across this desk!" (75). The promise of the frontier and the promise of *Oklahoma!* had been implicit in every day of Willy's working life: something will grow; my offspring will flourish; my claim on the land will bring a harvest of riches. He played by the rules; he opened new territories; he "knocked 'em dead," and now what about the promise? Willy is exhausted and depleted, and yet there is no place for replenishment. The soil is barren; his life is barren, and they are barren for the same reasons. Willy's lifeblood has been mined for its value like a strip-mined landscape. In *Death of a Salesman*, the personal psychobiological cost of American frontier as environmental ethos comes full circle. Willy's life is a litany of the failure of *Oklahoma!*'s promise. His death is the violent culmination of an ecological ethos that denies the permeability between people, culture and the natural world: a personal silent spring.

This glimpse of the multiple and complex ways that *Oklahoma!* and *Death of a Salesman* resonated with the politics and environmental issues in the decades of the 1940s and 1950s are examples of the ways that theater reflects popular understanding and deeply held convictions about the land, its uses, who belongs, and what livelihood means. These ideas about what land is for, and who it serves, are intertwined with implicit questions of social justice. The very basis of a so-called American relationship to land is founded on a presumed "right" to take that land from its current inhabitants and to enslave others in the project of turning the land into wealth. Miller's play attests to lived experience that what we do to the land, we do to ourselves. Ecologically dangerous as well as unjust, many of these ideas about what land is for have been represented on stage throughout the twentieth century. The stories told on stage participated in environmental debates, informing social convictions about the land and its diverse ecological communities (including human), and sometimes directly shaping public opinion and policy. The body of the land and the bodies of people are one whole living system, "land and bodies passing in and out of one another" in an imperative of kinship that can no longer be ignored (Berry [1966] 1977, 22). In the 1980s, the environmental justice movement would make the social justice connections that ecologists had long known: the most basic fields of exchange of air, food, and water betray the imagined boundaries between people and their ecological kin.

Notes

1 All subsequent quotations from the play are taken from this edition and are referenced by page numbers. For details on the adaptation of Riggs's play to *Oklahoma!*, see Max Wilk, *OK! The Story of Oklahoma!* (New York: Grove Press, 1993), 35–35. For analysis and historical context, see Jace Weaver's (2003) "Introduction to *Green Grow the Lilacs*"; see also Hanay Geiogamah's *Stories of Our Way* (1999). For more on Riggs's professional career and significance, including *Cherokee Night*, see Kirby Brown's *Stoking the Fire: Nationhood in Cherokee Writing 1907–1970* (University of Oklahoma Press, 2018), especially Chapter 4; see also Barry Witham's *A Sustainable Theatre: Jasper Deeter at Hedgerow* (New York: Palgrave Macmillan, 2013), especially Chapter 4.
2 *Oklahoma!* has enjoyed frequent New York revivals, including in 1951, 1970, 1980, and 1988. In 1999, Trevor Nunn directed a new film adaptation starring Hugh Jackman as Curly, which was followed quickly by a (2002) London stage production. Notable regional and productions have included Portland, Oregon's Portland Center Stage (2011), featuring a completely African American cast; Washington, D.C.'s, Arena Stage (2010), in which Aunt Eller and Laurey were played by African American actors; and Seattle's 5th Avenue Theatre (2012), in which an African American actor played the role of Jud. Most recently, Oregon Shakespeare Company earned critical recognition in 2018 for its interpretation of *Oklahoma!*, in which many key roles were cast to represent gay, bisexual lesbian, and trans-identified characters.
3 The phrase "age of anxiety" was oft-quoted by psychologists and cultural critics throughout the 1950s and 1960s, to such a degree that in 1961 *Time* ran "The Age of Anxiety" as a cover story. Auden's *Age of Anxiety* (1947) inspired Leonard Bernstein's "Symphony No. 2, The Age of Anxiety" (1948–1949) and a 1951 ballet by Jerome Robbins.
4 All subsequent quotations from *Death* are taken from this edition.
5 The Choctaw were indigenous to what became Mississippi; the Cherokee had inhabited parts of what are now called Georgia and Alabama; the Seminoles had lived in Florida. For more about the many other Native tribes who inhabited the area of the Louisiana Purchase and who were displaced by the Jacksonian-era Removal Policy, see: White (1991), especially Chapter 4 (86–115), Anderson (1991), Zinn (1995), especially Chapter 7 (124–146), and Globe (1980), Chapters 2 and 3.
6 Andrea Most discusses the assimilationist ideology of *Oklahoma!* the musical in "'We Know We Belong to the Land'" (1998). For example, she observes that Hakim's "Persian" identity may represent a euphemism for "Jewish" (77–116).
7 Arab-American theater scholar Michael Malek Najjar (2015) details the troubling representations of diverse Arab characters across theater and film in *Arab American Drama, Film and Performance: A Critical Study, 1908 to the Present*.
8 This text was originally published as Turner's *The Frontier in American History* in 1921.
9 Riggs, himself of mixed ancestry, wrote many plays with characters that represented the complex racial and cultural of Cherokee identity. For more on this topic, see Brown (2018), Chapter 3.

154 We know we belong to the land

10 To press the signification of Jud's phenotype and its link to land even further, eastern Oklahoma was a central cite of lead mining for the manufacture of bullets used in both World War I and II. Picher, Oklahoma, where a large surface mine left piles of tailings, or lead chat, is part of the Tar Creek Superfund Site mandated by the Environmental Protection Agency in 1980. See, for example, K.E. Juracek and K.D. Drake (2016), "Mining-Related Sediment and Soil Contamination in a Large Superfund Site: Characterization, Habitat Implications, and Remediation."
11 For more about the development of the U.S. cattle industry and its relationship with railroads and eastern investors, see White (1991, 222–227) and Worster (1992), especially Chapter 3 (34–52).
12 In the United States, the cattle boom spread rapidly from Texas and New Mexico into the grasslands of Oklahoma, Kansas, Nebraska, Colorado, Wyoming, Montana, and the Dakotas. Everywhere the bison had roamed, human-bred bovines were multiplying in great numbers. By 1880, White writes, there were "approximately 4 million cattle" (222); by the mid-1880s there were "an estimated 7.5 million head of cattle on the Great Plains north of Texas and New Mexico" (223). Foreign capital poured into the new industry that would become the icon of U.S. individualism and entrepreneurship.
13 In *The Fatal Environment*, Slotkin (1994) notes that the historical attention paid to Custer's battle at Little Big Horn far exceeds the importance of his military achievement in the incursions against tribes in the west. The story becomes, however, a guide to perception and behavior with moral and political imperatives. Slotkin points out that General George Armstrong Custer's Last Stand has become a U.S. ritual reenacted not only on stage and screen but also in the streets and on the battlefield (see pages 435–476 and 499–532).
14 A near obsession with frontier narratives dealing with nineteenth-century western life saturated popular culture of the 1950s, typified by television series such as *Roy Rodgers, Maverick, Rawhide, Bat Masterson, Wyatt Erp, The Lone Ranger, The Virginian*, and *Bonanza* (to name only a few). In *Gunfighter Nation* (1998), Slotkin argues that these narratives conveniently overlaid contemporary events such as the anticommunist era, and the Vietnam War (see, e.g., chaps. 15–17).
15 In 2012, the Fifth Avenue Theatre of Seattle mounted a production of *Oklahoma!*, casting an African American actor in the role of Jud. The production received strong pushback from audience and critics for racist casting. See *KIRO Radio*'s (2012) "Is 'Oklahoma!' racist?" for more context. See also Alice Kaderlan (2012). Relatedly, in 2011 Portland Center Stage produced an "all black" production to call attention to the presence of black cowboys, black homesteaders, and the many all-black communities in Oklahoma in 1907.
16 Strikers were often depicted as "savage" forces to be conquered and controlled. A case in point is the 1931 workers strike at Boulder Canyon, where one worker died on average every three days, some poisoned by gas in the diversion tunnels. (Congress had passed the Boulder Canyon Project Act in 1928, authorizing the corporate conglomerate, the Six Companies, to do the construction and to provide housing for workers.) The IWW had tried to organize the workers, but some workers did not want to be associated with Wobblies who were perceived as too socialist. But when the Six Companies cut wages for tunnel workers, project workers organized and presented a list of demands that included fair wages and health and safety measures. Six Companies refused to meet any of the workers' demands, and instead it began a series of layoffs, calling the workers "malcontents." See Rocha (1981), 217.
17 For more about how environmental activism was painted with an anticommunist brush, see Michael G. Barbour's (1994) "Ecological Fragmentation in the Fifties," 256–268.
18 The jingle was part of a 1953 General Motors advertisement for the new 1953 Chevrolet. It was written by Leo Corday and sung by Dinah Shore.
19 The subplot of Ado Annie and her cowboy suitor, Will Parker, also offers insights into the musical's gender politics and enforcement of heteronormativity. In particular, Ado Annie echoes popular and military entertainments in which women's sexuality was

exploited for the pleasure of male spectators. In this light, Ado Annie is a kind of pasteurized burlesque element in a musical that follows the example of Jack Kirkland's popular adaptation of Erskine Caldwell's *Tobacco Road*, which played at the Masque Theatre in New York from December 1933 to May 1941. Fearnow observes that *Tobacco Road* portrays the misery of southern sharecroppers as a "festival of sexual explications in a context of low humor" (107). He also notes that the adaptation "assembled a product that catered to the traditional public hunger for sexual excitement and, at the same time, performed the cultural work of reducing the tremendous 'poverty anxieties' of depression America" (107).

20 In December 1952, more than 4,000 people died from London's smog. British Parliament passed the Clean Air Act of 1956, and similar controls were enacted in Los Angeles in the 1950s. For more on this topic, see Clive Pointing's *A Green History of the World* (1991), 358.

21 My allusion here is to Walter Benjamin's (1996) comment on Paul Klees's 1921 painting, *Angelus Novus*, which Benjamin describes as "an angel looking as though he is about to move away from something he is fixedly contemplating. His eyes are staring, his mouth is open, his wings are spread. This is how one pictures the angel of history. His face is turned toward the past. Where we perceive a chain of events, he sees one single catastrophe which keeps piling wreckage upon wreckage and hurls it in front of his feet. The angel would like to stay, awaken the dead, and make whole what has been smashed. But a storm is blowing from Paradise; it has got caught in his wings with such violence that the angel can no longer close them. The storm irresistibly propels him into the future to which his back is turned, while the pile of debris before him grows skyward. This storm is what we call progress" (215–223).

22 The Doctrine of Discovery, derived from Anglo-Saxonism and Christian ideology discussed in Chapter 2, was ingrained through legal case law, dating back to the colonial era. It is still used to divest indigenous peoples of their lands. Its history, which preceded and shored up Manifest Destiny, gave legal precedent to the belief that conquest itself gave title and dominion to the conquer. See, for example, Blake A. Watson (2012). Watson argues that the 1823 case of *Johnson v. M'Intosh*, in which the Marshal court determined that Native people had no claim to land because they merely subsisted on the land but did not improve it, nor understand it as property, became the basis of Indian law going forward. In other words, its value derived from seeing it as a commodity rather than as a homeland. Robert J. Miller (2011) lays out the interconnection between the case law of the doctrine of discovery and the cultural ideology of Manifest Destiny: "When Euro-Americans planted flags and religious symbols in newly-discovered lands, they were not just thanking god for a safe voyage across the ocean; instead, they were undertaking the well-recognized legal procedures and rituals of the Doctrine designed to make their country's legal claim to the lands and peoples" (330).

References

"The Age of Anxiety." 1961. *Time*, March 31.
Anderson, William L., ed. 1991. *Cherokee Removal: Before and After*. Athens and London: University of Georgia Press.
Auden, W.H. 1947. *The Age of Anxiety: A Baroque Eclogue*. New York: Random House.
Barbour, Michael G. 1994. "Ecological Fragmentation in the Fifties." In *Uncommon Ground: Rethinking the Human Place in Nature*, edited by William Cronon, 223–255. New York: W.W. Norton & Company.
Barthes, Roland. 1972. *Mythologies*, edited by Annette Lavers. New York: Hill and Wang.
Bates, Barclay W. 1983. "The Lost Past in *Death of a Salesman*." In *Twentieth Century Interpretations of Death of a Salesman*, edited by Helene Wickham Koon. Englewood Cliffs, NJ: Prentice Hall, Inc., 60–70.

Benjamin, Walter. 1996. "Theses on the Philosophy of History." In *The Continental Philosophy Reader*, edited by Richard Kearney and Mara Rainwater, 215–223. New York: Routledge.

Berry, Wendell. (1966) 1977. *The Unsettling of America: Culture and Agriculture*. San Francisco, CA: Sierra Club.

Brown, Kirby. 2018. *Stoking the Fire: Nationhood in Cherokee Writing 1907–1970*. Norman, OK: University of Oklahoma Press.

Carter, Tim. 2007. *Oklahoma!: The Making of an American Musical*. New Haven, CT and London: Yale University Press.

Daly, Augustin. 1978. *Horizon*. In *Six Great American Plays*, edited by Allan G. Halline, 335–375. New York: Modern Library.

Debo, Angie. 1998. *And Still the Waters Run: The Betrayal of the Five Civilized Tribes*. Princeton, NJ: Princeton University Press.

Dippie, Brian W. 1998. *West Fever*. Los Angeles, CA: Autry Museum of the American West.

Dunbar-Ortiz, Roxanne. 2016. *An Indigenous People's History of the United States*. Boston, MA: Beacon Press.

Ellis, Jeffrey Charles. 1995. "When Green Was Pink: Environmental Dissent in Cold War America." PhD diss., University of California Davis.

Ellsworth, Scott. 1982. *Death in a Promised Land: The Tulsa Race Riot of 1921*. Baton Rouge, LA: Louisiana State University Press.

Evans, Larry James. 1990. "Rodgers and Hammerstein's *Oklahoma!*: The Integrated Musical." PhD diss., University of California.

Fearnow, Mark. 1997. *The American Stage and the Great Depression: A Cultural History of the Grotesque*. Cambridge: Cambridge University Press.

Friedman, Michael, and Alex Timbers. 2010. *Bloody Bloody Andrew Jackson*. New York: Music Theater International.

Geiogamah, Hanay, ed. 1999. *Stories of Our Way: An Anthropology of American Indian Plays*. Los Angeles, CA: UCLA American Indian Studies Center.

"General News: The Great Race." 1889. *West Side News*, April 27.

Globe, Danny. 1980. *Progressive Oklahoma: The Making of a New State*. Norman, OK: University of Oklahoma Press.

Gottlieb, Robert. 1993. *Forcing the Soring: The Transformation of the Environmental Movement*. Washington, DC: Island Press.

Griffith, D.W., dir. [1915] 2002. *The Birth of a Nation*. Griffith Masterworks. New York: Kino on Video.

Helburn, Theresa. 1960. *A Wayward Quest*. Boston, MA: Little, Brown and Company.

"Is '*Oklahoma!*' Racist?" 2012. *KIRO Radio*, February 17.

Juracek, K.E., and K.D. Drake. 2016. "Mining-Related Sediment and Soil Contamination in a Large Superfund Site: Characterization, Habitat Implications, and Remediation." *Environmental Management* 58 (4): 721–740. https://doi.org/10.1007/s00267-016-0729-8.

Kaderlan, Alice. 2012. "A Rousing '*Oklahoma!*' at the Fifth Avenue Theatre." *Seattle PI*, February 11.

Kirle, Bruce. 2003. "Reconciliation, Resolution, and the Political Role of '*Oklahoma!*' in American Consciousness." *Theatre Journal* 55 (2): 251–274.

Knapp, Raymond. 2005. *The American Musical and the Formation of National Identity*. Princeton, NJ: Princeton University Press.

Lockridge, Richard. 1931. "Lynn Rigg's Southwest: *Green Grow the Lilacs*." *New York Sun*, January 27.

"Magnificent Death." 1949. *Newsweek*, February 21, 78.
Manes, Sheila. 1982. "Pioneers and Survivors: Oklahoma's Landless Farmers." In *Oklahoma: New Views of the Forty-Sixth State*, edited by Anne Hodges Morgan and H. Wayne Morgan, 93–132. Norman, OK: University of Oklahoma Press.
May, Theresa J. 2000. "Earth Matters on Stage." PhD diss., University of Washington.
Mielziner, Joe. 1965. *Designing for the Theatre: A Memoir and a Portfolio*. New York: Atheneum.
Miller, Arthur. 1971. *Death of a Salesman*. In *The Portable Arthur Miller*, edited by Harold Clurman, 3–133. New York: The Viking Press.
Miller, Arthur. 1987. *Timebends: A Life*. New York: Grove Press.
Miller, Arthur. 2016. "The Crucible in History." In *Collected Essays*, edited by Arthur Miller and Susan C.W. Abbotson, 252–275. New York: Penguin Books.
Miller, Robert J. 2011. "American Indians, the Doctrine of Discovery, and Manifest Destiny." *Wyoming Law Review* 11 (2): 329–349. https://heinonline.org/HOL/P?h=hein.journals/wylr11&i=338
Most, Andrea. 1998. "'We Know We Belong to the Land': The Theatricality of Assimilation in Rodgers and Hammerstein's *Oklahoma!*" *MLA of America* 1113 (1): 77–116. https://doi.org/10.2307/463410
Nadel, Norman. 1969. *A Pictorial History of the Guild Theatre*. New York: Crown Publishers.
Najjar, Michael Malek. 2015. *Arab American Drama, Film and Performance: A Critical Study, 1908 to the Present*. Jefferson, NC: McFarland & Company.
Orr, David W. 1995. *Earth in Mind: On Education, Environment and the Human Perspective*. Washington, DC: Island Press.
Pointing, Clive. 1991. *A Green History of the World*. New York: Penguin.
Riggs, Lynn. 1931. *Green Grow the Lilacs: A Play*. New York: Samuel French.
Rocha, Guy Louis. 1981. "The I.W.W. and the Boulder Canyon Project: The Death Throes of American Syndicalism." In *At the Point of Production: The Local History of the I.W.W.*, edited by Joseph R. Conlin, 213–234. London: Greenwood Press.
Rogers, Richard, and Oscar Hammerstein. 1942. *Oklahoma!* New York: Random House.
Roszak, Theodore. 2002. *The Voice of the Earth: Exploration of Ecopsychology*. New York: Simon & Schuster.
Schlesinger, Arthur. 1963. *The Politics of Hope*. Boston, MA: Houghton Mifflin.
Sellars, Nigel Anthony. 1998. *Oil Wheat and Wobblies: The Industrial Workers of the World in Oklahoma, 1905–1930*. Norman, OK: University of Oklahoma Press.
Slotkin, Richard. 1994. *The Fatal Environment: The Myth of the Frontier in the Age of Industrialization, 1800–1890*. Norman, OK: University of Oklahoma Press.
Slotkin, Richard. 1998. *Gunfighter Nation: The Myth of the Frontier in Twentieth-Century America*. Norman, OK: University of Oklahoma Press.
Stein, Howard F., and Robert F. Hill. 1993. "The Culture of Oklahoma: Group Identity and Its Images." In *The Culture of Oklahoma*, 198–235. Norman, OK: University of Oklahoma Press.
Steinbeck, John. (1939) 1993. *The Grapes of Wrath*, edited by Brad Leithauser. London: David Campbell.
Thomashow, Mitchell. 1995. *Ecological Identity: Becoming a Reflective Environmentalist*. Cambridge, MA: MIT Press.
Thompson, Gary L. 1993. "Green on Red: Oklahoma Landscapes." In *The Culture of Oklahoma*, edited by Howard F. Stein and Robert F. Hill, 3–28. Norman, OK: University of Oklahoma Press.

Turner, Frederick Jackson. 1956. "The Significance of the Frontier in American History," in *The Turner Thesis: Concerning the Role of the Frontier in American History*, edited by George Rogers Taylor, 1–18. Revised Edition. Boston: D.C. Heath and Company.

Waldau, Roy S. 1972. *Vintage Years of the Theatre Guild: 1928–1939*. Cleveland, OH: Case Western Reserve University Press.

Watson, Blake A. 2012. *Buying America from the Indians: Johnson v. McIntosh and the History of Native Land Rights*. Norman, OK: University of Oklahoma Press.

Weaver, Jace. 1977. *That the People Might Live: Native American Literatures and Native American Community*. New York and Oxford: Oxford University Press.

Weaver, Jace. 2003. "Introduction to *Green Grow the Lilacs*." In *Cherokee Night and Other Plays*, by Lynn Riggs, 3–5. Norman, OK: University of Oklahoma Press..

White, Richard. 1991. *It's Your Misfortune and None of My Own*. Norman, OK: University of Oklahoma Press.

Wilk, Max. 1993. *OK! The Story of Oklahoma!* New York: Grove Press.

Wister, Owen. 1945. *The Virginian*. New York: Grosset & Dunlap.

Witham, Barry. 2013. *A Sustainable Theatre: Jasper Deeter at Hedgerow*. New York: Palgrave Macmillan.

Worster, Donald. 1979. *Dust Bowl: The Southern Plains in the 1930s*. Oxford: Oxford University Press.

Worster, Donald. 1992. *Under Western Skies: Nature and History in the American West*. New York: Oxford University Press.

Worster, Donald. 1994. *Nature's Economy: A History of Ecological Ideas*. 2nd ed. New York: Cambridge University Press.

Worth, Robert Miller. n.d. "Populist (People's) Party." Oklahoma Historical Society. *Encyclopedia of Oklahoma History and Culture*.

Zinn, Howard. 1995. *A People's History of the United States*, Revised ed. New York: Harper Perennial.

5
(RE)CLAIMING HOME AND HOMELANDS

Lorraine Hansberry's *A Raisin in the Sun*, Luis Valdez's *Bernabé*, and Sam Shepard's *Buried Child*

In the decades following World War II, two dramatically contrasting images galvanized environmental activism in the United States. Both the visage of annihilation from nuclear war in the atomic mushroom cloud and the iconic image of Earth from space gleaned from the 1968–1969 Apollo missions were visual products of the Cold War. Considered together, they represented a new collective imaginary that seeded an awareness of the fragility of "the blue planet"[1] and the (human and nonhuman) lives that depend upon that planet (Worster 1994). Wilderness preservationists found solidarity with a rising antinuclear movement in rallying public concern for the health of the land, its diverse environments, and its peoples. Post-World War II economic strategies – such as planned obsolescence, the citizen as consumer, the repurposing of chemicals developed as weapons into household and agricultural products, and the configuration of a national highway system that positioned the automobile at the center of American life – contributed to a mounting environmental crisis. Pollution events – including nuclear accidents, chemical spills, and killer smog events – were becoming everyday news, while predictions of population increase over the remainder of the century gave rise to alarming increase of drought, hunger, and migration. Environmental activism during this era prompted new state and federal legislation that responded to scientific warnings about the many new threats to the quality of life and lifeways. Many of the laws enacted during the 1960s and 1970s, passed by coalitions of Republican and Democratic lawmakers, formed the bulwark of environmental protections through the end of the century (see, e.g., Gottlieb [2005]). When Congress designated April 22, 1970, as the first Earth Day, many U.S. citizens believed the nation had turned a vital corner toward a more just and sustainable future.

Despite new popular concern for the health of the environment, local struggles for social and economic justice by historically marginalized communities

DOI: 10.4324/9781003028888-6

continued to cope with the legacies of colonialization and slavery which carry white supremacy, racism, and disadvantage forward. The environmental movement of the period was steeped in ideologies handed down from early preservationist and conservationist movements that were riddled with white supremacist ideas and institutional patterns, and yet the burden of environmental degradation was felt most keenly by women, children, and communities of color. In the streets and in the fields, black, brown, and indigenous people demanded a reckoning with history. Collectively, the many civil rights movements of the 1960s shared a demand for a new vision and a remaking of the United States that included all who now call this land home.

Theater was a vital force in these struggles for justice and social change, as dramatists in the 1960s and 1970s took up the specific lived experiences of communities that had been systematically brutalized by U.S. culture. These playwrights demanded a rescripting of "America" that acknowledged the histories, experiences, rights of people, and places that U.S. white supremacy and environmental imperialism had long ignored or erased. When considered through an ecocritical lens, many of the plays of the 1960s and 1970s that expressed both the frustration and vision of communities engaged in struggles for economic and social justice also shared a focus on environmental concerns, including the embodied health of workers, women, and children; food security; connections to place; and the right to a safe and healthy home.

Yet, at the time, environmental and civil rights activism seemed to run on separate rails. The second-wave environmental movement was steeped in popular understandings of nature and environment as something apart from human culture.[2] Rooted in images and ideology of wilderness of the early conservation and preservation movements, the environmental movement of the 1970s focused primarily on saving what is left of pristine environments rather than posing the question of how can we better treat this blue planet as our home. However, as I argue earlier in this book, this binary thinking separates scenic beauty from industrial utility as it sanctions and rationalizes environmental destruction as part and parcel of "progress." Such thinking tends to disassociate human health, poverty, and racism as separate and social issues rather than understanding them as elements of the larger workings of extractive capitalism that affect lives and land.

The arts, including theater, were fundamental to the civil rights struggles of the 1960s and 1970s in the United States. The significance of the Black Arts Movement, the new Native theater, and the many Chicana/o *teatros*, which grew into the hemispheric theater movement TENAZ, bear out the important role of theatrical artistic expression as part of larger social justice movements (see, e.g., Elam [1997]; and Hanay and Darby [2000]). Environmental activists also made use of theatrical tools, staging protests that included performative elements like puppets, die-ins, parades, and other spectacles. Grassroots theater groups attended to the politics of place as well as the links among environmental problems, industrialization, and militarism. The San Francisco Mime Troupe, for example, touched on environmental themes as part of their satirical critique of capitalist culture, and grassroots companies like the Dell'Arte Company devised performances around local environmental issues that concerned the communities they served. Yet even during the surge of environmental activism of the 1970s, mainstream plays (regional and Broadway) that dealt *explicitly* with environmental issues were few and far between (see, e.g., Davis [1975]; Mason [2013]; and Standing [2008]). The only play explicitly focused on environmental themes to find its way to Broadway was a musical revue titled *Mother Earth*.[3] Successful reception on the West Coast did not carry to New York, and *Mother Earth* closed on Broadway after only three performances.

Rendering environmental problems in stories for the mainstream stage is complicated by what Una Chaudhuri (1994) would later identify as the "programmatically anti-ecological" bias of theater itself (23). Drama has long been a literary mode construed as an expression of strongly opposed human forces, antagonists and protagonists, and meant to illuminate the human condition. On stage, representations of the environment have long been understood as a backdrop to human affairs. A cultural expression solely concerned with the human-centric side of the nature/culture binary, theater is the very epitome of artifice, firmly within the realm of culture. Consequently, Chaudhuri argues, stories about ecological interdependence are typically understood as metaphors for human concerns and as grist for the humanist mill. Even when dramatists took up environmental themes within the context of social justice theater, those themes went unrecognized.

As a historiographic lens, ecodramaturgy can (re)illuminate plays already part of the dramatic canon to show their ecological through lines, recognizing how they carried implicit meanings that were integral to the development of environmental sensibilities going forward. Lorraine Hansberry's *Raisin in the Sun* ([1958] 1994), for example, was concerned not only with the ways by which racism and sexism translate into social and economic inequity, but also with its impact on mental and physical health. These themes of environmental justice were abundant in theatrical work associated with the social justice struggles of the civil rights era (such as the Black Arts Movement, Chicano theater, feminist theater, and Native theater). Thirty years before the rise of the environmental justice movement, Hansberry's *Raisin in the Sun* depicted the impacts of poverty and environmental degradation on communities of color, women, and children. In this way, Hansberry helped to redefine the "environment" as encompassing all of the places where people live, work, play, and worship. In particular, *Raisin* made palpable on stage the ways in which environmental degradation affected the neighborhoods, communities, and bodies of certain Americans more severely than others. Insofar as these wider and ecologically-material meanings are occluded, the intertwined realities of social and environmental justice are also masked. Injury and abuse of people perpetuate the degradation of the land; as Barry Commoner argued in his 1971 treatise, *The Closing Circle*, there can be no environmental health without social justice, and ecological concerns must be understood to be intertwined with economic justice and gender and racial equity.[4] To make the permeability of human and environment visible is to strengthen the sense that environmental justice and ecological survival are intertwined and irreducible. As I have argued throughout this book, theater, in particular, is uniquely poised to call out connections between human and environment. In this chapter, I examine the ways in which environmental and social justice concerns came together in some of the important theatrical projects of the 1960s and 1970s, where they took shape in the lives of characters and the living exchange between audience and performers. These dramatists and activists understood the environment as coupled with human health and welfare; they told stories of home and homeplace that rhetorically (re)claimed homelands, language, and legitimacy and asserted visions of the future.

My analysis here centers on how theater became an apt site to expose what Rob Nixon (2013) has called the "the slow violence" of industrial capitalism's environmental degradation as it plays out on the land and in the bodies of workers and their families. Nixon reminds us that while news media cycles litanize the catastrophic environmental events – the oil spills and extinctions – environmental degradation shows up most cruelly and invisibly in the day-to-day lives of the world's poorest people. Insofar as longstanding patterns of U.S. imperialism fuel both systemic injustice and environmental degradation, plays that engage social and economic justice also, implicitly, engage environmental justice. Approaching these plays with a greener lens means looking specifically for how they represent the lived experience of environmental injustice, asking questions about the welfare of bodies, families, and communities affected by racism and economic injustice and how those arise from the same ideologies of domination that have caused ecological havoc across the land. Not only do these plays demonstrate that theater engaged ecological thought and environmental issues long before theorists recognized those themes and topics on stage, but they also point to the ways in which theater as an immediate, embodied, and communal forum helped to shift public attention to the impact of environmental injustice by telling stories that focused on the lived (and thus ecological) experience of people in their communities. As I argue in this chapter, much activist theater during the civil rights era illuminates the intractable connections between environmental concerns and social justice. A greener lens reveals that both mainstream theater of the period, as well as grassroots activist theater, implicitly (re)claim home, work, food, family, and spirituality as significant and inseparable aspects of ecological and cultural well-being.

(Re)claiming home: *A Raisin in the Sun*

Five years after the Supreme Court's decision in *Brown v. Board of Education*, which struck down the "separate but equal" standards of the U.S. south (see, e.g., Gates and West 2000, esp. 129, 261–262), Lorraine Hansberry's *A Raisin in the Sun*[5] exposed how historic and institutionalized racism in northern cities actively excluded black Americans from full participation in postwar economic prosperity. In his introduction to the published and complete version, Robert Nemiroff

(1994) details the long arc of critical reception that has followed Hansberry's play from its debut as a "first" play (first black playwright, first black woman, youngest woman) to receive critical acclaim as a "universal" story about Americans to the later scholarly and artistic acknowledgment of the play's politics (5). A black "humanist" drama that centers black experience in both historic and still-relevant ways, since its premiere *Raisin* has been recognized for its specificity in relation to institutionalized racism, economic class, and the rising black consciousness of the late 1950s and 1960s (12).

I contend that the white supremacy that the Youngers face when they plan to move to a new home not only constitutes systemic environmental racism but reflects and arises out of the historic abuse of land and bodies on which the U.S. extractive economies depends. My analysis traces the subtle ways those legacies infect and damage the daily lives of those who carry the disproportional burdens of that historic abuse. In my analysis, I examine clues as to the larger U.S. environmental trends to show how Hansberry's play makes a claim to homeplace in the milieu of the 1950s.

In the discussion that follows, I am also interested in the prescience of Hansberry's *Raisin*, specifically the way it foreshadows the environmental justice movement by 30 years. Through the story of the Younger family, Hansberry points concretely to the correlation and interdependence of racial inequities, poverty, and environmental degradation, thereby revealing how the felt experience of white supremacy plays out in the daily life of one family. Demonstrated through the material-ecological aspects of the characters' lives, *Raisin* thus paints an intimate portrait of environmental racism at work. For example, the impacts of systemic poverty, redlining housing practices, and the health effects of longstanding societal racism shape the lived experience of Hansberry's characters. As the Younger family struggles within a context of economic structures shaped by white supremacy, the lived and living aspects of that environment impact their capacity to thrive. The characters' hopes for a better future, including a homeplace that nurtures their collective material, emotional, and spiritual well-being, chaffs against the stress that comes from living in a compromised environment.

On the whole, my ecocritical reading of *A Raisin in the Sun* foregrounds clues to the ecologies represented on stage, including human bodies and habitat, to examine the environmental implications of the Youngers' struggle and racial and class-based oppressions and hierarchies the play represents. Considered from the vantage point of longstanding humanist interpretations of Hansberry's play, critical reception and analysis of *Raisin* have often disguised the larger social, economic, and environmental contexts within and against which the Younger family's struggles play out.[6] Attention to these contexts, however, is central to the task at hand in this chapter – namely, to better understand the ways in which violations and injustices stemming from the legacies of slavery (e.g. institutional racism, explicit racism, racial microaggressions, and state-sponsored, racially-motivated economic inequality) also expose and underscore environmentally unjust and destructive practices and ideologies.

Defined as the disproportional burden of impact of environmental degradation on people and communities of color, environmental racism deprives individuals of the very foundations of life – clean air, water, healthy foods, and shelter adequate to support healthful families.[7] Poverty caused by a system of privilege by which generation after generation of African Americans has been shut out of the opportunities considered the birthright of those who pass as white in a given sociohistorical period is a form of environmental racism. A lack of access to healthy food, clean air, and clean water in impoverished urban neighborhoods results in higher incidences of asthma and other childhood diseases (see, e.g., Bullard [1993, 2005]; Jaynes [2014]; and Sampson [2012]). Structures of exclusion in housing, for example, have been built and maintained through unscrupulous yet legal lending practices and policed by social and institutional expressions of white supremacy, as well as violence and the threat of violence against families and individuals challenging that system. Throughout the play, Hansberry maps the ways those impacts land heavily on the bodies of women and children, determining longevity, childbearing, and mental health – all of which are now recognized as impacts of environmental injustice.

A Raisin in the Sun is set in the South Side of Chicago, in a two-room apartment that is home to a multigenerational family that includes Mama Younger; her daughter, Beneatha; her son, Walter Lee, Jr.; Walter's wife, Ruth; and their young son, Travis. The early stage directions describe a place of great "weariness" in which "too many people have lived for too long" and where a "single window [is the] sole source of natural light" (24). In the opening scene, the family eagerly awaits a life insurance policy due to arrive by mail, from the death of Walter Lee, Sr. Mama Younger hopes to use the money to buy a new home for the family, but her son, Walter Lee, Jr., presses his mother to allow him to invest in a liquor store, and her daughter, Beneatha, needs funds to pay for college. The conflict among the central characters is generational in nature and points to the arc of African American history of the twentieth century. Hansberry's Lena and Walter Lee, Sr., were part of the Great Migration of the twentieth century, when black families moved in high numbers to the urban, industrial north, seeking jobs, economic opportunities, and freedom from segregation laws, unfair labor practices, the threat of lynching, and other forms of systemized violence against black Americans. Lena Younger has held fast to the aim of a homeplace in an environment free from racism in which she and her husband, and now her children and grandchildren, might thrive.

Meanwhile, her adult children's goals and desires are focused on finding their own places in the world, and yet in 1959 these futures are under threat from the very patterns of racism that their parents sought to escape a generation earlier. When the life insurance check finally arrives, the family struggles to support these several goals on the small sum that represents the lifelong labor of their husband and father. The universalist and humanist proclamations of critics underscored this familial struggle in which parents want a better future for their children and must weigh present needs (like food) against future ambitions as one faced by parents across demographics. However, as Hansberry herself embraced this recognition of the ways in which her play engenders pathways of empathy by which white audience members might recognize aspects of their own human struggle, it also illuminates the specific struggles of a hardworking economically disadvantaged black family and, in doing so, asks more of its audience than mere identification. *A Raisin in the Sun* asks for expanded conscious awareness on the part of white audience members (as well as others), including recognition of their own role in (and privilege derived from) the injustices inflicted by systemic racism.

Homeplace – an environmental right

I read *Raisin* as an indictment of the ways that black families were shut out of the pathways to prosperity available to whites in the 1950s and a testament to the effects of that exclusion in the daily lives and health of families like the Youngers. Lena has held fast to her dream of a home over a lifetime: "I remember the day me and Big Walter moved in here. Hadn't been married but two weeks and wasn't planning on living here no more than a year. We was going to set away little by little . . . and buy a little place out in Morgan Park" (44).[8]

Like many who came north to flee the Jim Crow South, the Youngers brought skills and commitment to build a future. As a bricklayer, Walter Lee, Sr. participated in the postwar development boom, although he died before he could realize a share of the dream that his labor had helped to build. Drawing strength from the memory of her husband, Mama tells Ruth that he "sure loved his children. Always wanted them to have something – be something" (45) – Lena's commitment to her family's future, to their well-being, stands on the shoulders of her husband's labor and her own, as she remains determined to claim that home after a lifetime of struggle:

> Been thinking that we maybe could meet the notes on a little old two story somewhere, with a yard where Travis could play in the summertime, if we use part of the insurance for a down payment and everybody kind of pitch in. I could maybe take on a little day work again.
>
> *(44)*

Ta-Nehisi Coates (2017) reminds us that conditions in the north continued, sometimes in subtler, but no less racist, ways, effectively excluding black families from the opportunity to provide for their families, accumulate wealth, and build a future for their children in the way that white families were more empowered to do, particularly after World War II (184–194). In "The Case for Reparations," Coates describes that in the 1950s, "85 percent of all black home buyers who bought in Chicago bought on contract," given that conventional collateral loans were unavailable to blacks (170). The growing mortgage industry sanctioned these highly predatory contracts as a matter of course, while white lenders profited. Holding a note for a black family almost guaranteed future repossession of the property when the high interest rates and inflated housing prices could not be paid. Since black homebuyers were systematically denied conventional mortgages, the chance that Lena Younger would buy on contract, becoming victim to such unfair lending practices, was high.

A safe home that supports thriving lives is an environmental right. Lena Younger explains, "I just tried to find the nicest place for the least amount of money for my family. . . . Them houses they put up for colored in them areas way out an all seem to cost twice as much as other houses" (91). Instead, they face a racist housing market that issues from inequities consistent with relations to land in the United States that shored up slavery, a system in which the exploitation of the bodies of African slaves made possible the extraction of wealth from the soil in the globally-traded cash crops of cotton and tobacco. Through the process of redlining, in which some neighborhoods were coded/mapped as more or less desirable based on racial demographics, the Federal Housing Administration, Coates (2017) writes, "exhorted segregation and enshrined it as public policy" (188). Redlining – a system that excluded blacks from certain housing markets and devalued properties already owned by blacks – is one tactic of ongoing settler colonialism aimed at the extraction of wealth from the land. In a public lecture I attended in 2017 at the University of Oregon, Coates observed that accumulation of wealth, and the black bodies in which that wealth was "secondarily housed," lays as the foundation of the U.S. nation-state as surely as Turner's lie of "free land" (see, e.g., Coates [2015], 103–111). Both stances functioned to mask colonialism's theft of land, lives, and labor as expressions of free enterprise. This system of sequestering wealth is brought home in *Raisin* with the visit of Mr. Linder from the Clybourne Park "neighborhood committee," who urges the Youngers *not* to move into the house, the house on which Lena has already made a deposit (114–115). In fact, he offers the family cash to do so – not only the equivalent of their down payment but also a bribe. Indeed, Linder's offer is a means to prevent blacks from owning property in neighborhoods where the home values are more reasonable.

Lena's commitment to securing a new home represents an act of fierce resistance in the face of very credible threats made palpable in the play. In Act 2, Scene 2, Lena's neighbor Mrs. Johnson stops by for a piece of pie and some conversation. As Mrs. Johnson learns of the Youngers' plan to move into a new home and into a home of their own, she reminds them of the risks: "I guess you all seen the news what's all over the color paper this week . . . 'bout them color people that was bombed out their place out there? . . . [G]etting so you think you right down in Mississippi!" (100). Mrs. Johnson's story echoed real-world events such as that of the Myers, which Coates describes. In August 1957, "Daisy and Bill Myers, the first black family to move into Levittown, Pennsylvania, were greeted with protests and a burning cross" (Coates 2017, 187.) The play does not specify that Mrs. Johnson was referring to this much-publicized incident or others like it. Nevertheless, through the presence of this character, Hansberry implicitly called out the racist acts of terrorism aimed to prevent families like the

Youngers not only from acquiring a home that would better serve their needs but also from exercising freedom generally. By the late 1950s, a litany of such incidents would have been familiar to the Youngers and to the audience as well (190–191). The real-life harassment of the Myers by white mobs lasted many weeks, yet the Myers stayed. Likewise, within the fictional narrative of *Raisin*, Lena Younger is equally determined to move into their new home in the face of a decade of violence by whites against black families. Packing boxes on stage in Act III represent a collective act of resistance and resilience that (re)claims the promise of freedom on their own terms, including their human right to a safe home in which they can thrive.

As Coates observes, the mining of black labor occurred not only by way of the institution of slavery from which the economic wealth of the still-young nation grew, but also through the exploitation of black Americans that continues in our present moment in discriminatory practices that have long institutionalized and perpetuated the economic power and privileges of those perceived as whites. A closer look at the unfolding dynamic over money between Mama and Walter Lee, Jr., reveals the extent to which this exclusion, compounded by the poverty it holds in place, has had a corrosive effect on the family. Lena's adult children, Walter Lee, Jr., and Beneatha, want what any young person might want – to invest in and own a business, to go to college, to start a career, to make a difference in the lives of their family and society. Yet for the Younger family, social and economic inequities have limited employment and educational opportunities in such ways that those basic, childhood goals exist in a constant state of jeopardy. Walter Lee, Jr., earns a living as a chauffeur, while his wife, Ruth, does domestic work for a white family. *Raisin* opens with a disagreement between Walter Lee, Jr., and Lena, centered on whether to give Travis 50 cents or whether he should carry groceries after school to earn the money himself. Walter Lee, Jr., wants more for his son, and he presses his case to use his father's insurance money to invest in a liquor store. Lena refuses, and Walter Lee, Jr., flies back at her:

> Well, *you* tell that to my boy tonight when you put in to sleep on the living-room couch. . . . and tell it to my wife, Mama, tomorrow when she has to go out of here to look after somebody else's kids . . . every time we need a new pair of curtains and I have to watch *you* go out and work in somebody's kitchen.
>
> *(71)*

His anger, like his father's, can be read as a palpable response to a world that has long exploited and deprived black Americans of ownership of land, the fruits of their labor, and the possibility of intragenerational wealth. In a later scene, Walter Lee, Jr., confides his sense of displacement: "Mama – sometimes when I'm downtown and I pass them cool, quiet-looking restaurants where them white boys are . . . sitting there turning deals worth millions of dollars . . . guys that don't look much older than me" (74). His observation is not merely the reflection of his desire for a place in, or a piece of, that marketplace but also an indictment of its racist exclusion and the impart of racism on her son's sense of self. Recognizing his need for self-determination, Lena finally allows Walter Lee, Jr., to invest a portion of the insurance money.

Self-determination, which would become one of the Principles of Environmental Justice, is a foundation for social and ecological well-being (see, e.g., Chiro [1996]). Yet an economic system that prays on disadvantaged communities thwarts Walter Lee, Jr.'s choices. Neither his father nor he can gain ground in a system designed to exclude their participation. At the end of Act II, Walter Lee's business partner Bobo reveals that their investment in the liquor store has been stolen by their third partner, who was to "go down to Springfield and spread some money 'round so we wouldn't have to wait so long for the liquor license." Bobo's bewilderment points to the underbelly of a system of graft and greed mapped across the state of Illinois. "Everybody said that was the way you had to do it" (126). As the men realize that this system is larger and deadlier than they dreamed, the stage directions indicate that Walter Lee, Jr., "starts to pound the floor with his fists, sobbing wildly" and cries out "THAT MONEY IS MADE OUT OF MY FATHER'S FLESH" (128). When Lena Younger realizes what has happened – the fact that all the plans they have made and her trust in her son have been lost. She invokes her husband's labor – labor now mined again by systemic theft in which her son has been caught up: "I seen him grow thin and old before he war forty . . . working and working and working like somebody's old horse . . . killing himself . . . and you – you tell me you give it all away in a day." In Lena's fury, the stage directions tell us, "she raises her arms to strike him again" but stops herself (129). The system that seduced her son into a venture that promised financial gain and the system that mined her husband's labor, never paying him enough to own a home or adequately feed his family, are one. In this moment, the play brings home the connection between institutionalized racism born of colonial systems of exploitation that continue to exclude blacks not only from mortgages but also from a host of economic avenues typically open to whites.

Bodies at risk

The economic system predicated on greed and fueled by graft, institutionalized theft, and manipulation of investments in the economic sectors of neighborhoods ultimately finds its way into the bodies and long-term health impacts affecting families like the Youngers, where poverty is a kind of poison. As Coates describes, racialized exclusion from the processes of wealth accumulation (like home and business ownership) is not merely a legacy of slavery; it is also the active perpetuation of racial exclusion that, in turn, results in toxic stress, reduced air and water quality, exposures to environmental contaminants, disease, and death. Mama Younger has already lost her husband and a child to these effects, and she is determined to resist and (re)claim a homeplace that provides this most basic ecological foundation.

That toxic stress (see, e.g., Shern et al. [2016]) – the long-term cumulative effect of poverty, insecurity, and racism – has caused losses that continue to harm the lives and bodies of the Younger family (see, e.g., Guest, Almgren, and Hussey [1998]; Holosko and Feit [1997]; and Hartman and Poverty [2006]). When it becomes apparent in Act I that Ruth is pregnant and planning an abortion – not an abortion that arises out of choice but one demanded by the lack of space to live healthy lives, to raise healthy children, Mama recalls the infant death of one of her own children and the toll it took on her husband:

> Big Walter would come in here some nights back then and slump down on that couch. . . . I'd know he was down then. . . . And then when I lost that baby – little Claude – I almost thought I was going to lose Big Walter too. Oh, that man grieved. . . . I guess that's how come that man finally worked hisself to death like he done. Like he was fighting his own war with this here world that took his baby from him.
>
> *(45)*

Lena confides these memories to Ruth, testifying to both women's embodied experiences of the inhospitable environment that limits their freedom, impacts their health, and endangers their children. The "world that took his baby from him" is one documented in studies on infant mortality in urban communities marked by a lack of healthy food, adequate medical care, space for children to play, and locally-owned businesses.[9] The losses bolster Lena's resolve to claim a homeplace in which her family might thrive: "We done give up one baby to poverty and that we ain't going to give up nary another one" (75). Lena's loss is an example of the slow violence of poverty that is perpetuated by institutional racism. Lena demands that her son bear witness to his wife and his mother's lived experience and implicitly asks the audience to acknowledge the psychological and physical burden of poverty and its effect on the health of children, families, and communities.

Similarly, when Ruth calls the apartment a "rat trap" (44) in Act I, Scene 1, her complaint is not a metaphor. Living conditions that require regular chemical pest control, rat poison, leaded paint, marginal sanitation, a lack of fresh air, and even light to grow a small houseplant constitute a kind of secondary violence represented in *Raisin*, a violence that lands most heavily on children, pregnant women, and the elderly – all of whom reside in the Youngers' apartment. Scene 2 opens as Mama and family members clean their small apartment. Stage directions describe that Beneatha, "with a handkerchief tied around her face, is spraying insecticide into the cracks in the walls" (54). Playfully, Beneatha chases her younger brother with a spray bottle full of roach killer, while Mama calls, "Look out there, girl, before you be spilling some of that stuff on that child!" (55). Rachel Carson's (1962) research would show what women of color across the country already knew from their lived experience: that the pesticides and herbicides, developed as weapons of war but increasingly used in civilian homes and gardens, impacted the health of society's most vulnerable populations. All over the nation, these newly created "miracle" chemicals promised pest-free farms, gardens, parks, countryside, and neighborhoods (Gottlieb 2005, 83–85). Many of these same chemicals were commonly used within the home, particularly in homes infested with insects as a result of poor construction and negligent management. In the scene described earlier, Beneatha retorts, "I can't imagine it would hurt him – it has never hurt the roaches . . . There's really only one way to get rid of them, Mama . . . Set fire to this building!" (55). In the context of the play, it is not only a chilling metaphor for the white neighborhood committee's effort to keep "pests" out of their privileged community, but also a material testament to environmental risk. Mama's response – "Well, little boys' hides ain't as tough as Southside roaches" (55) – foreshadows issues of eco-racism and environmental justice that would embattle many communities of color in coming decades.[10]

Scene 2 offers another demonstration of the Youngers' compromised habitat as Beneatha calls out the apartment window "TRAVIS! TRAVIS . . . WHAT ARE YOU DOING DOWN THERE? (*She sees*) Oh Lord, they're chasing a rat!" The moment comes just after Ruth, Travis's mother, has revealed that she is planning to terminate her pregnancy. Travis enters the scene here, eager to recount his adventure to his mother:

> Mama, you should have seen the rat. . . . Big as a cat, honest! . . . Bubber caught him with his heel and the janitor, Mr. Barnett, got him with a stick – and then they got him in a corner and – BAM! BAM! BAM! – and he was still jumping around and bleeding like everything too – there's rat blood all over the street – (*Ruth reaches out and grabs her son without even looking at him and clamps her hand over his mouth and holds him to her.*)
>
> (59)

The physical violence represented in the preceding scene is as unsettling as the presence of the rat itself. Ruth's fear and impulse to protect her son from both are palpable visages of the dangers of an out-of-balance ecosystem, where human children are at higher risk of health impacts from chemical pollutants within and outside their home, where they must fend against rats and roaches to sleep or to play.

As Coates (2017) reminds us, black bodies were stolen, possessed, traded, and mined under the structures of settler colonialism.[11] As those structures shaped governmental policies and programs, the dangers of exposures came directly from the government. In 1957, for example, the U.S. government sprayed Duxbury, Massachusetts, in an effort to control mosquito populations (Glotfelty 1996). Similar mass sprayings took place on Long Island and along roadways, forests, and urban neighborhoods around the country – anywhere that insects proliferated.

Americans paid, and continue to pay, a high price for their constructed environments. As Carson (1962) later detailed in *Silent Spring*, after repeated insecticide sprayings insect populations develop resistance to the chemicals meant to kill them, only to return often in exponentially greater numbers, requiring higher and higher doses of poison (see, e.g., "DDT" [2014]; Paull [2013]; and "Industry" [1962]). Carson initially published her research on the use of DDT and other chlorinated hydrocarbons as a three-part series in the *New Yorker*, in which she offered Americans a lesson in the fundamentals of ecology and chemistry (Rothman 2012). In sobering detail, Carson's early essays already underscore the inter-permeability between the chemistry of the human body and the larger ecological systems. By the 1970s, these chemicals were also responsible for the rapid decline of numerous species of birds, including the American eagle. Carson (1962) predicted that the widespread use of chemicals like DDT would also damage human cells, in turn causing a variety of diseases, including cancers and sterility. Yet even in the face of scientific evidence, in the 1950s the United States largely promoted a do-it-yourself garden culture that touted an array of petroleum-based chemical fertilizers, herbicides, and pesticides for home and garden as products ideal for suburban homeowners seeking green lawns and bug-free roses.[12]

Tending the roots of belonging

In Act II, Scene 3, Lena's family present her with gifts – garden tools, gloves, and sunbonnet – in anticipation of the gardening she will do at their new home. It is a celebratory scene in which her children honor her love of the earth. Positioned just after a visit by Mr. Linder, the scene is a kind of response to Linder's white supremacist message. Travis presents her with a hat, "like the ladies always have on in the magazines when they work in their garden" (124). But Lena's love of gardening should not be confused with middle-class desires for a suburban pastiche of the little house on the prairie. Rather, it is a signal of resistance and intentional autonomy consistent with the early civil rights activism and the rise of black nationalism, signifying social, political, and ecological resilience emerging from her own connection to the land in the south – an expression of Lena's commitment to herself and to her family.[13] In "Earthbound on Solid Ground," bell hooks (2002) reminds us of the "relationship to the earth" that verified for southern black folks that "white supremacy, with its systemic dehumanization of blackness, was not a form of absolute power" given that no one, not even plantation owners, could make it rain (69). Even as black bodies were commodified in service of the nation's first cash crop, slaves and then, later, sharecroppers tended small gardens of their own, providing for their own families, nourishing an intimacy with the soil that maintained a "concrete place of hope" in the face of white supremacy (70; see also hooks [2009]). In *Raisin*, Lena refers to a similarly-felt connection with the land, explaining that her desire for a garden like the ones she saw at "the back of the houses down home" (53) comes from her life in the south before moving to Chicago. Given the growing home and garden industry that was so prevalent at the time, Lena might be tempted to use the many advertised miracle-chemical products. On the other hand, she is already sensitive to the impact of roach spray on her grandson.

Hansberry's play (often produced and frequently taught in high school literature classes) carries the weight of its longstanding place in the U.S. dramatic canon. As such, the place-based, racialized, and economic particulars of the Younger family's struggle have too often been interpreted (and taught) as a black family's search to achieve what has been popularly understood as the American dream. Lena's small struggling houseplant has often been read as a metaphor for that ubiquitous dream. But while it clearly represents her resilience, when understood in terms of its ecological relatedness to the family, the little houseplant points to the white privilege underpinning that dream. It sits on a windowsill, soaking in what little light there is, as family members routinely tease Lena for the care she affords it: "You going to take that to the new house . . . that raggedy old thing?" Beneatha teases, yet Lena defends her attention by exclaiming, "It expresses ME!" (121). Lena's identification with the plant is more than symbolic; it is empathic and biochemical. She empathizes with the plant's struggle to survive with little light: "Lord, if this little plants don't get more sun than it's been getting it ain't never going to see spring again," she pronounces early in the play (40). However, the reciprocity goes further and becomes more material, signifying a relatedness that is also physiological by virtue of the biochemical exchange between people and plants. The family produces carbon dioxide for the plant, which, in turn, produces oxygen for exchange in human blood. The plant is thus kin, family, blood. In the final moment of the play, all the moving boxes are cleared, the plant sits on a table, and the children are calling from the street for her to hurry. Lena pauses, "a great heaving thing rises in her as she puts her fist to her mouth to stifle it." She puts on her coat and goes out, and the lights begin to dim. Then, remembering the plant that has long been the recipient of her care, she "comes back in, grabs her plant, and goes out" (151). Here, Lena engages in an act of defiance and resistance against the forces that nearly killed her little plant, as well as an act of resurgence and resilience as she claims this small piece of soil as one in which she too is rooted and has a right to thrive.

Raisin echoes the ideas from the history of black intellectual thought in the United States that reaches back to Marcus Garvey and forward to Malcolm X. Rising black nationalism is expressed through scenes that point back – over the arc of history in which the Younger's ancestors were stolen from their homelands to provide labor for extracting wealth from the land – but that also point forward, acknowledging, asserting, and celebrating African heritages. As a black intellectual, Hansberry engages the ideas that would inspire the Black Power Movement and the need for black people to break free of internalized [white] stereotypes. Walter Lee, Jr.'s drunken performance of an African "warrior" in Act II, Scene 1, throws open the doors of collective imagination to dream into and reclaim – specifically through performance – cultural ties to African homelands.[14] Beneatha appears in the Yoruba garment Asagai has given her and "natural" hair: "OCOMOGOSIAY!" she shouts, translating the meaning for her sister-in-law Ruth as "welcome," in the sense of welcoming "the hunters back to the village." Beneatha also dances and sings a traditional Nigerian folk song, "Alundi, Alundi" (77). The inclusion of Yoruba language in the scene resists the interpretation of the scene as mere parody, locating the garment, the dance, and song in a specific traditional homeland. Hansberry suggests that cultural knowledge, traditions, and ties to a homeland are potential tools with which people might decolonize their own thinking and claim a self-determined future. Scenes such as this served not only to remind audiences that African ties to homelands are real and legitimate but also assert resilience and power of cultural knowledge and tradition/identity. Walter Lee, Jr., joins in the scene's assertion of embodied connections to Africa. He jumps on the table, as "a leader of his people, a great chief, a descendant of Chaka," the stage directions tell us, saying, "Listen, my black brothers – ... – Do you hear the water rushing against the shores of the coastlands?" (78). Beneatha remains caught up in her brother's reverie, "OCOMOGOSIAY! ... We hear you Flaming Spear!" It is a meta moment on Hansberry's part that asserts black nationalism and African place-based identity and demonstrates the power of theater itself, the power of embodied enactment to empower and enliven. But then, in the opening of Act III, after Walter Lee has lost his father's money to local corruption, Beneatha challenges Asagai's vision of African renewal: "What about all the crooks and thieves and just plain idiots who come into power and steal and plunder the sale as before – only now they will be black" (133–134). Asagai's response to her rightful despair lays blame on systemic injustice, rather than the individual, by asking,

> [W]as it your money he gave away? ... Would you have had it at all if your father had not died? ... [I]sn't there something wrong in a house – in a world – where all dreams, good or bad, must depend on the death of a man?
>
> *(134–135)*

Asagai calls into question colonial history in which the acquisition of wealth on the part of some is invariably dependent on violence and loss. Decolonization, Asagai suggests, is a generations-long project in which he sees himself and Beneatha participating. His proposal envisions not a reversal of history but a new vision of home: "when it is all over ... come home with me ... home to Africa" (136). Through the character of Asagai, Hansberry asserts and reclaims a connection among identity, home, and homeland, between intellectual and cultural traditions, in a vigorous and unabashed vision of a replenished future that is at once African and American.

The claim of a right to homeplace and homelands that is at the center of Hansberry's play would be at the heart of the environmental justice movement 30 years later. The articulation of an ecological framework that is inclusive of people – envisioned as part of, and not separate from, nature – is one that recognizes justice as fundamental to ecological health and well-being. The fronts of struggle of the diverse civil rights movements of the 1960s and 1970s would lead in the years to come to the recognition that social justice and environmental health are intertwined. Before her death from breast cancer in 1964, Rachel Carson was the target of attacks by defenders of industry; she was branded "anti-American" and a "hysterical woman" (Gore 1962, xvi). Nonetheless, her indictment of herbicides and pesticides provided the foundation for the rallying cry of "Ecocide!" by protesters of Dow Chemical's Agent Orange, which was used as a defoliant during the Vietnam War. Years before the Cuyahoga River caught fire (1969; see, e.g., Hartig [2010]); before mothers protested the poisoning of schoolchildren at Love Canal (1977; see, e.g., Blum [2008]); before Three Mile Island (1979), the first major nuclear accident in U.S. history; and the largest oils spill in U.S. waters, the *Exxon Valdez* (1989), *Silent Spring* (1962) exposed the personal cost of the illusion that nature and culture are separate. On the occasion of the 50th anniversary of the publication of Carson's *Silent Spring*, the *New York Times* compared her book's impact to that of Harriet Beecher Stowe's *Uncle Tom's Cabin*, which fueled the antislavery movement in the early nineteenth century (Grizwald 2012).

The Environmental Protection Agency was established in 1970, and DDT was finally banned in the United States in 1972 under the Federal Environmental Pesticide Control Act. The Clean Air Act of 1970 and the Clean Water Act of 1972 began concurrent municipal and industrial regulation of wastes entering the environment. Yet for ideological as well as economic reasons, U.S. culture continued to embrace chemical production and use. Regulations and health standards were delayed or ignored. Certain chemicals, like DDT, were replaced but often with substances that are more toxic (Gore 1962, xxi). Twenty years later in 1992, then U.S. Vice President Al Gore observed that in a single year,

> 2.2 billion pounds of pesticides were used in this country – eight pounds for every man, woman and child. Many of the pesticides in use are known to be quite carcinogenic; other work by poisoning the nervous and immune systems of insects, and perhaps of humans. Although we no longer have the doubtful benefits of one household product that Carson described – "We can polish our floors with a wax guaranteed to kill any insect that walks over it" – today pesticides are being used on more than 900,000 farms and in 69 million homes.[15]
>
> *(xxii)*

By the time Gore wrote those words, the environmental justice movement was already calling attention to the ways in which communities of color – including families like the Youngers – were at higher risk of exposure not only to pesticides but also to all manner of environmental degradation, including air and water pollution. Whether considered through the vantage point of its moment of production, or from our present sociocultural moment, Hansberry's play serves as an enduring reminder that women, children, and communities of color bear the burden of environmental ills. Connecting the dots in this play, for example, between unfair housing laws, the siting of waste facilities, the ecological violence of air and water pollution, and human health demonstrate the subtle and personal ways that environmental racism is felt in bodies, families, and neighborhoods of those whose lives and labor were tools of extractive capitalism and the consumer culture that it fostered in the 1950s and beyond.

Although it would be a generation before the Principles of Environmental Justice would be formally adopted by the People of Color Environmental Summit in 1990, the connection between environmental and social justice visible in many personal and palpable ways in Hansberry's play was also already a central concern of civil rights leaders. In April 1968, Dr. Martin Luther King traveled to Memphis to support the sanitation workers' fight against low wages and unhealthy working conditions. Meanwhile, that same year, César Chávez and Dolores Huerta of the United Farmworkers Movement (UFW) testified before the U.S. Congress about the impact of pesticides on Mexican and immigrant laborers. Using the Equal Protection Clause of the 14th Amendment, as well as Title VI of the 1964 Civil Rights Act, environmental justice activists would begin to turn the attention of the environmental movement toward an idea that had long been obvious to scientists – that what we do to the land we do to ourselves.[16]

El Teatro Campesino on the grounds for justice

The 1960s and 1970s were characterized by multiple and diverse civil rights movements. Despite differences in foci and participants, each unique movement arose from collective lived experiences of injustice, racism, and inequities. These sometimes-converging social struggles stemmed (at least in part) from long-standing environmental ideologies and frameworks on which extractive industrial practices depended. As a result of a national exploitative relationship to the land, U.S. urban and rural communities of color faced (and still face today) a barrage of parallel and interrelated issues, including unjust and nonliving wages, inadequate and unhealthy living conditions, routine pesticide and herbicide exposure, food insecurity, a lack of access to education, and systemic and institutionalized racism. In *Environmentalism and Economic Justice*, Laura Pulido (1996) argues that "mainstream environmental groups . . . sought to intervene in the spraying of pesticides in national parks or in their use in mosquito control." Pulido also notes that the effects of these exposures on childhood health are "quality-of-life issues in which the social actors are fairly removed from the actual threat," while for farmworkers, resisting pesticide exposure became fundamental to "a major power struggle to alter the conditions of farmworkers' powerlessness" (58). Symbolically conflating human bodies with the land, the U.S. agribusiness empire also mined the farmworker him- or herself for profit. Pulido argues that the programmatic application of pesticides by California growers disregarded the safety of Mexican American farmworkers and their homespaces precisely because of the subaltern status of the migrant worker (59).[17] As a result of the early efforts and successes of the UFW in California, a broader Mexican American sociopolitical activist branch emerged in the U.S. Southwest; *El Movimiento*, also known as the Chicano Movement in English, began organizing to fight for Chicana/o civil rights in every sphere of national life.[18] While the *Movimiento* and the localized fight to unionize Mexican American, Filipino, and other immigrant workers through the UFW are not synonymous, these intersecting fronts of Chicano/a civil rights activism were grounded in a reclaimed sense of transnational cultural heritage and identity.

Throughout the 1960s and 1970s, many Chicana/o activists and artists returned to and reclaimed the Aztec conceptualization of Aztlán, the indigenous homeland that remembers the land of the U.S. Southwest as a contiguous indigenous landscape that existed before – and has continued to exist after – colonization.[19] The ancestral homeland of Aztlán became a galvanizing cultural image and historical touchstone that resists dominant U.S. Anglo power structures, in direct counter valence to historical whitewashing and deterritorialization of the peoples native to what we now call the United States and Mexico (Anzaldúa 1987, 23–35). Aztlán provided a foundation for the *Movimiento* in the form of shared cultural identity grounded in the land. Throughout the *Movimiento*, asserting the precolonial geography of Aztlán has subverted Anglo cultural messaging of illegitimacy by representing Mexican origin people, regardless of citizenship, as owning an identity in and connection to the land that predated current national boundaries. Aztlán empowered twentieth-century Chicano/as struggled for sociopolitical and economic equity, and its power continues today, providing a sense of identity rooted in indigenous homelands that precede the current nation-states of the United States and Mexico (see, e.g., Mesa-Baines [2001]; and Ontiveros [2014]).

The indigenous founders of the ancient city of Tenochtitlán, the site of contemporary Mexico City, came from *El Norte* – the region that encompasses what is now called the southwestern United States. After colonialization by the Spanish, these northern regions were taken/stolen and annexed, in turn, as part of the new nation-state of Mexico through the Mexican–American War, known in Spanish as the War of Northern Aggression (1846–1848). In 1848, Mexico ceded to the United States the territories now called Texas, New Mexico, Arizona, California, and parts of Utah and Colorado in the Treaty of Guadalupe Hidalgo. The treaties promised that the people autonomous to those regions – indigenous, Mexican – and those who owned farms and ranches would become naturalized citizens, retaining all the rights, privileges, and properties of U.S. citizens.[20] All the treaties were subsequently violated, broken, and then "erased" from official memory.

In her 1987 groundbreaking and oft-cited bilingual work, *Borderlands/La Frontera: The New Mestiza*, Chicana scholar, activist, teacher, and poet Gloria Anzaldúa reminds her readers – Chicana/o and Anglo – that the U.S. government failed to honor the terms of the 1848 Treaty of Guadalupe Hidalgo (see Anzaldúa [1987, 23–35]). Indeed, illegal land seizures, graft and a climate of racism made many Mexican-people-turned-U.S.-inhabitants into second-class citizens in a country constructed around their homeland and in the midst of their communities.

Aztlán is both an origin story and a spiritual geography in and of itself. Alicia Arrizón (2000) identifies the dual significance of Aztlán as both spatial/geographic homeland and spiritual identity: "According to myth, Aztlán is the ancestral homeland in the north that the Aztecs left in 1168 when they journeyed southward to found the promised land, Tenochtitlán (Mexico City), in 1325." This siting of cultural heritage in the geographic lands that are now part of the United States represented for Chicano/as "the spiritual power of unity among a people who see in their common pre-Hispanic heritage and indigenous past a source of cultural affirmation in the present" (23). In Arrizón's words, Aztlán's physical geography "justified contemporary efforts to reclaim this lost land," while its galvanizing spiritual power "helps combat racism and exploitation" (23). Similarly, in *The Road to Aztlán: Art From a Mythic Homeland*, Chicana scholar and activist Amalia Mesa-Baines (2001) describes the powerful symbolism of Aztlán as "an ancestral homeland emanated from the deep Chicana/o sense of dislocation and deterritorialization experienced in the aftermath of the Mexican-American war." Mesa-Baines cites "this long history of frustration and powerlessness that gave rise to demands for equality and social justice" as the inspiration for articulations of Aztlán as "a narrative of origin," particularly for young activists of the *Movimiento* in the 1960s (333).

For many Chicano/a artists and scholars, the poetics of Aztlán as a conceptual framework and vision assert place-full-ness and place-based legitimacy in a society in which Chicana/os have been displaced, disregarded, and devalued. Aztlán recognizes the contiguous culture, language, ancestry, and environments of what are now geopolitically considered parts of both Mexico and the United States. For activists within the *Movimiento*, Aztlán connects contemporary identities to ancestral histories, as it asserts the land-based legitimacy. It was – and remains today – a spiritual concept of a material and mythic past that have empowered and united Mestiza/o peoples of the U.S. Southwest from the 1950s into our present moment in a struggle for social, economic, and environmental justice. Aztlán and the histories, peoples, and homeland that the term connotes continue to serve as a source of inspiration and an assertion of belonging for diverse communities of present-day Chicana/os and Mexican Americans.

(Re)claiming Aztlán as the ancestral homeland of the Chicano people, a homeland that was both pre- and transnational, became a potent source of inspiration and provided a landscape into which Chicana/os could carve out an alternative possible future, as Anzaldúa (1987) poeticizes:

> This land was Mexican once,
> Was Indian always
> And is.
> And will be again.
> (23)

In a material/geopolitical and mythic/metaphoric sense, Aztlán was (and is) the grounds for socioeconomic justice and self-determination. Moreover, Aztlán also forms the basis of an ecological sensibility unique to Chicana/os of the southwestern U.S. Aztlán not only connotes a Chicana/o homeplace but also asserts the ecological congruency of the land and its rivers, deserts, species migration, and all the ways it has provided human sustenance. This invocation resists geopolitical nationalism in favor of lived, cultural-ecological connection to the land.

Early in the *Movimiento*, in the first recitation of what has since become a central and widely-known manifesto, Chicano poet-leader Alurista's *El Plan Espiritual de Aztlán* brought the meanings of Aztlán to bear on the material geopolitical lands of the U.S. Southwest. In his original, video-recorded public recitation of the spoken-word poem, Alurista (1969) demanded justice and liberation

> called for by our house, our land, the sweat of our brows, and by our hearts. Aztlán belongs to those who plant the seeds, water the fields, gather the crops . . . with our heart in our hands and our hands in the soil, we declare the independence of our Mestizo nation.[21]

Here, Alurista asserts that the embodied connections between the land's fecundity, the body's labor, and the community's historic presence on and ancestry in the land is the literal and metaphoric ground on which the *Movimiento* proceeds.

The power of performance and iconography of Aztlán as the central image for the *Movimiento* also carry an implicit Chicano/a environmentalism. In "Green Aztlán," for example, Ontiveros (2014) details how

> "El plan espirtual de Aztlán" . . . outlines a model of sustainability that united social justice and environmental protection through an emphasis on labor. . . . Alurista envisions an alternative economy built on the recognition of a shared moral obligation to "our lands" and "those who plant the seeds, water the fields, and gather the crops."
> (97)

As a counternarrative to U.S. Anglo historical discourse discussed in earlier chapters, the diverse and temporally specific conceptualizations of Aztlán articulate a sense of place and rootedness in the land for Chicano/as. It is a strikingly different ethos than that of industrial agriculture as practiced in the San Joaquín Valley and throughout the United States. Indeed, environmental historian Devon Peña (2010) observes that practices now recognized as "sustainable," such as dry-land farming, were at the center of Mexican agricultural practices for generations before the Treaty of Guadalupe Hidalgo and are currently still practiced by Mexican American farmers and ranchers in many regions of the U.S. Southwest. As Peña (2010) has also observed, the ethos of Aztlán shares common cause with ideas of sustainability, including land-use practices informed by an ethic of reciprocity and evidenced in cultural traditions, family ties to the land, generations-long habitation, and cultivation. The ecology of Aztlán carries a sense of kinship.

As part of the political mobilization of the *Movimiento*, Chicano/a artist-activists sought to animate the histories of Aztlán through storytelling, improvisation, agitprop theater, visual art, cultural festivals, and music. Chicana/o *teatros*, in particular, aimed to educate and empower Mexican Americans, as well as to leverage a collective indigenous past to mobilize Chicano/as in the fight for legitimacy and justice in the present (Worthen 1997). In his firsthand observations of the *teatros* of the Chicano Movement, Chicano theater scholar Jorge Huerta describes a cultural front that sought to recover, reclaim, and validate Chicano/a history by connecting and tracing precontact indigenous cultures to then current struggles in the fields and the *barrios* of the late twentieth century. Such didactic and improvisational performances called attention to the impacts of political policies and economic practices on the lived experience of Mexican Americans and immigrants. More recent scholarship of Peña (2010) and Pulido (1996) have shown how social and environmental justice are intertwined and interdependent and that this awareness is crucial to these movements today. Indeed, three decades of Chicana/o theater – through which dramatists amplified the connections between farmworker rights and health, community well-being and sustainability, the health of the land, and the ecological significance of spiritual connections to the land – prompted a conscious shift toward environmental justice within the field of ecocriticism in the late 1990s (see, e.g., Buell [2005]; and Deming and Savoy [2011]).

In what follows, I argue that the mythos of Aztlán, which figured prominently in Chicana/o artistic expression and theater and performance in particular, carries an implicit ecological ethic that functions as a counternarrative to white Anglo expansionist frontierism. In the first half of this chapter, I discussed the connection to home/homeplace/homeland (e.g., as a home in Clybourne Park, as a homeland in Africa) that figure mightily as an empowering visage in Hansberry's *A Raisin in the Sun*. In the analysis that follows, I explore the theatrical invocation of Aztlán that has given power and legitimacy to Chicana/o presence in the United States, just as it has asserted a connection to homeland that originates prior to the nation-states that are now called Mexico and the United States. In order to broaden our understanding of how Chicana/o theater engaged and continues to engage environmental frames of thought, I look closely at the ways in which animating Aztlán suggests an ecological ethic consistent with the values of the environmental justice movement.

El Teatro Campesino: staging Aztlán

Much of the theater and performance related to the *Movimiento* was centrally concerned with the ways in which the histories of people and land are irremovably connected. These themes are particularly visible in the work of El Teatro Campesino (ETC) – the Chicano farmworkers' theater – which inspired many Chicana/o teatros[22] of the *Movimiento*. Founded in 1965 by Luis Valdez, ETC served as the artistic arm of the farmworkers' struggle in the Central Valley of California (Broyles-González 1994, 129).[23] In both popular and scholarly discourse, Luis Valdez is often described as the father and singular founder of the Chicano theater movement. In *El Teatro Campesino: Theater in the Chicano Movement*, Yolanda Broyles-González notes that

> [s]peaking generally, the history of El Teatro Campesino since its founding in 1965 has been canonized as the history of the life and times of Luis Valdez. Like history in general, Teatro history has largely been reduced to a chronology of the doings of one individual, its director.
>
> *(130)*

The work of ETC as an ensemble undisputedly inspired Chicana/o theater across the Southwest from the 1960s through the 1980s. Yet the contributions and leadership of women – as artists, as subjects, as activists – have been omitted consistently from dominant Anglo and Chicano accounts of ETC.

Beginning in the early 1960s under the grassroots leadership of founders and organizers Dolores Huerta and César Chávez, the UFW sought to challenge the power of U.S. agribusiness through unionizing. The union fought for fair wages and occupational health standards consistent with standards for workers in other U.S. business sectors. Dolores Huerta and Chávez worked through legal and legislative means, as well as organizing public boycotts, worker protests, and strikes (Pulido 1996, 58–124). During the 1960s and 1970s, ETC supported the collective organizing in the San Joaquin Valley of California through grassroots theater performed by, for, and about Mexican American farmworkers, laborers, and communities. The role of theatrical performance within the farmworkers' movement not only helped to articulate what was (and still is) happening to human beings who labor(ed) in California's agricultural industry but also made visible the relationships of power that allowed an abusive system to continue and to exercise farmworkers' active participation in identifying, understanding, and intervening within and beyond that system.[24] ETC was a fiercely activist, creatively controversial, worker-centered *teatro*. The performances and theatrical demonstrations staged by ETC intended to educate and raise consciousness as well as to entertain and support the morale and resolve of those engaged in the struggle. Working in common cause with the political organizing of Chávez and Dolores Huerta, ETC's collectively sketched, improvised, and (later) scripted works empowered farmworkers to educate themselves and their communities by bringing performances right to the fields, neighborhoods, and urban centers where Chicana/o people worked, lived, and led their lives.

In its early years, ETC's community-based theater took the form of short, improvisational *actos*, which were meant to inspire action through didactic and accessible community-based theater. Jorge Huerta (1982) describes the *acto* as a "short, *comedia*-style sketch, broad and often farcical, political in its subject matter," and underscoring the need for a farmworkers' union (14). The *actos* sought to intervene in the abusive practices of industrial agriculture, thereby empowering farmworkers and promoting the causes of the UFW. As Broyles-González (1994) notes, the often satirical *actos* of ETC in the 1960s underscored the poor working and living conditions of farmworkers in California (xii). Some of the *actos* portrayed the conditions of everyday life, legitimizing union demands for fair wages, adequate and healthy housing, and protection against pesticide use; others called out conditions of racism and stereotyping of Mexican American farmworkers; still others allowed farmworkers themselves to step in to the roles of growers or middlemen; others celebrated an explicitly Chicano/a history, heritage, and identity. In all cases, these early performances and roles helped to galvanize the *Movimiento* and foster deeper understanding of the economic systems of large-scale agribusiness in which Mexican American labor and bodies functioned. As farmworkers took part in theatrical performances and demonstrations depicting the ways in which industrial agriculture worked to commodify and exploit their labor and the land, they gained an invigorated sense of agency and solidarity.

By 1967, ETC had established a more permanent home base in San Juan Bautista near Delano, California – the epicenter of the UFW and the first grape boycott, located squarely in the heart of California's vast agribusiness landscape. Jorge Huerta (1982) observes that land, both as setting and co-player, is central to the politics and the power of ETC's work. ETC came to the farmworkers where they worked and lived. Frequently staged on the back of a truck, performances generally took place in the space between the harvest fields and the matrix of dirt roads that carried the fruits of the labor to markets that profited the Anglo growers. Jorge Huerta also notes that the physical presence of the land as setting and co-player, stretching as far as the eye could see surrounding players and workers, carried material and symbolic meaning: "The farmworker *actos* enjoyed an incredible realism . . . Because the action was usually related to the fields, references to the vast expanse of agricultural wealth had ironic impact on the strikers, who could see it all around them" (19). Performances also took place in community centers and granges that became union halls in which farmworkers gathered. In these ways and others, the ETC participated directly in the radicalizing and community-building that was part of the UFW's struggle.

In addition to issues directly affecting farmworkers, ETC's work began to explore what Valdez called the *mitos* – theatrical works that represented the spiritual aspects of Chicano/a identity rooted in the histories and cosmologies of an indigenous heritage, as well as works that celebrated the heroes of more recent histories including the Mexican Revolution. Jorge Huerta (2000) describes the *mitos* as a "Chicano/a imaginary," a body of mythic, heroic, and indigenous tales that provided an alternate narrative and one that empowered Chicano/s with a history of their own, distinct from the dominant narratives of frontier America (19). Indigenous ancestry provided the burgeoning artistic and sociopolitical *Movimiento* with a spiritual reorientation and an ecological standpoint. While the *actos* were implicitly political, exposing the racist stereotypes of Mexican Americans perpetuated by Anglos and the U.S. discourse as comical "over-the-top" satirical vignettes, the *mitos* expressed and celebrated the historical, spiritual, and mythic underpinnings of the *Movimiento*.[25] *Mitos*, Jorge Huerta (1982) observes, were meant to infuse "a sense of the sacred [into the] commonplace" (51), such as a backyard, a vacant lot, a railroad easement, a barrio, or a street corner. In the *actos*, the earth was represented allegorically through characters that represented the seasons, sun, rain, or wind; the *mitos*, however, brought to life the indigenous gods and goddesses, linking allegory to longstanding history on and connection to the land underfoot. The *mitos* assert that this land, however subjugated and degraded by the mechanisms and chemicals of agribusiness, *is* Aztlán. The central mythos of Aztlán so integral to many of the *mitos* performed by ETC, ultimately illuminates how the land is both a material location in which workers bodies are used and abused and an imaginative space through which solidarity forms around a common sense of homeland.

The problem of *Bernabé*

With this context as background, I turn now to Valdez's *Bernabé*, a *mitos* in which the embodied connection between the soil and the body of the farmworker represents a spiritual and cultural identity and a connection to ancestral homelands. Following on his beginnings with ETC, Luis Valdez began a second phase of his career as a solo dramatist.[26] My analysis seeks to discover if this play about a farmworker who falls in love with La Tierra, the earth, carries an ecological message through its invocation of the mythos of Aztlán.[27] Through *Bernabé* (as well as Cherríe Moraga's *Heroes and Saints* discussed in the following chapter), I argue that, among other interpretations, Aztlán constitutes a counternarrative that centers peoples' historic spiritual, social, economic, and ecological relationship as the grounds for social and environmental justice in the present.

First produced in 1970, *Bernabé* asserts an earth-centered spirituality and land-based identity born of the mythos of Aztlán. Its parable of a farmworker who falls in love with La Tierra, the earth, argues that a sensual, embodied engagement with the lands of indigenous heritage offers empowerment and renewal. For Chicano/as, Jorge Huerta (2000) argues, indigenous spiritual traditions function as part of lived experience in a "syncretism of the mythical with the everyday" (42). He describes *Bernabé*'s protagonist as an "archetypal Chicano seeking his connection with Mother Earth" (195), who would be "[f]amiliar to any barrio audience . . . people do not hide their abnormal children" (97). Elizabeth Ramírez (2000) describes Bernabé as "the village idiot, who through innocence seeks to marry La Tierra, the earth" (88). In the first scene, Bernabé digs a hole in the ground through which to commune with his lover, La Tierra, and there enjoys the erotic pleasures of his connection to the earth. Mocked by the men of the village as a fantastical and deranged obsession, a mere unmet sexual hunger, La Tierra is a real woman to Bernabé, who insists that he loves and wants to protect her. His mother, a one-dimensional character who expresses her disgust and moral condemnation of her son's sexuality, represents not only the Catholic Church but also the internalized oppression of women and subjugation of bodies of all genders and sexual expression, and perhaps signals society's disconnection from and domination of both the earth and sensorial experience. To convince Bernabé to abandon La Tierra, his cousin, Primo, tries to entice him with a prostitute named Consuelo.

On stage, Consuelo and La Tierra are played by the same actress. On the surface of things, this doubling points to the binary thinking about both women and land as "virgin" or "whore," a vestige of settler colonialism, and one that has been internalized in the thinking of men and women, both Chicano and Anglo. Yet as Bernabé is pressed on the one hand by Primo to show his manhood by consuming Consuelo's services, he is condemned on the other by his mother for this love of La Tierra: "Come out of that dark ugly hole! . . . You think I don't know what dirty things you do in there? . . . one of these nights the moon is going to come down and swallow you alive" (144–145) she chastises. In the next scene, the older men set Bernabé up with Consuelo. They taunt him by saying, "She's ready, she's willing, and she's able, carnal. So get in there. Go get it!" (151). In Scene 5, Bernabé's confusion and sense of isolation are palpable as he becomes agitated, disoriented, and wants only to escape Consuelo's room: "I don't know what happened. *Está loco, ¿qué no ves?*" (155), Consuelo concludes. In this way, Bernabé remains trapped between two forces of colonialism: white settler colonization of the land, women's bodies, and sexuality have been commodified; meanwhile, under Catholicism, the body, sexuality, and women are subjugated as well.

That subjugation, and the ways in which it goes unquestioned within the extractive economic system of industrial agricultural, are made clear in an earlier scene when the men also tease Bernabé, telling him the "gabachos" are going to take La Tierra away from him: "[L]ook how the ranchers treat her, hombre. They sell her whenever they feel like it," Torres chides. Then he presses the metaphor of the industrial prostitution of the land further by exclaiming, "Give me a few bucks, and I'll let you have her – for the night!" (149). The binary thinking on which *Bernabé* depends recapitulates the narrative of "virgin land" and male agency that was central to the project of Manifest Destiny: the land as waiting and ready for masculine use, exploitation, penetration, and violation. As I argue in my second chapter, in this feminized nature land serves the project of settler colonialism by erasing indigenous presence, history, and culture, as well as rationalizing the license to extract (i.e., rape, conquer, pick your trope) profits from the land's bounty. Industrial agriculture is predicated on the idea that the earth is inert material for use. This ideology allows the commodification not only of the land, the natural world, but also of the labor and bodies of workers.

Bernabé's worldview and his refusal to participate are read as *loco* by his community. But his refusal also calls out the dichotomy, the collision of cultures, between a Western/industrial view of land as a material resource, an inert matter to be shaped and used for profit, and an indigenous view in which La Tierra is a living, vital, and intelligent being with whom human beings are engaged in a reciprocal relationship. Rooted in Mayan and Aztec spiritual knowledge and practices, this view understands the land as a living life partner, as vital and sensual, and is born out in the theatrical performance. As Ramírez (2000) has observed, "Bernabé is seen in two spheres: the realistic setting in which Bernabé lives and the sphere in which he encounters the world of the Aztec gods" in which La Tierra "appears as an activist, powerful enough to make Bernabé want to fight for her, die for her and even kill for her" (88). In a 1975 production directed by Ramírez, La Tierra appears as "La Adelita, a woman soldier of the [Mexican] revolution" (88) and later as "maíz, or an ear of corn grown from the ground" (89). Using the space of the stage to give material presence to the spiritual/ancestral world, Ramírez's production underscored what is not readily apparent on the page – that this is a story about engaging with indigenous presence in the land itself, presence and ancestry that predate the harsh realities of the farmworker's living and working conditions and lay claim to an identity in and of the land.

Bernabé is troubling, however, and may even be disturbing to contemporary readers for its depiction of disability as well as gender stereotypes. Indeed, the work is saturated with racist and sexist ideas about land, and these have their source in a settler-colonial frame of thought. Viewed within a conceptual binary that sees the earth as passive, feminine, brown, and exploitable, while agency is understood as masculine and white, Bernabé's connection to La Tierra is, at best, a symptom of his disability and, at worst, a kind of rape in which Primo and the other men urge him to participate. Bernabé's protestations against the buying and selling of his lover may seem ludicrous, and the ways in which Valdez equates the body of the land and female bodies are often more problematic than progressive. Yet, when Bernabé pledges to protect her against the crop-dusting planes, his relationship with La Tierra does signal certain ethical considerations about how the beloved is to be treated and, for those who labor and live on the land, how farmworkers are to be treated.

Nonetheless, it is important to consider Pulido's socioanalytical perspective about the work, in which the farmworker occupies a subaltern position as a result of the social and economic structures instituted by and intended to serve extraction (and thus grow colonial power). Understood through this lens, the play becomes a parable in which the disabled character of Bernabé occupies the subaltern position within his own community, thus revealing the very ways in which Anglo frames of thought have informed and distorted a sense of kinship with the earth. In one of the largest, most productive industrial agriculture regions in the United States, California farmworkers – most of whom were Mexican origin – functioned within an international commodity market that depended on the subjugation of their labor (Pulido 1996, 58–60). Pulido points out that the extraction of labor and land was made "possible only through groundwater extraction or surface diversion" – large public works projects that sequestered water for industrial use. Moreover, the intersection of economic and social positions within a global labor market meant that the "class structure and racial and ethnic lines were drawn fairly sharply, with the Anglos occupying positions of authority as community leaders and growers and the Mexican-origin population serves as highly marginal, socially stigmatized, impoverished work force" (60). This social position, Pulido argues, is "subaltern," that is, located at the nearly invisible and often silenced margins of sociopolitical and economic power structures. Consequently, activism by these groups necessarily seeks "to change the distribution of power and resources to benefit the less powerful . . . seeking greater social justice" (24).

In the play itself, the character of Bernabé represents both the material embodiment of, and a metaphor for, this subaltern position. Bernabé finds power neither in Primo's sexism, recapitulated from Anglo society, nor in the structures of his mother's Catholicism (nor in the church as a social institution), but rather in his common cause with the earth itself. While the play is fraught with troubling images of women – the one-dimensional guises of the virgin, the mother, and the whore, representations that blatantly subjugate, typecast, and rob women of their agency and their voices – Pulido's framework of the subaltern makes plain the way in which these stereotypes function from/within the Anglo domination of the land.

In *Borderlands/La Frontera*, Anzaldúa (1987) interrogates the myth of "machismo" in Mexico and in Latina/o and Chicana/o cultures, exposing it as a rhetorical invention of Anglo culture that perpetuates a system in which Chicano/a bodies are subjugated and mined for their labor. In a section titled "*Que no se nos olviden de los hombres*" ("Let us not forget the men"), Anzaldúa writes that

> [t]he modern meaning of the word "machismo," as well as the concept, is actually an Anglo invention. For men like my father, being "macho" meant being strong enough to protect and support my mother and us, yet being able to show love.
>
> *(105)*

She asks what happens to a man when he becomes part of a system that does not pay him a fair wage for his labor, steals his vitality to turn into profit in a global agricultural market, and endangers his body and his family through pesticide exposure and inadequate housing. Anzaldúa argues that so-called machismo is an adaptation to oppression and poverty and low self-esteem, the result of hierarchical male dominance that issues from the condition of multiple layers and generations of colonization:

> The Anglo, feeling inadequate and inferior and powerless, displaces or transfers these feelings to the Chicano by shaming him. In the Gringo world, the Chicano suffers from excessive humility and self-effacement, shame of self and self-deprecation. Around Latinos he suffers from a sense of language inadequacy and its accompanying discomfort; with Native Americans he suffers from a racial amnesia which ignores our common blood, and from guilt because the Spanish part of him took their land and oppressed them.
>
> *(105)*

Here and elsewhere, Anzaldúa invites a reframing of "macho" as "excessive compensatory hubris" (83), which must also be understood as an expression of resistance and resilience.

As Bernabé holds to his dream of unity with his lover, he not only asserts a worldview in which La Tierra is a living being with whom he shares intimacy and acceptance, he actualizes/materializes the indigenous Aztec heritage that still – despite its centuries of silencing by colonialization – lives in the land. First, however, Bernabé must undergo a cosmic shift in his own thinking.[28] In Scene 6, in particular, La Tierra enjoins him to fight for her, when she asks him,

> [H]ow am I yours, Bernabé? Where and when have you stood up for me? All our life you've worked in the fields like a dog – and for what? So others can get rich on your seat, while other men lay claim to me? . . . I was never meant to be the property of any man – not even you.
>
> *(160)*

In the dream sequence that follows, Bernabé must contend with the cosmic forces El Sol and La Luna. El Sol confronts Bernabé with a vision of himself and challenges Bernabé to step into it:

> I know who you are, Bernabé. Eres el último y el primero. . . . The last of a great noble lineage of men I once knew in ancient times, and the first in a new raza cósmica that shall inherit the earth. Your face is a cosmic memory.
>
> *(164; ellipses in original)*

Here, Valdez invokes the concept coined and developed in 1925 by Mexican philosopher José Vasconcelos (Vasconcelos and Jaén 1997) as "*La Raza Cósmica*," meaning the "cosmic people" and signifying not a "race" but rather a community of people connected across geography and across time. For Vasconcelos, Valdez, and others, the mestizo/a – born of both colonial Spanish and indigenous heritage – represents the possibility of a new civilization with the potential to transcend racial and ethnic identities.

Returning to the play, the previously cited reference to Bernabé's face invites him – along with the farmworkers who view the play – to see himself newly, not as a farmworker but as a representative of the new Chicano with deep personal and sociocultural roots in Aztlán.[29] Like Bernabé, real-life Chicano poet Francisco X. Alarcón recognizes history in his own flesh when he articulates that

> this image in the mirror is me, and I am a mestizo. I am the physical proof of violent transformation suffered by Native peoples on this continent in the last five hundred years. My face, my body, my soul are in constant turmoil.
>
> *(29)*

This visage of the concept of "In Lak'ech," or other self/*otro yo*, is central to Valdez's *Pensamiento Serpentino* and encapsulates the dual self that Bernabé claims in the final tableau of the play. Bernabé and La Tierra rise up; he is naked, wearing only a loincloth: "He is Coatlicue, Mother Earth, the Aztec Goddess of Life, Death, and Rebirth." While the play ends in what the stage directions describe as "a cosmic embrace" (Valdez 1990, 167), the moment of recognition is better understood as an ongoing spiritual struggle to come to terms with one's relation to past and present, to land and spirit.

In *Borderlands*, Anzaldúa (1987) writes about a state of being she calls the Coatlicue State, which occurs when "[w]e are not living up to our potentialities and thereby impeding the evolution of the soul – or worse, *Coatlicue*, the Earth, opens and plunges us into its maw, devours us." This moment she describes can serve as an aperture, as an opportunity for those Chicana/os who witness it. In this section, Anzaldúa continues that "*Coatlicue* is the mountain, the Earth Mother who conceived all celestial beings out of her cavernous womb . . . the incarnation of the cosmic processes" and as such is "life-in-death and death-in-life" (69). In this way, what Anzaldúa calls "the *Coatlicue* state" is the generative space suggested here. Thus, Bernabé stands in the indeterminate space of possibilities, a space of potentiality that holds within it seeds of a different future, a future available, the play suggests, to all those who have suffered under the mechanisms of industrial agriculture.

In the decades following the production and publication of *Bernabé*, Latina feminists and their Anglo ecofeminist counterparts have wrestled with the ways in which Western values of domination have impacted women, men, and (human conceptualizations and treatment of) the land. Chicana dramatists, in particular, have continued to respond to, reject, and reenvision Valdez's representations in rich and nuanced works that deepen and complicate the poetic as well as embodied connections between women and the land.[30] As I explore in the context of Cherríe Moraga's *Heroes and Saints* in Chapter 6, these Chicana activists-cum-dramatists have exposed the ways by which industrial agriculture commodifies human labor as it simultaneously oppresses men as well as women through an ethos of exploitation and profit (see, e.g., Ramírez [2000], 80–124). *Bernabé* suggests a way of relating to the land that is relational rather than commercial. The land is understood in the continuum of kinship (in this case lover), an idea that has ancestral, cultural, spiritual, and erotic power to heal and connect.

As Chicano/a dramatists worked to claim and reclaim home/homeland, some Anglo dramatists, such as Lanford Wilson, Sam Shepard, and Steven Dietz, also wrote plays that exposed the broken, bankrupt, and toxic legacies of extractive capitalism's quest for profits at the expense of land and lives. In the 1970s and early 1980s, Wilson set a number of plays in locations that called attention to current environmental questions about land heritage and use, exposures of workers and communities of color, and the impact of resource profiteering on people and communities. Wilson's *The Mound Builders* (1975; first published 1976), for example, is concerned with a highway and resort development on lands culturally significant to North American indigenous cultures, while *Angels Fall* (1983) is set within the circumstances of a radioactive spill from a uranium mine on Navajo land in the southwest. Meanwhile, Sam Shepard (1979) wrestled with the legacies of frontierism, its implicit instruction of American masculinities, and degradation of familial relationships to land. While not ecodramas in a strict sense – that is, they do not have a clear ecological sensibility or environmental focus – at the center of these plays might be seen at the very least as an attempt to struggle with complex social costs of a mainstream society that continued to turn a blind eye to the intertwined nature of ecological degradation and human injustice. Yet, in the humanist listening that governed mainstream theater, the ecological implications were largely ignored, and the indigenous thematic elements of the plays were also brushed over in favor of the angst of the largely white characters. In the following section, I close this chapter with a brief demonstration of how the material meanings of nature on stage might be rejoined with the metaphoric in Sam Shepard's 1979 Pulitzer Prize winner.

Buried Child's face in the mirror

Many American dramatists of the twentieth century have critiqued the social and environmental impact of progress at any cost; several also confront the deeply personal cost of frontier masculinity, wrestling in particular with the destructive generational impact of this construction for American men. Sam Shepard's (1979) Pulitzer Prize-winning play *Buried Child*, for example, laid bare some of the deeply personal legacies of frontier values and the ecological violence of settler colonialism, which expresses as impacts not only to the environment but to families. When the central metaphor of *Buried Child* is rematerialized, it embodies a searing critique of the impact – both personal and environmental – of a century of land abuse. A multigenerational family gathers on the family farm in the final days of the patriarch's life. The original homestead of a great-great-grandfather, the farm is two steps short of economic catastrophe as it strains under drought, debt, and neglect. The family is riddled with alcoholism, cycles of generational abuse, and dysfunction, bearing witness to the personal costs of the lies buried in the land. Dodge, the raving drunk father, insists the drought and chemically-ravaged soil has not grown a thing for years, yet other family members harvest armloads of carrots and fill up the living room floor with carrots and corn, a fecundity rooted in buried acts of violence. Breakage is apparent: in relationships, in material structures, in bodies, and in the land – a result of layers on layers of lies. Tilden and Bradley are physically and emotionally disfigured (Bradley lost a leg to a chainsaw). When the grandson, Vince, arrives, neither his grandfather, Dodge, nor his father, Tilden, recognize him. Shepard forces a confrontation with the legacies of violence, as embodied history, as Vince sees – in his own face, his own flesh – the men, the generations of men that have come before him. The irony of father and grandfather not recognizing their progeny begs the question, What is it in the multigenerational juggernaut of the American Dream that mainstream society does not recognize or cannot face?

The distorted interactions between characters peel back the husks of those lies to reveal that the patriarch murdered a child born incestuously to his wife and eldest son. He buried the infant in the backyard. Thus, the family's garden is also a gravesite: a poignant metaphor for the layers of violence, illegitimacy, and its denial that lie silent in American soil. Ultimately, it is Vince who sees into his own heritage a legacy that he carries in his bones. In a vision of the grandson in which Vince confronts the legacies of his family history, Shepard presses settler descendants to face the image of the past that is coming toward us from the future as Vince speaks:

> I was gonna run last night. I was gonna run and keep right on running. . . . I could see myself in the windshield. My face. My eyes. I studied my face. Studied everything about it. As though I was looking at another man. As though I could see a whole race behind him. Like a mummy's face. I saw him dead and alive at the same time. In the same breath. In the windshield, I watched him breathe as though he was frozen in time . . . and then his face changed. His face became his father's face. Same bones. Same eyes. Same nose. Same breath. And his father's face changed to his grandfather's face. And it went on like that. Changing. Clear on back to faces I'd never seen before but still recognized. Still recognized the bones underneath. The eyes. The breath. The mouth. I followed my family clear into Iowa. Every last one. Straight into the Corn Belt and further. Straight back as far as they'd take me.
>
> *(64)*

Vince implicitly sees the legacy of his ancestor's settlement, use, and abuse of the land, and his reverie is an invitation to settler descendants to face our ancestors and to take account, to take responsibility for the stories and the lies that live in our bones. In a hauntingly disturbing final image, Tilden, Vince's father, carries his sister/daughter into the light for his own son to witness. He appears "dripping with mud from the knees down. His arms and hands are covered with mud. In his hand he carries the corpse of a small child." Meanwhile, almost echoing lyrics from the musical *Oklahoma!*, Hallie, mother to both Tilden and the murdered child, raves from upstairs, "I've never seen such corn. . . . Tall as a man already. . . . Carrots too. Potatoes. Peas. It's like a paradise out there" (66). Bounty, in this play, grows from the blood and bones, not of proud pioneer ancestors as the frontier mythos suggests but in the bones of the family's own (illegitimate) future. As it populates the stage in great heaps, the corn itself seems to shout back to Aztlán, to the long indigenous history in which the corn, now so bountiful, grows. Personal, generational, and environmental losses are unified in one moment, in one body, as Tilden brings the girl-child's corpse in from the land, dripping with mud. This is a parable that asks at what cost, and on what lies, does American bounty grow? Who are we (as a family, as a nation) if our abundance arises from killings and takings, from illegitimacy, and if our present bounty comes at the expense of future generations?

Amid the dysfunctional chaos of alcoholism, old familial rage, resentments, and betrayal, the men continue to harvest armloads of vegetables from the field behind the house, where Dodge insists that nothing has grown since the Dust Bowl. But clearly something is growing in the soil made rich by the blood and bones of the family's incestuous child. Reviewers at the time, and writing of the play's several revivals (1996 and 2016), understood the child as a metaphor for the weakness and dysfunction of the patriarch, a metaphor for the nation's disillusionment with the American Dream. But if we look at the child, and the vegetables themselves, in their materiality – that food, nourishment, is being grown from the bodies of the next generation, the body of the child is an indictment of that dream. While Vince does not see, or at least he does not speak about, the long history of plunder represented by the faces of his own settler-ancestors in the mirror, he does see the horrors of this family's history. As these are revealed in the play, we are led to wonder if Vince understands their implications. The murdered child under the soil is rich with ecological implications: the present bounty is quite materially nourished by murdering the future.

The buried child is a metaphor for all the ways in which the future – children's future – is put at risk by greed. Present prosperity is predicated on cycles of violence, and cycles of denial of that violence, which, in turn, infect even the most personal of our self-concepts, particularly for men. As a product of incest, the buried child is both a metaphor for and a product of the abuses of the patriarch and the systems of exploitation made possible by his privilege and power. Yet the child *is also a child, a person, a body*, one who suffered, who lost their body. As both a flesh-and-blood being and a metaphor, the child cannot be disassociated from the lived experience of the thousands of real-world counterparts whose lives and liberties are imperiled by increasing environmental degradation, the result of a long history of exploitation. Furthermore, when the meaning of the buried child is extended into this chapter's previous discussion of works by Valdez and Hansberry, its materiality centers the ecological violence that resulted in the deaths of Mama Younger's babies and the poisoned and disabled children of farmworkers in California's Central Valley. Although clearly quite distinct from the plays by Hansberry and Valdez, Shepard's work is relevant here for the ways in which it exposes the violence inherent in ongoing settler colonialism. The body of the murdered and injured child as a signifier of the social and environmental injustice would appear in plays by Latino/a and Euro-American dramatists in ways that also demand a focus on their material meanings.

In 1991, Robert Schenkkan's *The Kentucky Cycle* (published in 1994) would seem to answer the question that Tilden's vision conjures, in this case in a play that closely examines the multigenerational legacies of the frontier ethos, its entitlement, and its exceptionalism. In Schenkkan's play, an ancestral child's body is also unearthed, but the child's parents are both settler and indigenous. However, neither *Buried Child*, nor, as we will see in the discussion in the next chapter, *The Kentucky Cycle*, go beyond acknowledging that history of violence and exploitation, land-takings and body-takings. To reckon with history must also mean taking responsibility for it in ways that shift the narrative in the direction of social and ecological justice.[31] In 1992, Chicana feminist Cherríe Moraga's *Heroes and Saints* (published in 1994) also seemed to speak back to Shepard's buried child; however, as I discuss in the next chapter, Moraga's children refuse burial. Instead, their bodies are the material evidence of environmental injustice writ large upon the landscapes of Central California.

In the 1960s and 1970s, middle-class white Americans remained significantly shielded from the lived consequences of environmental degradation. It was this segment of American society that largely drove the second-wave environmental movement, an extension of the earlier conservationist movement with a more pressing and pervasive call to protect species and "wild" spaces, to conserve resources and combat pollution. Informed by global fears of nuclear war and global visions of the earth as a fragile planet, the theater that arose from the second-wave environmental movement telescoped out to the environmental health of the United States and the world as a whole, as the 1972 production of *Mother Earth* illustrated. Meanwhile, dramatists of color located the struggle for social justice within the community and a lived relationship with place. Foregrounding the ways that environmental justice advocates would later press for a redefinition of environment, these dramatists explore places as sites of intimate connection – whether homeplace, workplace, playground, neighborhood, or sacred site – in which people experience the interconnection of human and environment as the daily circumstances of their struggle for civil and human rights. These ecodramas portray the experiences of people on the front lines of environmental degradation; they expose the exploitative relationship of the United States to American

192 (Re)claiming home and homelands

land and give increased attention to indigenous perspectives. For these dramatists, the body is a central and material signifier of meaning. Communities and dramatists of color were also cognizant of global environmental issues. Plays like ETC's *Vietnam Campesino* indicate that farmworkers in the southwestern United States were painfully aware that they shared common cause with farmers and their families whose lands and lives were being eradicated by U.S. defoliants and chemical weapons (Valdez 1990). In the opening scene of Hansberry's ([1958] 1994) *A Raisin in the Sun*, Walter Lee, Jr. reads the morning paper and comments to Ruth, "they dropped another bomb" (26). Yet while the nation rehearses ending life on earth, the Youngers struggle to make and to claim a home that supports life. Later in the play, in a prescient connection between warm weather and atomic test, Beneatha quips, "Warm isn't it? . . . Everybody say it's got to do with them bombs and things they keep setting off" (82). In the decades to follow, dramatists would take up the myriad ways in which global environmental issues affect diverse local communities and the places they called home.

Notes

1 The phrase "the blue planet" is credited to astronomer Carl Sagan, known for his 1980 television program *Cosmos*. The 2001 mini-series *The Blue Planet* brought increasing popular attention to the uniqueness of Earth in the solar system. See, for example, Sagan's "Cosmos" (1980) and *Pale Blue Dot* (1994).
2 Second-wave environmentalism refers to the environmental movement of the late twentieth century, which was characterized by an increasing awareness of the magnitude of human impact on earth's ecological systems, as well as of the interdependency of nature and culture. See, for example, Gottlieb (2005). Comparatively, first-wave environmentalism refers to the early twentieth-century wilderness movement, which is associated with John Muir and Gifford Pinchot, among others, and which precipitated the conservation movement of early and mid-century. For more on this topic, see David Orr's (1977) "Environmentalism." See also Harry Justin Elam (1997), and Geiogamah Hanay and Jaye T. Darby (2000).
3 *Mother Earth*, by Thronson and Toni Shearer, premiered at South Coast Repertory Theatre in Costa Mesa, California, in January 1971. For more on the South Coast Repertory, see Vanderknyff and Herman (1988).
4 For more on environmental health and social justice, see Daniel Lewisoct's "Scientist, Candidate and Planet Earth's Lifeguard" (2012).
5 The original publication of Hansberry's *Raisin* (1958) was by New York: Random House. The play opened at the Ethyl Barrymore Theatre on March 11, 1959, received the New York Drama Critics Award that year, and has been frequently produced since then, including three Broadway revivals (1973, 2004, 2014). In the "Introduction" to *Raisin*, Robert Nemiroff (1994) writes that the 1989 film version (with Danny Glover playing Walter Lee, Jr.), which restored scenes considered too controversial in 1959, is a "luminous embodiment of the state play . . . not altered for the camera" (12).
6 For a discussion of the critical reception of *A Raisin in the Sun*, and the ways in which white critics at the time used tropes of universality to mitigate the play's activism, see Robin Bernstein's "Inventing a Fishbowl" (1999). See also Nemiroff's (1994) "Introduction" to *Raisin*, in which he discusses scenes of the play that were not included in the 1959 Broadway production and that were restored in later productions and publications.
7 For more on this topic, see "The Principles of Environmental Justice" as articulated by Giovanna Di Chiro in "Nature as Community" (1996).

8 Located on the South Side of the greater Chicago area, Morgan Park became a burgeoning African American community beginning in the early twentieth century. As a result of the construction of Interstate 57, the neighborhood became increasingly isolated and impoverished between 1969 and 1974. For more on this history, see Ta-Nehisi Coates's (2017) "The Case for Reparations" and (2017) *We Were Eight Years in Power*, especially 163–208. According to the online *Encyclopedia of Chicago History*, greater Morgan Park demographics in 2013 were recorded as approximately 57 percent African American and 35 percent white.
9 The reduction of infant mortality rates in the United States between the 1950s and the early twenty-first century has not benefited blacks and whites equally. See, for example, Gopal Singh and Stella Yu (1995) and Kenneth Y. Chay and Michael Greenstone (2000). In the present, infant mortality rates for blacks in the United States continue to be recorded as 50 percent higher than for whites (see, e.g. Jessica Firger [2017]). Attesting to racism as a kind of toxicity, John Yang similarly reported in a December 27 *PBS NewsHour* program (2018) that "it's not just physical conditions that influence their [mothers'] health. In Cincinnati, and nationwide, high infant mortality is driven by the fact that black infants are more than twice as likely to die as white babies, regardless of the mother's income and education."
10 For more on the way in which the environmental justice movement transformed the messaging and rhetoric of mainstream environmental organizations, as well as activist priorities, in the United States, see Gottlieb (2005) Chapters 6 and 7, and particularly 240–250.
11 For an in-depth discussion of the impact of settler colonialism on black bodies, see Coates (2015), Part II, in which Coates reflects on the implications of the extractive system of commodification of black bodies as labor for contemporary society (73–132).
12 This trend was epitomized in magazines like *Sunset*, the western "lifestyle" magazine that was first published in 1898 as Southern Pacific Railroad promotion.
13 As one response to the condition of urban poverty, Lena's desire for a garden perhaps foreshadows the food security movement of the 1990s, when urban community gardens in vacant lots and schoolyards became sites of resistance in a larger struggle. Gardens enabled poor communities to take collective food security and community health into their own hands. For a discussion of food security, which became an important arm of environmental justice, see Erika Mundel's (2011) "Theorizing Community Food Security as a Multi-institutional Social Movement."
14 For further discussion and analysis of Hansberry political activism and its reflection in *Raisin*, see John Liver Killens's (2014) "Lorraine Hansberry: On Time!," especially 103–109.
15 In his introduction to *Silent Spring*, Gore (1962) further explains that 70 million tons of the herbicide atrazine, classified as a human carcinogen, are used on the cornfields of Mississippi each year; as a direct result, 1.5 million pounds of runoff from the cornfields flow into public drinking water. Even more troubling, atrazine is not removed by water treatment plants (xxiii).
16 Environmental justice activism and scholarship began in the mid-1980s with the work of Robert Bullard, who documented patterns of environmental impact in studies such as *Dumping in Dixie* (1994). For more on the origins of the environmental justice movement, see Luke W. Cole and Sheila R. Foster's (2001) *From the Ground Up*, especially the introduction and Chapter 1, as well as Gottlieb's (2005) *Forcing the Spring*, especially Chapters 5 and 7. Subsequent studies have stressed the significant risk factors of gender. See, for example, Rachel Stein (2004), *New Perspectives on Environmental Justice: Gender, Sexuality, and Activism*.
17 Pulido, like Devon Peña (2005), uses the term *Mexican-origin*, rather than Mexican American or Mexican, in order to include those who may have been U.S. citizens, or documented, along with undocumented workers (59).
18 *El Movimiento* refers to the larger Chicano/a movement in the U.S. Southwest arising in the 1960s and 1970s and continuing today in the twenty-first century across the United

194 (Re)claiming home and homelands

States. This sociopolitical movement has included political, social, and cultural activism; cultural and literary production; and community and global consciousness-raising. See, for example, Laura Pulido, *Environmentalism and Economic Justice: Two Chicano Struggles in the Southwest* (1996).
19 Some scholars locate historic Aztlán in northern Mexico, while others place it in what is now considered the U.S. state of New Mexico. Yet the precise geographic location matters less than the implications of the legendary homeland as a means, and as a shared history, around which to galvanize an identity and a movement. See, for example, Virginia M. Fields and Victor Zamudio-Taylor's *The Road to Aztlán: Art from a Mythic Homeland* (2001).
20 For more on this history of Mexico and the United States, see the 2017 *PBS* series "Latino Americans," Parts I and II.
21 Alurista's *Plan espiritual de Aztlán* was later published in *Aztlán: Essays on the Chicano Homeland* (Anaya and Lomelí 1991).
22 *Teatros* refers to theater troupes and groups, rather than to plays. See Huerta (1973, 1982, 2002) for more on this lexicon within a Chicana/o theater context.
23 In particular, Broyles-González (1994) maps women's contributions to and struggles within ETC.
24 In "The Theatrical Politics of Chicana/Chicano Identity" Elizabeth Jacob (2007) describes ETC of the 1960s as "resistance theater," a focus which later shifted when Luis Valdez took an alternative creative direction with his staging and publication of *Zoot Suit* in the 1970s. For more on this topic, see Ontiveros (2014), particularly Chapter 3, in which he argues that ETC empowered farmworkers by repositioning the idea of citizenship not in birthplace but "in the ideal of human dignity" (147).
25 The term *mitos*, in this context, was coined by Valdez in 1967 to describe his anti-Vietnam War play *Dark Root of a Scream*. See Jorge Huerta's introduction to Valdez's *Zoot Suit and Other Plays* (1992, 8).
26 Among Luis Valdez's solo plays perhaps the most well known is *Zoot Suit*, which premiered at the Mark Taper Forum in 1979 and moved to Broadway in 1979, establishing Chicano/a theater firmly in the mainstream (Huerta 1992).
27 Peña (1998) conceptualizes Aztlán as an ecological framework in his discussion of the ecologies and civic organization of rancheros in the Mexican West. His analysis fuses land and labor into a fabric of reciprocity that sustains culture, community, and environment.
28 For a poetic articulation and analysis of this cosmic shift in Chicano consciousness, see Valdez's (1990) *Pensamiento serpentino*.
29 Augustín Palacios (2017) observes that the appropriation – toward the end of a generalized Hispanic multiculturalism by scholars in a U.S. context – of both "mestizaje in general, and Vasconcelos's raza cósmica in particular, has proven to be an extremely malleable ideology that has been appropriated in different historical periods and places to meet diverse ideological and political needs" (418). Whether Valdez is guilty of such appropriation, in part by linking the early concept to the Chicano/a identity, is an important question worth considering.
30 One such dramatist is Cherríe Moraga, whose *Heroes and Saints* I discuss in Chapter 6; others include María Irene Fornes, Milcha Sánchez Scott, and Josefina López, to name only a few. For more on Chicana/os who have undertaken environmental writing, see Priscilla Solis Ybarra's "Walden Pond in Aztlán" (2006, 19).
31 In 1992, Jose Rivera's *Marisol* (published in 1997) also called on its audience to attend to the deaths of infants born to the crack-addicted poor of New York City. Chris Westgate has argued eloquently and in detail that *Marisol* constituted a clear and direct challenge to the social policies of the Koch mayoral administration in "Toward a Rhetoric of Social Spatial Theatre" (2007). Produced during a time when Mayor Edwin Koch enacted policies to rid the streets of homeless people, Rivera's Lenny reminds us: "People are buried here. It looks like a sidewalk. But it's not. It's a tomb . . . these are babies born on the street" (61).

References

Alarcón, Francisco X. 2011. "Reclaiming Ourselves, Reclaiming America." In *The Colors of Nature*, edited by Alison H. Denning and Lauret E. Savoy, 24–48. Minneapolis, MN: Milkweed Editions.

Alurista. 1969. "El Plan Espiritual de Aztlán." *M.E.C.H.A.: Movimiento estudiantil chicanx de Aztlán*, March.

Anaya, Rudolfo A., and Francisco A. Lomelí, eds. 1991. *Aztlán: Essays on the Chicano Homeland*. Albuquerque, NM: University of New Mexico Press.

Anzaldúa, Gloria. 1987. *Borderlands/La frontera: The New Mestiza*. San Francisco, CA: Spinsters/Aunt Lute.

Arrizón, Alicia. 2000. "Mythical Performativity: Relocating Aztlán in Chicana Feminist Cultural Productions." *Theatre Journal* 52 (1): 23–49. https://doi.org/10.1353/tj.2000.0001.

Bacon, J.M. 2018. "Producing, Maintaining and Resisting Colonial Ecological Violence: Three Considerations of Settler Colonialism as Eco-social Structure." PhD diss., University of Oregon.

Bernstein, Robert. 1999. "Inventing a Fishbowl: White Supremacy and the Critical Reception of Lorraine Hansberry's *A Raisin in the Sun*." *Modern Drama* 42 (1): 16–27. https://doi.org/10.1353/mdr.1999.0011

Blum, Elizabeth D. 2008. *Love Canal Revisited: Race, Class, and Gender in Environmental Activism*. Lawrence, KS: University Press of Kansas.

Broyles-González, Yolanda. 1994. *El Teatro Campesino: Theater in the Chicano Movement*. Austin, TX: University of Texas Press.

Buell, Lawrence. 2005. *The Future of Environmental Criticism: Environmental Crisis and Literary Imagination*. Malden, MA: Blackwell Publisher.

Bullard, Robert D. 1993. *Confronting Environmental Racism: Voices from the Grassroots*. Boston, MA: South End Press.

Bullard, Robert D. 1994. *Dumping in Dixie: Race, Class, and Environmental Quality*. Boulder, CO: Westview Press.

Bullard, Robert D. 2005. *The Quest for Environmental Justice: Human Rights and the Politics of Pollution*. San Francisco, CA: Sierra Club Books.

Carson, Rachel. 1962. *Silent Spring*. Boston, MA: Houghton Mifflin; Cambridge, MA: Riverside Press.

Chaudhuri, Una. 1994. "'There Must Be A Lot of Fish in that Lake': Toward and Ecological Theater." *Theater* 25 (1): 23–31. https://doi-org.libproxy.uoregon.edu/10.1215/01610775-25-1-23

Chay, Kenneth Y., and Michael Greenstone. 2000. "The Convergence in Black-White Infant Mortality Rates during the 1960s." *The American Economic Review* 90 (2): 326–332. www-jstor-org.libproxy.uoregon.edu/stable/117245

Chiro, Giovanna Di. 1996. "Nature as Community: The Convergence of Environment and Social Justice." In *Uncommon Ground: Rethinking the Human Place in Nature*, edited by William Cronon, 298–320. New York: W.W. Norton & Company.

Coates, Ta-Nehisi. 2015. *Between the World and Me*. New York Spiegel & Grau.

Coates, Ta-Nehisi. 2017. "The Case for Reparations." In *We Were Eight Years in Power: An American Tragedy*, 151–208. New York: One World Publishing Company.

Cole, Luke W., and Sheila R. Foster. 2001. *From the Ground Up: Environmental Racism and the Rise of the Environmental Justice Movement*. New York and London: New York University Press.

Commoner, Barry. 1972. *The Closing Circle: Nature, Man, and Technology.* New York: Bantam Books.
Davis, Ronald G. 1975. *The San Francisco Mime Troupe: The First Ten Years.* Palo Alto, CA: Ramparts Press.
Dawson, Susan E., and Gary E. Madsen. 1997. "American Indian Uranium Millworkers: A Study of Perceived Effects of Occupational Exposure." In *Health and Poverty*, edited by Michael J. Holosko and Marvin D. Feit, 223–237. New York: Haworth Press.
"DDT and Environmental Toxicology." 2014. In *Banned: A History of Pesticides and the Science of Toxicology*, 38–71. New Haven, CT and London: Yale University Press.
Deming, Alison Hawthorne, and Lauret E. Savoy. 2011. *The Colors of Nature: Culture, Identity, and the Natural World.* Revised ed. Minneapolis, MN: Milkweed Editions.
Dunbar-Ortiz, Roxanne. 2015. *An Indigenous Peoples' History of the United States.* Boston, MA: Beacon Press.
Elam, Harry Justin. 1997. *Taking It to the Streets: The Social Protest Theater of Luis Valdez and Amiri Baraka.* Ann Arbor, MI: University of Michigan Press.
Fields, Virginia M., and Victor Zamudio-Taylor, eds. 2001. *The Road to Aztlán: Art from a Mythic Homeland.* Los Angeles, CA: Los Angeles County Museum of Art.
Firger, Jessica. 2017. "Black and White Infant Mortality Rates Show Wide Racial Disparities Still Exist." *Newsweek*, July 3.
Gates, Jr., Henry Louis, and Cornell West. 2000. *The African-American Century: How Black Americans Have Shaped Our Country.* New York/London: Simon & Schuster.
Geiogamah, Hanay, and Jaye T. Darby, eds. 1999. *Stories of Our Way: An Anthology of American Indian Plays.* Los Angeles, CA: UCLA American Indian Studies Center.
Glotfelty, Cheryll. 1996. "Carson, Rachel (1907–1964)." 1: 151–171.
Gore, Jr., Albert. 1962. "Introduction." In *Silent Spring*, by Rachel Carson, xv–xxvi. Boston, MA: Houghton Mifflin; Cambridge, MA: Riverside Press.
Gottlieb, Robert. 2005. *Forcing the Spring: The Transformation of the American Environmental Movement.* Washington, DC: Island Press.
Grizwald, Eliza. 2012. "How Silent Spring Ignited the Environmental Movement." *New York Times*, September 21.
Guest, Avery M., Gunnar Almgren, and Jon M. Hussey. 1998. "The Ecology of Race and Socioeconomic Distress: Infant and Working-Age Mortality in Chicago." *Demography* 35 (1): 23–34. https://doi.org/10.2307/3004024
Gussow, Mel. 1975. "Theater: Wilson's 'Mound Builders'." *New York Times*, February 3.
Hanay, Geiogamah, and Jaye T. Darby, eds. 2000. *American Indian Theater in Performance: A Reader.* Los Angeles, CA: UCLA American Indian Studies Center.
Hansberry, Lorraine. (1958) 1994. *A Raisin in the Sun.* New York: Vintage Books.
Hartig, John H. 2010. *Burning Rivers: Revival of Four Urban-Industrial Rivers That Caught on Fire.* Burlington, Ontario, Canada: Aquatic Ecosystem Health and Management Society.
Hartman, Chester W., and Poverty & Race Research Action Council. 2006. *Poverty & Race in America: The Emerging Agendas.* Lanham, MD: Lexington Books.
Holosko, Michael J., and Marvin D. Feit, eds. 1997. *Health and Poverty.* New York: Haworth Press.
hooks, bell. 2002. "Earthbound on Solid Ground." In *The Colors of Nature: Culture, Identity and the Natural World*, edited by Alison H. Deming and Lauret E. Savoy, 67–71. Minneapolis, MN: Milkweed Editions.
hooks, bell. 2009. *Belonging: A Culture of Place.* New York: Routledge.

Huerta, Jorge A. 1973. "Concerning Teatro Chicano." *Latin American Theatre Review* 6 (2): 13–20.
Huerta, Jorge A. 1982. *Chicano Theater: Themes and Forms*. Bilingual Press/Editorial Bilingüe.
Huerta, Jorge A. 1992. "Introduction." In *Zoot Suit and Other Plays*, 7–20. Houston, TX: Arte Público Press.
Huerta, Jorge A. 2000. *Chicano Drama: Performance, Society & Myth*. New York: Cambridge University Press.
Huerta, Jorge A. 2002. "When Sleeping Giants Awaken: Chicano Theatre in the 1960s." *Theatre Survey* 43 (1): 23–35. https://doi.org/10.1017/S0040557402000030
"Industry Maps Defense to Pesticide Criticisms: Government Also Responds to Charges against Pesticides Contained in Rachel Carson's New Yorker Articles and Forthcoming Book." 1962. *Chemical & Engineering News* 40 (33): 23–25.
Jacob, Elizabeth. 2007. "The Theatrical Politics of Chicana/Chicano Identity: From Valdez to Moraga." *New Theatre Quarterly* 23 (1): 25–34. https://doi.org/10.1017/S0266464X0600061
Jaynes, Gerald. 2014. "On Neighborhoods and Neighborhood Effects." *Du Bois Review: Social Science Research on Race* 11 (2): 465–475. https://doi.org/10.1017/S1742058X14000253
Johansen, Bruce E. 2013. *Encyclopedia of the American Indian Movement*. Santa Barbara, CA: Greenwood.
Johnson, Troy. 1996. *The Occupation of Alcatraz Island*. Chicago, IL: University of Illinois Press.
Killens, John Oliver. (2014) "Lorraine Hansberry: On Time!" In *SOS—Calling All Black People: A Black Arts Movement Reader*, edited by John H. Bracey, Sonia Sanchez, and James Edward Smethurst, 103–109. Amherst, MA: University of Massachusetts Press.
LaDuke, Winona. 2009. "Uranium Mining, Native Resistance, and the Greener Path." *Orion*, January 22.
Lewisoct, Daniel. 2012. "Scientist, Candidate and Planet Earth's Lifeguard." *New York Times*, October 1.
Mason, Susan Vaneta. 2013. *The San Francisco Mime Troupe Reader*. Ann Arbor, MA: University of Michigan Press.
Mesa-Baines, Amalia. 2001. "Spiritual Geographies." In *The Road to Aztlán: Art from a Mythic Homeland*, edited by Virginia M. Fields and Victor Zamudio-Taylor, 132–141. Los Angeles, CA: Los Angeles County Museum of Art.
Mundel, Erika. 2011. "Theorizing Community Food Security as a Multi-Institutional Social Movement." *Appetite* 56 (2): 538–539. https://doi.org/10.1016/j.appet.2010.11.239
Nemiroff, Robert. 1994. "Introduction." In *A Raisin in the Sun*, by Lorraine Hansberry, 5–14. New York: Vintage Books.
Nixon, Rob. 2013. *Slow Violence and the Environmentalism of the Poor*. Boston, MA: Harvard University Press.
Ontiveros, Randy J. 2014. "Green Aztlán: Environmentalism and the Chicano Visual Arts." *In the Spirit of a New People: The Cultural Politics of the Chicano Movement*, 86–130. New York and London: New York University Press.
Orr, David. 1977. "Environmentalism: The Second Wave." *Public Administration Review* 37 (6): 730–735. https://doi.org/10.2307/975344
Palacios, Augustín. 2017. "Multicultural Vasconcelos: The Optimistic, and at Times Willful, Misreading of la raza cósmica." *Latino Studies* 15 (4): 416–438. https://doi.org/10.1057/s41276-017-0095-6

Paull, John. 2013. "The Rachel Carson Letters and the Making of *Silent Spring*." *SAGE Open* 3 (3): 1–13. https://doi.org/10.1177/2158244013494861
Peña, Devon Gerardo. 1998. *Chicano Culture, Ecology, Politics: Subversive Kin. Society, Environment, and Place*. Tucson, AZ: University of Arizona Press.
Peña, Devon Gerardo. 2005. *Mexican Americans and the Environment: Tierra Y Vida. Mexican American Experience*. Tucson, AZ: University of Arizona Press.
Peña, Devon Gerardo. 2010. "Environmental Justice and the Future of Chicana/o Studies." *Aztlán: A Journal of Chicano Studies* 35 (2): 149–157.
Pulido, Laura. 1996. *Environmentalism and Economic Justice*. Tucson, AZ: University of Arizona Press.
Ramírez, Elizabeth C. 2000. *Chicana/Latina in American Theatre: A History of Performance*. Bloomington, IN: Indiana University Press.
Rich, Frank. 1982. "Play: 'Angels Fall,' Lanford Wilson's Apocalypse." *New York Times*, October 18.
Rivera, José. 1997. "*Marisol*." In *Marisol and Other Plays*, 1–68. New York: Theatre Communications Group.
Rothman, Joshua. 2012. "Rachel Carson's Natural Histories." *New Yorker*, September 27.
Sagan, Carl. 1980. *Cosmos*. New York: Random House.
Sagan, Carl. 1994. *Pale Blue Dot: A Vision of the Human Future in Space*. New York: Random House.
Sampson, Robert J. 2012. *The Great American City: Chicago and the Enduring Neighborhood Effect*. Chicago, IL and London: University of Chicago Press.
Schenkkan, Robert. 1994. *The Kentucky Cycle*. New York: Dramatists Play Service.
Shepard, Sam. 1979. *Buried Child*. New York: Dramatist Play Service.
Shern, David L., Andrea K. Blanch, and Sarah M. Steverman. 2016. "Toxic Stress, Behavioral Health, and the Next Major Era in Public Health." *American Journal of Orthopsychiatry* 86 (2): 109–123.
Singh, Gopal, and Stella Yu. 1995. "Infant Mortality in the United States: Trends, Differentials, and Projections, 1950 through 2010." *American Journal of Public Health* 85 (7): 957–964.
Standing, Sarah. 2008. "Human/Nature: Eco-Theatre Politics and Performance." PhD diss., New York University.
Stanford University Libraries. 1998. *Sunset Magazine: A Century of Western Living, 1898–1998: Historical Portraits and a Chronological Bibliography of Selected Topics*. Stanford, CA: Stanford University Libraries.
Stein, Rachel. 2004. *New Perspectives on Environmental Justice: Gender, Sexuality, and Activism*. New Brunswick, NJ: Rutgers University Press.
Steinman, Erich. 2012. "Settler Colonial Power and the American Indian Sovereignty Movement: Forms of Domination, Strategies of Transformation 1." *American Journal of Sociology* 117 (4): 1073–1130. https://doi.org/10.1086/662708
Valdez, Luis. 1990. *Early Works: Actos, Bernabé, Pensamiento serpentino*. Houston, TX: Arte Público Press.
Vanderknyff, Rick, and Jan Herman. 1988. "South Coast Repertory: A Chronology." *Los Angeles Times*, September 4.
Vasconcelos, José, and Didier Tisdel Jaén. 1997. *The Cosmic Race: A Bilingual Edition*. Baltimore, MD: Johns Hopkins University Press.
Westgate, Chris J. 2007. "Toward a Rhetoric of Sociospatial Theatre: José Rivera's 'Marisol'." *Theatre Journal* 59 (1): 21–37. https://doi.org/10.1353/tj.2007.0076
Wilson, Lanford. 1976. *The Mound Builders: A Play*. New York: Hill and Wang.

Wilson, Lanford. 1983. *Angels Fall*. New York: Hill and Wang.
Worster, Donald. 1994. "Healing the Planet." In *Nature's Economy: A History of Ecological Ideas*, 342–387. Cambridge: Cambridge University Press.
Worthen, William. 1997. "Staging America: The Subject of History in Chicano/a Theatre." *Theatre Journal* 49 (2): 101–120. https://doi.org/10.1353/tj.1997.0058
Ybarra, Priscilla Solis. 2006. "Walden Pond in Aztlán? A Literary History of Chicana/o Environmental Writing Since 1848." PhD diss., Rice University.

6
STORIES IN THE LAND/ LEGACIES IN THE BODY

Robert Schenkkan's *The Kentucky Cycle*, Cherríe Moraga's *Heroes and Saints*, and Anne Galjour's *Alligator Tales*

In the late 1980s, scientists helped to shift the tide of public opinion and the policies of nation-states to prevent a potentially catastrophic change in Earth's atmosphere. Throughout the decade, scientists had suspected that the earth's ozone – a protective layer of the outer atmosphere – was thinning. In 1987, chemical analysis and direct observation by high altitude aircraft confirmed that ozone depletion was being caused by a particular class of gases called chlorofluorocarbons, or CFCs, that break apart the ozone molecule when they reach the stratosphere (see, e.g., Gribbin [1988]).[1] Ubiquitous to modern life, CFCs were commonly used in air-conditioning, refrigeration, polystyrene products (such as to-go cups), and propellants of all kinds, including hairspray and asthma inhalers. The "ozone hole," as it was soon called, was the most global, stunning, and imminently dangerous threat to the planet, second only to thermonuclear war. Scientists around the world warned their governments to take action.[2] If the ozone continued to deplete at the current rate, the science indicated, life as we knew it on earth would be irrevocably harmed, perhaps extinguished. In 1987, the United States joined Canada and 12 European nations in negotiating the Montreal Protocol on Substances that Deplete the Ozone Layer (Stevens 1995).[3] Despite the protests of multinational corporations like DuPont, one of the largest producers of CFCs, participating nations agreed to phase out CFCs (Weisskopf 1992).

During the same time period, National Aeronautics and Space Administration (NASA) scientists and their international counterparts reported a new phenomenon: the earth's mean temperature was increasing at a rate that exceeded normal rates of geologic change. Now understood as "the greenhouse effect," these scientists had begun to document the earth's rising temperature since the industrial era, noting that warming had accelerated markedly beginning in the 1950s. Once more, believing that their research was crucial to the national security of nations around the world, scientists reported the growing body of conclusive

DOI: 10.4324/9781003028888-7

evidence to their governments. In 1988, U.S. climatologist James Hansen, of the NASA Goddard Institute for Space Studies, testified before the Senate about the dangers of anthropogenic (human-caused) climate change (Hansen 2009, iv). In 1988, the United Nations formed the Intergovernmental Panel on Climate Change (IPCC) to bring together the most conclusive and scientifically-vetted information about global warming to assist U.N. member nations in developing policies and protections. Meanwhile, a January 2, 1989 issue of *Time* magazine pictured the planet in chains with the descriptor "Planet of the Year: Endangered Earth." The special "planet of the year" section translated decades of scientific study into prose accessible to the general public. Dangers described ranged from the dangers of CFCs and carbon emissions to the atmosphere to the acidification and pollution of the oceans, to the loss of biodiversity around the world. *Time* functioned as a mini-IPCC, gathering the best science and summarizing it for the layperson. This 1989 issue also included a closing essay titled "Planet of the Year: What the U.S. Should Do." (Note the absence of a question mark.)

The 1980s also saw the rise of the global AIDS epidemic. In the early years, medical research was slow and not adequately supported by the U.S. government. Treatment options were few, and communities that were most severely impacted were often shunned, blamed, or ignored. During these years, theater-makers across the United States participated in the call to action for increased research funding, public education, compassionate health care treatments based on sound science (not implicit bias and ignorance), and community support (see, e.g., Fee and Fox [1992]). As a result, dozens of new plays gave voice to the lived experience of AIDs in the 1980s. By decade's end, Tony Kushner's two-part epic *Angels in America* (1993–1994) had radicalized and galvanized a generation of theater-makers, changing the landscape of American theater forever (see, e.g. Worthen [1995, 2011]).

By the early 1990s, many theater-makers began to wonder how a form of cultural production that had been so much a part of the social justice movements of the 1960s and 1970s, one which had rallied with such verve and creative innovation during the AIDs struggle, had failed nonetheless to respond in any significant measure to the many and increasingly alarming environmental crises. Throughout this decade, then, artists and theorists at last began to articulate what Una Chaudhuri (1994) has called an "ecological theater," suggesting that theater might inspire not only environmental awareness but also cultural transformation (23).

As I argue in Chapter 5, theater linked to communities or connected to ongoing social justice movements continued to engage powerfully with environmental concerns, revealing the interlocking systems of racism, sexism, and environmental imperialism. This "grassroots theater," as Downing Cless (1996) affirms, was infused with environmental themes because it reflected the voices of people and communities experiencing loss, injury, and struggle as the result of environmental destruction. Dramatists continued to amplify the call for environmental justice by representing the concerns and challenges still facing the migrant labor force – both immigrant and U.S.-born – in the industrial agriculture of California and the Southwest and in the cities and neighborhoods. Native theater artists continued to call attention to the interweaving of land and identity, exploring the ways in which loss of land was coupled with loss of culture, language, and often health. They used theater to reclaim the places and assert the diverse voices and sovereignty of the indigenous people of this continent.

Ecofeminist theater and performance art of the 1980s captured, critiqued, and ritualized the ways in which the desecration of the earth and women's bodies were aspects of the same patriarchal social values. Rachel Rosenthal (2001), for example, imagined the outrage of the earth itself in works such as *Gaia Mon Amour* and *L.O.W. Gaia*. Rosenthal was among the first solo female performance artists to call specifically for a synthesis of feminism and rising ecological consciousness.

Many artists (myself included) simply left the confines of theatrical spaces (see, e.g., May [1999a]). The work of María Irene Fornes, Peter Schuman, and Richard Schechner defined a new "environmental theatre," one which made use of spaces and places beyond traditional theater venues.[4] Outside the confines of the black box, creating site-specific works seemed an avenue to engage environmental issues implicitly, as the place itself becomes a creative collaborator. In *Environmental Theatre* (1994), Schechner articulates this concept of environmental theater design as

> [t]he fullness of space, the endless ways space can be transformed, articulated, animated.... It is also the source of environmental performer training. If the audience is one medium in which the performance takes place, the living space is another. The living space includes all the space in the theatre, not just what is called the stage.
>
> *(1–2)*

In *Staging Place*, Chaudhuri (1995) maintains that "[t]he premise of environmental theater is, then, an assault on the traditional divisions, taxonomies, and codings of theater space, and a commitment to overriding, erasing, or even destroying them" (23). What became known as environmental theater cannot be conflated with ecotheater (theater focused on environmental themes). Nevertheless, the inquiries and experiments of environmental theater gave rise to questions about how theater makes meaning and, thus, how it might inspire the sense of place that is foundational to ecological values.

Operating outside of the theatrical mechanisms upon which artists depend to produce aesthetic meaning (e.g., lighting, scenography, seating arrangements), environmental and site-specific performance foster multifocal attention and the proliferation of meaning. Audience members encounter the felt environment around them in corporeal ways – one audience member might run their hand over a piece of furniture while another notices the warmth of light across a path. The bright boundaries between audience and performers dissolve, allowing for a shared experience more expansive and unpredictable than in a darkened theater.

Connection to others, to community, is also a focal part of the experience: audience members bear witness to one another's experiences as well as the performer's. Choreographer Anna Halprin's outdoor participatory dance events, to cite only one example, made use of these unique qualities of sited performance to engender a sense of connection to land and community, among other ceremonial events, in her *Planetary Dance* (see, e.g., Worth and Poynor [2004]; and Merriman [2010]). Meanwhile, environmental organizations made increasing use of theatrical methods – from guerrilla theater to large outdoor spectacles – in order to attract media attention and communicate their messages (Standing 2008, 2012). Yet, by the last decade of the twentieth century, environmental awareness was only beginning to gain traction in regional theater and on Broadway. Though rarely the focus, dramatists did sometimes touch on environmental themes, often as part of the backstory of characters and conflicts.

There were exceptions, however, to the relative lack of environmental themes explored in the 1990s. Steve Tesich's *The Speed of Darkness* (1991), for example, follows two Vietnam veterans who were exposed to chemical agents during the war take out their postwar rage on U.S. soil by dumping chemical wastes in the hills and polluting the groundwater under a future housing development; years later, the men are haunted by the deed and by one another as they struggle with taking responsibility for past actions. In Robert Schenkkan's 1992 *The Kentucky Cycle* (published in 1993), the environmental and social impacts of frontier ideology are explored and exposed. In José Rivera's 1993 *Marisol* (published in 1994), humans find themselves trapped in an ecologically dystopian future caused by a revolution of angels. These plays personalized environmental issues by exploring the ethical questions and the decision-making processes that precede human action. Yet, with the exception of *The Kentucky Cycle*, reviewers did not comment on the ecological questions they raised. Why were critics, audiences, and scholars unable or unwilling to recognize environmental themes when they did take the stage?

Two important transformations in thinking, one within the discourse of the field, and one within the larger discourse of environmentalism, gave rise to an ecological turn in U.S. theater during the 1990s. Together, they changed the terms of debate both in theater and in the environmental movement. Theorists and artists began to recognize that as the very epitome of cultural artifice, theater was deeply allied with the ideologies of modernity, including industrialization and Western European humanism. In a special summer 1994 issue of *Theater* on "Ecological Theater," Erika Munk observed that "our playwrights' silence on the environment as a political issue and our critics' neglect of the ecological implications of theatrical form are rather astonishing" (5). Chaudhuri (1994) likewise theorized that theater is "programmatically anti-ecological" and allied with a Euro-American humanist paradigm, resulting in a long-standing tendency for artists, critics, and audiences to interpret *any* representation of ecological circumstances as the mere backstory to human action, or as metaphors for human struggles. This humanist habit of mind means that even when a play makes direct reference to ecological themes, ideas, or circumstances, audiences and critics miss them. Ironically, metaphors that originate in lived experience can become, in the theater, mechanisms that erase the living, palpable world around us. Thus, Chaudhuri argued for "a new materialist-ecological theater practice that refuses the universalization of metaphorization of nature"; she encouraged theater artists not only to take up ecological themes but also to use theatrical forms in ways that resist humanist interpretation (24). This bending of the material form of theatrical expression to return the audience's attention to the sensible world can be seen in the plays discussed in this chapter.

At the same time, a growing environmental justice movement challenged second-wave environmentalism, which was historically dominated by white leadership and privilege. Environmentalism, advocates argued, must include the people and communities that bear the brunt of environmental degradation. The discourse of environmental justice also represented a tear-down of the binary thinking at the heart of modernity (i.e., the separation "nature" and "culture"). Human beings, after all, are part of and not separate from the wider ecological fabric of the planet. The environmental justice movement transformed the environmental debate and redefined its moral agenda by reframing how we understand the environment.[5] As I argue in Chapter 5, dramatists who recounted and represented the intractable interdependence of communities, families, workers, and the environments with which they are intertwined, opened up new theatrical space. This shift in focus from the singular, universal, human story to stories of collective experience signaled a rising ecodramaturgy that told stories of the intertwining of human and nonhuman, peoples, and land.

Turning a corner toward ecological theater

The creative work, organizing, and thinking of artists and scholars in the 1990s form the ground from which ecodramaturgy and the current burgeoning climate change theater movement grew. As earlier chapters have illustrated, environmental themes in theater did not begin in the 1990s, but this period did mark a threshold of new thinking, informed by the environmental justice movement that would inspire theater-making, historiography, and criticism going forward. As mentioned earlier, Chaudhuri's (1994) eco-critical analysis illuminates how artists might own theater's complicity in our interdependence with, and our injuries to, the natural world. In addition, Munk's (1994) recognition that, from the perspective of ecological thought, there exists "vast open fields of histories to be rewritten and contexts to re reexplored" (5) not only inspired theater-making but also implied that environmental messages, themes, and implications are always and already present in any and all plays, past and present.

Many of the challenges that artists and companies face now, and the innovative approaches they employ, were informed by the "green theater" movement in the 1990s. In a 1992 *American Theater* cover story, Lynn Jacobson investigated the many examples of ecologically-minded productions, festivals, and conferences offered by regional and community-based companies. These outcroppings were examples of "'thinking globally, acting locally,' sifting through the toxins in their own backyards" by developing projects about the environmental/social concerns impacting their respective communities (17). For example, the Dell'Arte Company in Blue Lake, California, staged its commedia-style "Scar Face Trilogy",[6] which was focused around the politics of timber in the Pacific Northwest; Merrimack Repertory in Lowell, Massachusetts, staged an adaptation of the granddaddy of ecodrama, Henrik Ibsen's *Enemy of the People*, calling attention to the pollution of the Merrimack River by local industry; and in New Orleans, where, even before the hurricanes of the twenty-first century, the water table rose so high that "bodies can't be buried underground; anything you put down into the earth floats right back up," a community collaboration by the Contemporary Arts Center called out the disproportional impact of chemical emissions in Louisiana's "cancer alley" on low-income and African American communities. Although two of these three examples used preexisting plays written by European playwrights, Jacobson also identified an emergent "eco-canon" that expanded out from Ibsen to include plays and performances that rose out of local community concerns (1992, 20).

Meanwhile, new works began to spring up across the country, from the high-tech *Nature Is Leaving Us* at the Goodman Theatre in Chicago to "household appliances" used by the Center of Puppetry Arts in Atlanta for a piece titled *My Civilization* about nuclear power and overpopulation (Jacobson 1992). Other theaters developed educational outreach programs, creating ecotheater for and with young audiences.[7] Underground Railway Theater of Boston created and produced two eco-cabarets (see, e.g., Sanders and Wise 2017), as well as several other environmentally-themed productions. In 1991, Life on the Water Theatre in San Francisco organized an eco-writing workshop series called Earth Drama Lab, and an eco-rap festival. Meanwhile, New York's Theatre for the New City produced a series of ecotheater festivals.[8]

Each of the plays from the 1990s discussed in the pages that follow represents trends in an emergent ecodrama and varied responses to some of the questions that were on artists' minds across the country. If the habitual reading of theatrical expression as programmatically human-centered was not problem enough, dramatists also struggled with temporal scope: How does one tell the story of the land when environmental problems occur over extended historic periods? How are human bodies, human lives, and histories intertwined with the land? How can stories personalize ecological issues and yet inspire collective responsibility? How can theater represent the voices and experiences of communities that have been and continue to be most at risk, including women, children, workers, and communities of color? And how might theater do any of the above while it can only do so through artifice, which makes use of both human bodies and natural resources?[9] The plays and productions discussed in this chapter approach these challenges differently, but all serve as models for ecological theater in the twenty-first century.

In 1991, the Intiman Theatre of Seattle produced Robert Schenkkan's *The Kentucky Cycle* (1993), an epic revisionist history that delivers a searing critique of the social and environmental impacts of settler colonialism and extractive capitalism and wrestles with the values and images of the nation's deeply ingrained frontier narrative. The play deals with many of the kinds of environmental justice issues that activists in communities across the nation were raising at the time, including the theft of indigenous land by early settlers, the irreparable harm to land and ecosystems, and the human health impact of extractive industries. *The Kentucky Cycle* went on to receive critical acclaim at other west coast regional theaters, was awarded the Pulitzer Prize for Drama for 1992 prior to opening on Broadway, and has much to reveal about how dramatists might cope with the challenges of representing environmental issues across generations. However, the play also demonstrates the artistic risk of perpetuating the very narratives and ideologies one hopes to critique.

Environmental justice thinking and activism not only inspired a plethora of regional and community-based works, signaling a new attentiveness among theater artists to environmental issues, but also a willingness on the part of critics and scholars to recognize and theorize ecological meanings in ways that had been invisible before. Furthermore, these new perspectives pointed toward a growing recognition that theater has always and already been part of our collective conversation and imaginings about the natural world. Taken as a critical lens, environmental justice discourse reveals that plays thought to be about social justice also inherently deal with environmental justice. As I demonstrate in Chapter 5, plays such as *A Raisin in the Sun* (1995) are richly threaded with themes that speak to ecologies of neighborhood and home, and calling out the burden of human habitat degradation on women, children, workers, and communities of color. During the 1980s and 1990s, women of color, whose activism may have begun as part of the liberation movements of their communities, turned their energies to naming and calling out environmental injustice where it occurred. They demanded accountability not only from polluters but also from the white-dominated environmental movement. And in more recent decades, women have led transnational environmentalists in the fight against corporate globalization, the subjugation of the poor, and the plundering of the resources of the planet by profiteering forces.

Cherríe Moraga's *Heroes and Saints* (1994) stands out among the plays from the 1990s that would become central to an eco-canon going forward. Moraga's story centers on a fictionalized community in California's San Joaquin Valley; it was inspired by real-world counterparts in the town of McFarland, California, where birth defects linked to pesticide exposures continued to increase throughout the 1980s. Moraga's play not only theatricalizes the body as permeable and one in the flesh with the land but also pays tribute to courage and intractable activism in the face of incomparable loss. Theatricalizing the reciprocity between bodies and the land, Moraga roots the activism of her characters in their homeplaces and bodies, where the slow violence of environmental and economic policy becomes visible and lived.

The emergence of environmental justice discourse as a prominent aspect of environmental thought and activism inspired many artists to explore the ways in which person and place are permeable. The final play discussed in this chapter, *Alligator Tales* by Anne Galjour (1997), utilizes the simplest of theatricalities – the body of the actor – as a means to explore the ways in which our bodies and our homelands are intertwined and interdependent, sharing both aliveness and vulnerability. These examples from the rising tide of ecodrama in the 1990s provide some of the key ongoing questions for ecodramaturgy as theater-making, shedding particular light on how theater might engage the long trajectories of environmental history, give voice to communities affected by environmental risk, and how the artifice of theater, including the human body, might be key to how ecological meaning is made in the theater.

Stories in the land – *The Kentucky Cycle*

In *The Kentucky Cycle*, Robert Schenkkan expands on this review of settler colonial history to examine how that history has impacted land and people. Schenkkan's nine-hour epic drama traces the history of the Cumberland Plateau through seven generations of three Kentucky families from 1775 to 1975, mapping the relationship between people and the land over time: land from which they have drawn life and profit, on which they have spilled one another's blood, and in which they have buried both enemy and kin. *The Kentucky Cycle* reminded (and still reminds) its audiences that the sum of collective memory is embedded in the land and the places that people have called home. Ecological identity is central to the psychological, social, economic, and political aspects of human affairs and history. Composed of nine one-act plays – each telling the story of key moments of Cumberland history – *The Kentucky Cycle* begins with an Irish settler named Michael Rowen, who coerces land from the Cherokee, takes a Cherokee wife by force, and then fathers, murders, and buries their child in the land he stole. Modeled on the classic Greek tragedy of the House of Atreus, Rowen's actions begin a cycle of revenge that continues through multiple generations. In Part One, acts of genocide against indigenous people, land theft, slavery, and violence against women are interlocked in this play as part of the mechanisms of greed that destroy the land. As Rowen's wealth grows, he becomes a slaveholding landowner, fathering Jesse Biggs with one of his slaves, forming the third family whose history is central to the play. As the cycles of violence and revenge continue, the Rowen land is stolen again, this time by the Talbert family, the local arm of a New York mining company. In Part Two, later generations of Rowens and Biggs labor in the mines, their land destroyed and their health imperiled by black lung disease and typhoid. The struggle of miners to unionize becomes the center of "Fire in the Hole." In the final play, "The War on Poverty," Joshua Rowen, the great-great-great grandson of the original Michael Rowen, visits the homestead where, together with the descendants of the Talberts and the Biggs, he surveys the land that has born the history of generations of abuse.[10]

As a whole, Schenkkan's nine one-act plays portray how white settler colonists' choices have shaped and reshaped the land. The ethos of frontierism unfolding over time has resulted in environmental destruction, community instability, and human poverty. As economic forces carve up the land, ecologies are destroyed, human communities are fractured; the land is rendered unrecognizable. The play is an indictment of the way in which settler colonial violence has resulted in environmental devastation and human suffering, and how its doctrine of discovery and rationales for extraction carry that suffering forward generation by generation. Performed over the course of a full day or two consecutive evenings of theater, *The Kentucky Cycle* offered a relentless confrontation with human choices that have led contemporary society to the brink of environmental devastation. The sheer length of the play throws a spotlight on history's repeating patterns. Reminiscent of Vince's vision from Sam Shepard's *Buried Child* (1979), in which bones and faces carry signatures of recognition from the past into the present, an ensemble of actors play multiple roles through multiple generations. Entering the stage from four directions under work lights, they introduce the play, set the stage, and dress those among them who will play the first scene. Actors not playing characters in the first scene sit "in full view of the audience" on benches along the sides of the playing area (4). This staging arrangement served to remind the audience throughout the performance that the characters are not the actors but rather that the actors lend their presence to a story that implicates all of us. The same human bodies (of the actors) cycling through the narrative in different roles embeds the play with a haunting sense of patterns repeating through time, making the point that the consequences of our actions surface *in the body* of both earth and children. Actors sat silent at the edge of the stage, bearing witness along with the audience to 200 years of U.S. history as it unfolded on stage. Their presence was a palpable reminder that we all carry the stories of our ancestors in our cells, bones, and sensibilities. Sometimes, after a scene was played and those actors moved to their seats at the edge of the stage, their breath or tears continued audibly even as the next scene began. In this way, history – and its ethical implications – was given flesh.

At the time of its production, Schenkkan's play challenged the audience's willingness to examine what he calls a "mythos" that prevents contemporary environmental disasters from being linked to the longstanding ideologies and economic systems that have come down from the Enlightenment (Schenkkan 1991). In the opening scene, Michael Rowen steals a gold watch from the corpse of Earl Tod, a Scottish trapper, whom he has killed. The watch becomes an heirloom and moves through the play from father to son a symbol of the Enlightenment's legacies, an ideology inherited from the European continent that coupled itself to colonialism, capitalism, and the domination of nature symbolized by a clockwork universe. In "The Concept of the Enlightenment," philosophers Max Horkheimer and Theodor Adorno (1996) write that "the program of the Enlightenment was the disenchantment of the world . . . What men want to learn from nature is how to use it in order wholly to dominate it and other men. That is the only aim" (199). Frontier ideology, sired in the Enlightenment, gave colonials permission to take from the land whatever, whenever, and wherever their economic ambition required, making nature the servant of mankind. The historic reach of *The Kentucky Cycle* is an effort in the direction of the missing self-consciousness, an interrogation of that domination, an illumination of the pillage carried out by ongoing settler colonialism – a structure through which both black bodies and indigenous lands were stolen and abused. To decipher how and whether Schenkkan's play was successful in this project requires not only examination of the stories it tells and the author's intent but also his methods and the field of listening into which the play played.

Speaking in Seattle at the Theater in an Ecological Age conference in 1991, shortly after *The Kentucky Cycle*'s West Coast premiere at the Intiman Theatre, Schenkkan (1991) described three implicit lies, which he called myths, perpetrated by the frontier narrative: "the myth of the pioneer, the myth of abundance, and the myth of escape." The first, he remarked, was Turner's notion that the "'free land' to which the pioneers moved was available for the taking and that American progress began with a regenerative retreat to the primitive, followed by a recapitulation of the stages of civilization." Schenkkan aimed to indict Turner's thesis and western expansion's ethos of taking by showing the perpetual cycles of violence and pillage that plague the fictional families Rowen, Talbert, and Biggs. In his keynote remarks, Schenkkan then debunked the second lie, the "myth of abundance," as the mistaken, anthropocentric notion of the "inexhaustible bounty of the land," which perpetuates the deception that "these resources are so vast they will never end. You can't possibly use them up."[11] To unmask the ways in which the "myth of escape" normalizes environmental destruction, Part Two of *The Kentucky Cycle* tracks the rise of coal company dominance in the region and maps the cost through the destruction of the social and ecological fabric of life. In "Tall Tales," the Rowens are swindled out of the land they farm. The rise of union organizing among coal miners is the core of "Fire in the Hole." Finally, "War on Poverty" delivers a visage of the impact of 200 years of environmental abuse: people, land, and hope are used up.

Collectively, the nine one-act plays that compose *The Kentucky Cycle* recouple ancestral values and actions to present consequences, interrupting the exceptionalist "idea that what one did in the past doesn't matter, that a man can endlessly reinvent himself and begin anew." That myth of escape, Schenkkan (1991) argued, is predicated on the myth of abundance:

> The myth of the frontier says, if you don't like where you are literally or metaphorically, well pick up stakes and move. There's plenty of land

out there. Change your name, change your address, change your history. Because what you did yesterday doesn't matter.

Both escape and abundance are implicit in Turner's notion of "free land" discussed in Chapter 1. This deception provides cover, the play suggests, for those executing plunder by relieving them of responsibility and culpability. Each act carries implications and responsibility forward, working hard to provide settler viewers an illumination that puts past and present into the conversation.

The structure of "Tall Tales," for example, which is the first act in Part Two, sets up this reflective mode. Affirming the ways those histories are carried in land, in bodies, and in stories, "Tall Tales" is framed by a prologue in which an older Mary Anne Rowen reflects on the visit of a certain young man who would change her family's life forever. As a young girl, she tells the audience directly, she marked her place in the world by the seasonal, sensible, sensuous counterpart to her living – a land that she remembers as intimately as "a lover's promise" (171) but one that was broken by the economic forces of U.S. industrialization. Mary Anne equates the industrial juggernaut with nature itself, likening it to the force of water. "Fella once told me a story, said these ain't no real mountains here at all . . . it was all just one big mound that had been crisscrossed and cut up into so many hills and valley by the spring runoff" (172).

As her story unfolds, Mary Anne's younger self enters and meets JT, a traveler who works for a New York coal company and has come to buy the mineral rights to the Rowens' farm land. Innocent to his intentions but charmed by his polish, Mary Anne flirts with him. She describes the old oak tree under which, according to her father, her grandfather made a treaty with the Cherokee:

> I used to think that tree was all that kept the sky off my head. And if that tree ever fell down, the whole thing, moon and stars and all, would just come crashin' down. I think sometimes how that tree was here way before I was born and how it'll be here way after I'm gone and that always makes me feel safe.
>
> *(175)*

The oak tree embodies both a sense of place and a sense of self for Maryanne; it is living kin, a character in the ancestral stories she has heard – stories that, as Part One of the play has already revealed, are lies.

Enchanted by his worldliness in comparison to her own family, Mary Anne takes JT home. There he poses as a spinner of yarns to earn an evening's company and a warm meal, ultimately using stories as weapons to rob the Rowen family of their mineral rights. Through a series of tales that draw on Shakespeare and the Old Testament, JT sets loose the geologic force of stories that put changes in motion that will make the land unrecognizable to those who called it home. Mary Anne is charmed; Tommy, Mary Anne's boyfriend, fumes silently, while her father, Jed Rowen, is swindled and his ego puffed up by JT's comradery. In the lie at the heart of the deal, Rowen is paid one dollar an acre for the mineral rights to his land. Mary Ann's mother, Lallie, protests:

LALLIE: This land been in your family back before anybody can remember, and I don't think you oughta be sellin' it.
JED: You heard him Lallie – I ain't sellin' the land, I'm just sellin' the mineral rights.
LALLIE: I don't think you oughta be sellin' any part of it, even them rocks.

(193)

JT plays on the gender roles he knows exist and chides Jed for allowing his wife to tell him what to do. "Hush up, now! JT and I are talkin' business," Jed snaps (194). The negotiation over price that follows is a charade in which JT exploits Jed's blustering pride and settles for one dollar per acre. With Jed's mark (for he cannot read or write) on the promissory note, the land exchanges hands – a commodity transferred from a settler family to an extractive corporation headquartered in New York.

Mary Anne is as innocent to JT's motives as her father is gullible, but she drives a different bargain. As she walks him to back to the creek where they first met, she begs JT to kiss her and take her with him to New York. When JT relents and kisses her, Tommy jumps from the woods and attacks him with a knife. Mary Anne intervenes and saves JT. Yet in this adrenaline-charged threat to his life JT panics, drops his veneer and tells the truth:

> Your poor old pa, thinking he's slick as goose shit – *a dollar an acre!* . . . there he is, sitting on top of maybe fifteen, twenty thousand tons of coal an acre! . . . Listen to me, god damn it! First, they cut down all your trees. Then they cut into the land, deep – start huntin' those deep veins, diggin' 'em out o in the deep mines, dumpin' the crap they can't use in your streams, your wells, your fields, whatever. . . . Leaving your land colder and deader'n that moon up there.

JT confessed his stories were a weapon of plunder to extract land and defraud the Rowens. When Mary Anne refuses to believe he is telling the truth, JT presses her to recognize the ways stories shore up dominant power structures.

> Truth? Hell, Woman there ain't no such thing. All there is is *stories!* . . . Your daddy got his stories. Civil War hero, right? . . . He's the son of thieves, who came here and slaughtered the Indians and took their land!
>
> *(203)*

He describes a process of erasure that she and her family face. Those who take the coal, who cut down her oak tree, JT claims, will merely make "a fine banker's desk and swivel-back chair for Mr. Rockefeller himself" and will not remember her story (204). "You a hillbilly just like my daddy, just like me!" Mary Anne reminds him. "At the end of the day . . . who are you?" she demands (204). She reminds him of his own connection and moral obligation to the land on which he stands – he, too, is a Kentuckian, but who left the Cumberland to seek his fortune in New York. JT's reversal, arising from his vulnerability in the moment of being injured, also suggests that individual and collective identities are collaborations with place, a process of marking and being marked.[12] Faced with the consequences of his actions on people he knows, places he, too, has lived, he gives the promissory note back to her, saying "Tear it up. Tell Jed to tear his bank note up, too" (205). Here, as in other sections of the play, Schenkkan suggests that individual action and individual choices are always matters of conscience.

However, the scene also argues that an economic system founded on extraction and entitlement has a forward momentum that cannot be stopped. When Mary Anne reveals JT's duplicity to her father, Jed refuses to listen, saying that JT just wants to cut out of the deal because Jed got the best of him. The scene leads the audience to ponder Jed's hubris – rather than corporate aggression – as the cause for the destruction of any remaining ecological wholeness in their land and lives.

More than land has been lost by the time Mary Anne describes her homeland after the mining companies take possession. In the epilogue to "Tall Tales," the settler-colonial story that she understood as her family's story of origin is revealed as bankrupt. The landscape of the past is washed away and with it, Mary Ann's identity and history, including the oak she loved: "[W]hen they chopped it down, everything fell in: moon 'n starts 'an all"; the sensible world collapsed; there was nothing to hold "up the sky" and nothing to "hold the land down. . . . What you get is a just a whole lotta rain, movin' a whole lotta mud." The character of the land and the home she knew have been erased; for Mary Anne the past is "just a story." Even her children don't believe her; "Mama's telling stories again," they say (206), encouraging caution about the incomplete histories that have been mainstreamed as fact.

In "Tall Tales" and other scenes, characters are portrayed as loving land they call home, and this love exists in cognitive dissonance with a system that may require the exploitation of that home when a person, family, or community is faced with financial need. Here the play anticipates those who argue that the emerging epoch would more correctly be called the Capitalocene. In "The Rise of Cheap Nature," for example, Jason W. Moore (2016) argues that the Anthropocene "obscures, and relates to context, the actually existing relations through which women and men make history with the rest of nature: the relations of power, (re)production, and wealth in the web of life" (83). Moore argues that it is not humans generally who have caused climate change but the capitalist system, which drives an extractivist ethos at the expense of people and environments.

Returning to *The Kentucky Cycle*, even as Schenkkan's epic aims to expose a national narrative that supports an ethos of exploitation, it also argues for a new ethos of habitation, or homemaking, that celebrates a connection to place. Key moments tie humans to place by showing that people are bound physically and spiritually to the ecological worlds of plants, animals, land, air, and water. Like the character of Mary Anne, in "The Homecoming," second-generation Patrick Rowen describes his experience of hunting:

> When I hunt, I don't 'pretend' I'm a deer or nothin' I just am. I'm out here in the woods and things just get real . . . still . . . or somethin' . . . When I reach that place, when I just am, there, with the forest, then it's like I can call the deer or somethin'. I call 'em and they come. Like I was still waters and green pastures, 'stead of hunger and lead.
>
> *(51, ellipses in original)*

Generations later, one of his descendants expresses a similar experience of "oneness." In the play's final scene, Joshua recalls that even the miners drew their identity from the mountains they cut open, saying, "It was *all one thing* – all of us and them mountains" (322, italics in original).

Native characters appear only in the first two scenes, signifying to its audience the past tense of indigenous America as a point of origin but not a present presence. In the final scene of the last Act, stage directions tell us that the "land is slowly regenerating itself" (315), and Joshua Rowen returns to his family homestead walking among the hard rock and broom sedge. He comes upon grave robbers who have unearthed a piece of buckskin; he fires his rifle and scares them off. He is joined by the descendants of the Biggs and Talberts; the men reminisce as they talk through the challenges of the War on Poverty of the 1970s. One of them discovers the buckskin dropped by the thieves, calling it "a collector's item" (328), and as they peer into the hole, they recognize a bundle – the wrapped bones of the Morning Star's child, murdered by his great-great-great-grandfather. Echoing Tilden's similar moment of unearthing kin in *Buried Child*, Joshua holds the bundle in his arms, cognizant of the layered history of colonialism buried in the land, willing – or so the tableau might lead the audience to believe – to take responsibility for the history of which he is a product.

The other men continue to remark on what seems like a rare anthropological find that would fetch a high price from collectors – a kind of mummy preserved down to her "tiny fingernails" (329). Rowen instead wraps the baby in the buckskin, in the same way that the child's mother, Morning Star, had done 200 years earlier. Then he takes off his own coat and wraps it, too, around his ancestral kin. As he lays the child back in her grave, he takes off the gold watch that his ancestor stole from the Scottish trapper and tucks it into the coat, burying it with the long-ago murdered child. Whether and how this act signifies reparations, healing, or a change in how Rowen might live going forward is difficult to decipher, and the play provides few clues. As Rowen does this, the stage directions describe that "the ground erupts. Pushing up through the soil are the figures of those we have watched live and struggle and die over this ground." From Michael and Morning Star, through Mary Anne and the miners, to Joshua's present moment, all the characters rise up from land in a final tableau. Then a wolf howls, and Joshua fires his rifle in the air, shouting, "RUN YOU SON OF

A BITCH! RUN! RUNNNNNN!" in "sheer joy and exultation" (332). While the moment may be moving for settler-descendant audience members, it does not suggest a way forward, and while it provokes reflection – a kind of settler grief in the face of history – little is restored or repaid. Indeed, while the play's final act acknowledges that poverty and unemployment followed in the wake of closed coal mines, it could also be construed as a jobs-focused argument for the continuance of big coal rather than an indictment of it.

Contemporary theater artists can learn much from *The Kentucky Cycle*'s critical reception, and the play's long-term lessons may indicate that how we remember is as important as what we remember. Early productions in Seattle (spring 1991), Los Angeles (winter 1992), and Washington DC (spring 1993) all enjoyed favorable reception. In a *New Yorker* article published during the play's Washington, D.C., run (before it arrived on Broadway), Bobbie Ann Mason (1993) interviewed Kentucky scholar and activist Gurney Norman. Schenkkan's depiction of Kentuckians and their history, Norman claimed, had "run roughshod over a whole culture" and "the play's vision blames the victim" while "the absentee owners [of the coal mines] are ignored" (58, 59). Kentucky scholars and cultural leaders argued that it demonstrated cultural appropriation and stereotypical representations of Kentuckians. By taking the stories of Appalachia as a metaphor for national hubris, the production reinforced national stereotypes of mountain people. By packaging the complex cultural and economic history of the Cumberland Plateau into revisionist history, they argued, Schenkkan had acted as cultural colonialist, a story pirate, repeating the very patterns of colonial theft that his play sought to critique. Norman claimed that Schenkkan had appropriated regional stories and that *The Kentucky Cycle* was "ambition masquerading as art" (Mason, 58). The insult that Kentuckians felt inspired a national conversation about the politics of representation in a collection of essays titled *Back Talk from Appalachia*, edited by Norman (Billings, Norman, and Ledford 2001). Schenkkan's (1991) project seemed clear enough – to face what we have long denied by confronting the myth of the frontier as "extremely seductive and ultimately very dangerous" (Mason, 58). Schenkkan attempted to make consequential connections between human action, the social and environmental devastation, and the grief which he perceived as a symptom of those losses:

> One of our big problems is how much we're in denial about our past, and how unwilling we are to examine our past and to come to terms with it. There's so much loss in this country, so much grief that we're in denial about. There's a river of loss that runs through the bedrock of this country.
> *("Keynote" 1991)*

Yet, as Richard White (1995) reminds us, when a historian, storyteller, or dramatist depends on mobilizing familiar images, they run the risk of erasing a "larger and more confusing and tangled cultural story." Experience is "more varied and complicated than Buffalo Bill," he writes. To "tell so many stories of this kind is to cut off the telling of other stories, other narratives, other imaginings" (54). By the time the New York production opened at the Royale Theatre on November 14, 1993, the critical skies had darkened. Jack Kroll (1993) concluded in *Newsweek* that Schenkkan's characters rang false, that he "writes with a bludgeon," and that the production was "a simulacrum of true theatrical power" (77). In the face of this criticism, the Broadway production closed after two weeks.

The Kentucky Cycle's fate, however, was also linked to larger cultural denial.[13] At the time of the play's premiere, environmental historians had been trying to define, debunk, and deconstruct the frontier in American culture for several decades without success. These revisionist histories met with considerable resistance among the public.[14] Environmental historian Patricia Nelson Limerick (1995) observes that

> trying to grasp the enormous human complexity of the American West is not easy under any circumstances, and the effort to reduce a tangle of many-sided encounters to a world defined by a frontier line only makes a tough task even tougher.
>
> *(72)*

Limerick describes the frontier as the "mental flypaper" of the American mind, one that historians (and dramatists it seems) are hard-pressed to dislodge: "[n]icknamed the 'f-word' and pummeled for its ethnocentrism and vagueness," the frontier is elastic, mutable, both invisible and opaque – a quagmire of meaning (72). She likewise warns that the frontier is so ubiquitous in U.S. culture as to be a "habit of mind" that "uses historians before they can use it"; moreover, she underscores, "the popular understanding of the word 'frontier' and the scholarly effort to reckon with the complex history of cultural encounters in colonialization share almost no common ground" (79). As scholars thrash about in its sticky meanings and discomforting implications, Limerick maintains that

> the public is paying absolutely no attention . . . it's perfectly evident that the public has a very clear understanding of the word 'frontier,' and that understanding has no relation at all to the definitional struggling of contemporary historians.
>
> *(79)*

Perhaps it is no wonder that Schenkkan's undertaking snagged, and yet there are important takeaways from the proverbial dust it kicked up.

Stories do not exist in a vacuum, but in an echo chamber that reverberates to the rhythm of the market-based theater economy and the master narratives that are part of it. The fate of *The Kentucky Cycle* should remind artists to ask, "Who's listening?" Artists must be cognizant and reflexive about the ways our own varied privileges infuse any work we undertake. If we can understand how

Kentuckians felt seeing their history represented in stereotypes that perpetuated assumptions about who they are, seeing their story told by a dramatist and a theater industry of outsiders trading on their stories, then surely we can understand why Native artists, allies, and audiences were similarly outraged by the representations of indigenous people in *Bloody Bloody Andrew Jackson*, which I discussed in Chapter 1. The politics of embodied representation have much to do with how effectively, or not, theater responds to the environmental crisis. Questions such as "Whose story is it?" "Who gets to tell it?", and "Who is listening?" must become central to any theatrical project. The injuries that are evident in the land – in this case, *"mountaintop removal"* strip-mining on the Cumberland Plateau – have a history felt in the lived experience of generations of people – including people who are alive and struggling today.

The Kentucky Cycle illustrates how quickly artists can be used by the very ideas they hope to critique. Its reception asks settler-descendant theater-makers as well as critics to examine the difference between a deconstructionist project and a decolonizing one, for the mechanisms of colonization easily play out like history on automatic repeat, even in a cultural product that aims to critique or satirize. Acknowledging acts of genocide and histories of plunder is merely a first step. Our work must move beyond settler grief and regret, working with those who have endured that history toward a reclaimed, restorative future. In *Decolonizing Methodologies*, Linda Tuhiwai Smith (1999) emphasizes this point when she describes that

> [t]aking apart the story, revealing underlying texts, and giving voice to things that are often known intuitively does not help people to improve their current conditions. It provides words, perhaps, an insight that explains certain experience – but it does not prevent someone from dying.
>
> *(3)*

As the discourse of environmental justice and indigenous sovereignty increasingly finds expression in artistic work, Tuhiwai Smith's "indigenizing projects" resonate with the efforts of dramatists not only to map the history of colonialization and the effects of environmental injustice but also to leverage the lively, present, immediate, and communal art of theater towards envisioning and embodying ecological justice and resilience on stage.

The 1991 People of Color Environmental Leadership Summit, which articulated the Principles of Environmental Justice now widely used as the basis for environmental justice discourse and activism, inspired a handful of dramatists and theater-makers to shift the focus of their work. Among them, the Underground Railway Theater (URT) of Boston, the Cornerstone Theatre of Los Angeles, and Brava! Women's Theater Collective of San Francisco sought to call attention to social and environmental justice concerns of specific communities.[15] These companies represented a theater-making methodology that centered community engagement. Since the 1970s, URT has sought to define its work through its relevance to social concerns by incorporating a unique mix of material practice that included elaborate puppetry with live actors.

Despite conceptual and topical differences, these productions were developed as collaborations between artists and communities toward the end of amplifying the voices of those who live at the frontlines of environmental racism and injustice were developed as collaborations between artists and the communities.

Stories in the body – *Heroes and Saints*

Bodies matter in the theater: stories are told and meaning is made in and through the bodies of actors; the collective endeavor of telling those stories both mirrors and enacts the body politic; and, finally, the audience bears witness with, in, and through individual and collective embodied presence. As dramatists and theater companies took up the cause and stories of environmental injustice in the 1980s and 1990s, the body – both the actual bodies of performers and the bodies of those whose lived experience is represented – became central to the complex ecological and political meanings of plays-in-performance. Inspired by the work of El Teatro Campesino (ETC), in particular, and aligned with the social justice and economic aims of the Chicano movement of the 1960s and 1970s, the national and hemispheric artistic movement of TENAZ – El Teatro Nacional de Aztlán – inspired future generations of Latino/a dramatists.[16] Among them, Cherríe Moraga's play *Heroes and Saints* (1994) leverages the presence and absence of the body on stage to call attention to the bodies at risk in the fields of the San Joaquin Valley and give testament to the UFW's central concerns with pesticide poisoning in the second grape boycott of the late 1980s (see, e.g., Pulido [1996]).

Farmworkers have long understood that the body is permeable, a reciprocal interface with the land; that what we do to the land, we do to ourselves. After the election of George Deukmejian as governor of California in 1982, funds for research into the effects of pesticides on farmworker health were cut, and his administration stopped enforcing state labor laws, many of which were a result of UFW protests in the 1960s and 1970s. In response, in the mid-1980s César Chávez, Dolores Huerta, and the UFW called for a second grape boycott to call attention to high rates of cancer and other pesticide-related health effects in farmworkers and their families. The toxic soup that results from the mixing of chemicals causes a cumulative toxicity from multiple chemicals, adding health burdens in farmworker communities. In 1986, the UFW produced a short documentary titled *The Wrath of Grapes*, which warned of the dangers of pesticides (including known carcinogens captan, dinoseb, methyl bromide, parathion, and phosdrin) to farmworkers and their families and documented the stories of children born without limbs or vital organs or suffering from cancer (see, e.g., Das, Steege, Baron, Beckman, and Harrison [2001]). Speaking in 1989, Chávez called attention to the community of McFarland, California's Central Valley, where studies indicated childhood cancer rates were 400 percent above normal (Chávez 1989).

Produced in 1992 by Brava!, a women's theater collective of San Francisco, Cherríe Moraga's (1994) *Heroes and Saints* theatricalizes bodies and land bound together in a communion that is material and spiritual. Work, food, family, water, air, and soil form a body politic infused with Catholic and indigenous spirits of resistance. Linda Margarita Greenberg (2009) argues that "these layers of the verifiably real in a fictional form enable the play to perform rich and multivalent understandings of community and social protest" (178). At the time of its production in 1992 by Brava! Women for the Arts in San Francisco, Moraga writes, unfolding events of 1988 "brought growing visibility to the United Farmworkers' grape boycott in protest against pesticide poisoning." Those events included the 36-day fast by César Chávez, and a month later, Dolores Huerta "was brutally beaten by a San Francisco policeman while holding a press conference protesting George Bush's refusal to honor the boycott" (Moraga 1994, 89).

Moraga (1994) was moved to imagine the lives of the women and children living with the embodied effects of pesticide poisoning:

> Behind the scenes of these events are the people whose personal tragedy inspired a national political response. In the town of McFarland . . . an image remained in my mind – a child with no arms or legs, born of a farm worker mother. The mother had been picking in pesticide-sprayed fields while her baby was still in the womb. This child became Cerezita, a character who came to me when I wondered of the child's future as we turn into the next century.
>
> *(89)*

In Moraga's story, set in an imaginary town of McLaughlin, a multigenerational farmworker family struggles with a daughter, Cerezita, born without a body. She is described as "a head" and appears atop a box, or *raite*, which she moves by touching her tongue to various buttons.[17] She lives with her mother, Dolores; sister, Yolanda; and brother, Mario. Her mother, feeling ashamed, hides Cerezita from the community – El Pueblo – made up of the children and mothers of McLaughlin. Moraga's description of El Pueblo is a reminder of the community-based genesis and values of the play. El Pueblo is made up of community, protesters, and the audience, "an ensemble of people from the local Latino community" (90). Other characters include a Latina news reporter, Father Juan, described as "a half-breed leftist priest," and Amparo and Don Gilberto, both community activists working for farmworker rights (90). English and Spanish are interchanged throughout the dialogue. As characters code-switch, an Anglophone audience member might experience a small measure of the lack of access for Spanish speakers in a nation where English is the norm.

Cerezita's body is the site not only of injury but also of resistance. She resists her mother's desire to protect and to hide her and, by the end of the play, casts off her cloister and becomes a vocal community leader, organizing a mass protest in the vineyards and putting what body she has in harm's way for a justice that understands that land and worker are of one flesh. Cerezita herself embodies one of the human costs of a culture in which food production and consumption are separated by industrial agriculture and market forces, a system that disregards the bodies of those who grow the food and those who eat it.

In *This Bridge Called My Back: Writings by Women of Color*, Cherríe Moraga and Gloria Anzaldúa (1983) have articulated a "theory in the flesh" as one that arises from "the physical realities of our lives – our skin color, the land or concrete we grew up on, our sexual longings [which] all fuse to create a politic born out of necessity" (23). In the character of Cerezita, Moraga has given theatrical form to this central idea of the body as the site from which identity and activism arise. In Cerezita, the body is the intersection of biological and cultural aspects of human ecology, and her lack of a body suggests the deadliness of severing the cultural from the ecological. In this, Moraga gives flesh to Gloria Anzaldúa's (1987) call to "stop importing Greek myths and the Western Cartesian split point of view and root ourselves in the mythological soil and soul of this continent" (90). The presence-of-absence of Cerezita's body is multivalent: it is not only stark material evidence of the impact of toxins on the health and well-being of workers and their families but also a metaphor for a society that lives in its head by privileging the abstract over the material and, as Western philosophy has tended to do since the Enlightenment, severing nature from culture.

In "Ecodramaturgy of Contemporary Women Playwrights," Wendy Arons and I (2013) argue that:

> Cerezita possesses all of the attributes philosophers have traditionally mustered in support of their claims for human exceptionalism – language, abstract intelligence, spiritual feeling, empathy, imagination, creativity – her bodilessness signals precisely how disastrous that conception of the human relationship to the nonhuman world has proved to be.
>
> *(183)*

Cerezita, and Yolanda and her newborn infant, are the immediate victims of environmental injustice in the play. Yet as their struggle becomes palpable, it registers a sickness in the larger body politic. Cerezita is a living warning not only to her community, signaling its loss of its future (its children) and also potentially the future of humanity as a whole. If people continue to treat the natural world merely as an exploitable resource for, and backdrop to, human activity, the body of the larger community may not survive.

The victims of environmental injustice are usually the poor, the disenfranchised, and women – especially minority women – make up a large percentage of those categories. In Moraga's (1994) play, it is women's bodies that bear the brunt of the effects of the toxic chemicals sprayed on the fields, and it is primarily women who must care for the dying children they bring into the world. Cerezita's sister, Yolanda, has given birth to a baby already ill from the poisons sprayed on the fields that have leached into drinking water and soaked the air and land around them. The crop duster is a nearly constant presence throughout the play. "Why don't you just drop a bomb, cabrones! It'd be faster that way!" Dolores yells as the sound of the crop duster passes overhead (126).

Later, in a scene of radical empathy and feminist solidarity that reaches into the heart of human intimacy with the earth, Yolanda grieves the death of her infant, a grief that expresses itself in her body as her breasts leak milk: "Every time I think of her, they run. Nobody told my body my baby is dead. I still hear her crying and my breasts bleed fucking milk. I remember the smell of her skin and they bleed again. . . . They feel like tombstones on my chest!"; Cerezita responds, "Sister! I wish I had arms to hold you" (142). Then, knowing that Yolanda's milk carries the same toxins that killed her baby, Cerezita suckles her sister's bereaved breasts to relieve the pain. Echoing the famous ending to Steinbeck's *Grapes of Wrath*, in which Rose of Sharon breastfeeds a starving man after her own infant dies, this scene marks the turning point in Cerezita's activism. Both Yolanda and Steinbeck's Rose of Sharon have lost a child to systems economic and environmental exploitation. In Moraga's play, Cerezita decides that she will not hide any longer; she will take up the cause and suffering of her own family, her people. Cerezita and Father Juan have planned a direct-action protest to parade the bodies of the dead children on crosses, to burn the fields. When Father Juan does not follow through, Cerezita cries out,

> If I had your arms and legs, if I had your dick for chrissake, you know what I'd do? I'd burn this motherless town down and all the poisoned fields around it. I'd give healthy babies to each and every childless woman who wanted one and I'd even stick around to watch those babies grow up! . . . You're a waste of a body.
>
> *(144)*

She calls out all those who profess, as Father Juan does, to Christian values and to the idea of environmental justice but do not translate that faith into action.

Jorge Huerta (*Chicano Drama* 2000) writes that Moraga's play must be understood as a kind of "miracle play" in which ordinary lives are infused with an extraordinary sense of communion among the living and the dead, the body, land, and the spirit (64–70). In a horrifying gesture of protest, the workers hang the bodies of their children, who are believed to have died as a result of pesticides, on crosses and place them in the fields among the trellised grapevines. The grapevine-like bodies of dead children draped on crosses and paraded in protest through the vineyards, an image rips into the imagination, forcing a rupture between the discrete spheres of self and environment, describing a solidarity of the flesh among children, workers, and land. As Greenberg (2009) observes, within the action of the play this public display of dead children "highlights the problems of environmental abuse of the Chicano/a farm-working community that the pastoral vision of California agriculture masks" (166). Yet the visual parallel between the crosses bearing the bodies of dead children and those bearing the pesticide-sprayed grapevines also invokes Catholic religious iconography in a purposeful way. This material allusion to faith forwards the play's activism and resists a display of death that merely underscores victimhood. Within the context of Catholicism, the death on the cross is a prelude to resurrection – itself a kind of revolution in the face of material death. Moreover, the *calavera* masks the children wear in their protest suggest the celebration of *Día de Muertos*, in which the borderlands between the dead and the living is permeable and fluid. Finally, the crucified children affirm the ecology of the sacrament communion at the heart of Catholic practice while lobbying the church to take a stronger stance on behalf of farmworkers.

The idea of "transubstantiation" within Catholic practice means that bread and wine are transformed through the ritual of the Mass into the actual body and blood of Christ. For Catholics, this is not a metaphor; it is a material reality that demonstrates a miraculous gesture of friendship and solidarity through time. According to the Catholic Mass liturgy, during his last Passover meal with his companions, Jesus held up bread and wine and proclaimed "this is my body" and "this is my blood" and instructed those present to "do this in remembrance of me." This ceremonial basis of communion acknowledges both the community that food always implies and the way in which what we consume enters into our bodies and becomes our own flesh. Through iconography of positioning the children as Christ figures, Moraga sets up a syllogism in the logic of Catholic belief that indicts the church (personified in the character of Fr. Juan), even as it reclaims the ecological foundation of this central sacrament. Grapes grown with pesticides become wine and in the Mass become the blood of Christ. If those same grapes have been the source of poisons, that blood of Christ is also poison, and the most sacred aspect of Catholic practice has been defiled. Moreover, the protest fuses the imagery of the Christ Child with the Crucified Christ in the symbolically crucified children, who reduplicate the crucifixion of Jesus at the hands of the Roman Empire. In this way, the play calls the Catholic Church – the church of many farmworkers – to task as a consumer of grapes (as wine) and as a social and civic as well as spiritual institution. But, more importantly, this visual language is a reminder that we all participate in an ecological communion. The body of the earth and the bodies of the workers (and, by implication, all those who consume grapes or any other megacrop of California's industrial agriculture) are part of one connected, interdependent, and complicit flesh.

In the final scenes of *Heroes*, Cerezita is transformed and transfigured, "draped in the blue-starred veil of La Virgen de Guadalupe" (144), who is believed by Catholics to have appeared to Juan Griego on the mountain of Tepeyac in Mexico in 1531. Luminous and fully embracing her role as leader of her community, Cerezita leaves the cloister of her mother's shame to lead the protests (144–146). While Moraga's Catholic-centered choice might seem antithetical to her self-identified queer, Xicana identity,[18] the iconography of Catholicism here reasserts Aztec and Mayan deities and the spiritual and historic connections to the land they represent. One of the many visages of the Mary, the biblical mother of Jesus, Our Lady of Guadalupe is herself the transmutation of Coatlaxopeuh, also known as the Earth Mother Tonantzín.[19] The use of the hybrid Catholic/indigenous deity also implicates the church as a civic institution within the hemisphere. Liberation theology had grown out of the Latin American bishops' conference in 1968, which took up the cause of the poor and indicted industrialized nations for growing rich on the labor and resources of developing countries (Gutiérrez 1973). The movement urged clergy and laypersons to become engaged in struggles for social and ecological justice worldwide, including the United States, where the work of César Chávez and Dolores Huerta were central examples (Floyd-Thomas and Pinn 2010). In the 1990s, factions of conservatism were often at odds with liberation theology and creation spirituality, which pressed the church to take a leadership role as a force for social and environmental justice. During that decade, the Vatican thus tried to rein in the liberation theology movement.

In the final scene of *Heroes and Saints*, Cerezita's call to action is reminiscent of liberation theology's hemispheric activism, as she speaks to the people and the cameras that have gathered to witness:

> [W]e live in a land of plenty. The fruits that pass through your fingers are too many to count – luscious red in the strawberry wonder, the deep purple of the grape inviting, the tomatoes perfectly shaped and translucent. And yet, you suffer at the same hands. . . . You are Guatemala, El Salvador. You are the Kuna y Tarahumara . . . today, this day, that red memory will spill out from inside you and flood this valley con coraje. And you will be free. Free to name this land Madre. Madre Tierra. Madre Sagrada.
>
> *(148)*

Invoking indigenous peoples of Central America, Cerezita's protest moves out beyond her community, beyond McLaughlin, California, taking up a hemispheric and global cause of the environmental justice movement at the dawn of the millennium.

Heroes and Saints and URT's eco-cabarets are only two of numerous examples of ecotheater that aimed to wake local and national audiences from what Moraga and Herrera Rodriguez (2009) call "duplicitous and strategic amnesia, resultant of a five-hundred-year history of invasion, beginning with the first European colonizer-settlers and their wholesale theft of native lands" (26). As Cerezita's protest bears witness to the bodies of the children whose lives became, for growers, part of the cost of doing agribusiness, she turns our attention outward to a hemisphere of struggle in which migrations of indigenous and mestizo families flee the political and social upheavals in Central America, including Guatemala's protracted civil war. The play concludes with a message thrown forward, that the pesticides sprayed on farmworkers like Cerezita's family and the tear gas sprayed at migrant families at the U.S. border are expressions, indeed the weapons, of settler colonialism. The structures of settler colonialism morph from one generation to the next, covered over by racialized rhetoric that masks the ways black and brown lives are subjugated by structures intent on extracting their land and labor while devaluing their contributions. Its current formation in a globalized corporate marketplace, Moraga argues, makes consumers the "speechless accomplices in an indefensible globalization, which continues to threaten the cultural integrity and economic stability of most of the Third World" (27). To call the land "Madre Tierra," the character of Cerezita suggests, requires recognition and to embrace what Moraga describes as "an incredible opportunity for the collective character of this country to distinguish itself from that of its bandit-leaders and make reparations for injustice done." This affirmation, Moraga asserts, is "a politic of mutually interdependent economic and ecological responsibility," a body politic that is cognizant of the shared and inexorably linked vulnerabilities of migrant and citizen, peoples and land, human and nonhuman (26).

Toward a transcorporeal self in *Alligator Tales*

The transmissions of stories (as I have sought to demonstrate in earlier chapters) carry ideology as well as trauma. On stage, as we saw in *The Kentucky Cycle*, the embodied aspect of performance is a central conduit of meaning. As actors cycled through scenes of depicting one generation after another of Cherokee whose lands were taken, Africans whose bodies were taken from their homelands, European settlers who did the taking, and they themselves became victims of extraction, they gave embodied material form to structures and patterns of settler colonialism handed down through seven generations.

In *Staging Place*, Chaudhuri (1995) writes that certain plays reflect "deeply ingrained convictions about the mutually constructive relations between people and place. Who one is and who one can be are . . . a function of *where* one is and how one experiences that place" (xii, italics in original). A sense of place *is* a sense of self in Anne Galjour's *Alligator Tales*,[20] a solo performance drawn from Galjour's childhood in the Louisiana bayou, in which Galjour embodied a vision of identity as interwoven fabric of touch-and-be-touched (see, e.g., May [2005]). Produced at Berkeley Repertory Theatre in 1996, and then at Seattle Repertory Theatre the following season, Galjour's story of self-as-place was originally developed as part of Life On the Water's Earth Drama Lab (Jacobson 1992; and Cless 1996 "Eco-Theatre"). Through the corporeality of her embodied performance on stage, Galjour pushed back against the long-ingrained theatrical habit of nature as background, backdrop, and backstory to human agency.

An actor's task is to live into the world of a play, to find in their own bodies and sensibilities a resonance – call it empathy – with the lived experience of others. During the 1990s, many individual actors took up stories of place, exploring whole constituencies and sets of relations through their singular but fluid self. Through these solo-performances actors explore the stories of the communities in which they live and whose environmental justice concerns they share.

serving to demonstrate another way in which the lively art of theater can serve as a democratizing force, a site of civic and ecological engagement. In this final discussion of emergent ecodrama of the 1990s, I argue that Galjour's play-in-performance reflects ecofeminist ideas that posited an ecological self in which human and environment are constituted by shared vulnerabilities and permeable boundaries. Galjour thus anticipates some of the central ideas of material feminisms articulated in the next millennium.

Taking place in the tidal flows of the bayou, *Alligator Tales* blurs distinctions between human and animal, land and water, nature and culture. Through the singular body of the actor, *Alligator Tales* envisions a mode of being in which human and environments are not defined by the border of skin but rather envisioned as a cacophonous riot of exchange. As Galjour's performance becomes a series of unexpected border crossings, individuals are infused with the terrain around them; identity and community are collaborations. A man is caught on a woman's fishing line and freed by a dog; alligators sleep on porches and must be shooed off with broomsticks in the morning; a cow saves a man's life; seeing through the eyes of a fish, a woman catches red snapper for dinner; a child is born of a hurricane; a woman is caught in a gill net and led back to breathable air by a fish.

Act I begins as Rosetta tells the audience, "Two men from Shell Oil came all the way from Houston to talk to my husband Grady about drilling for oil behind our house. . . . I told those men I had my neighbors to worry about." Thus, the story's point of departure is one of relationship: "I knew Inez and Sherelle Dantin would like waking up to a big truck carrying pipes and equipment. Then I got Urus Arceneaux on the other side of me, worrying about any little thing upsetting his cows." Grady, Rosetta, and their neighbors, are part of a community that has subsisted for generations by fishing and trapping, supplemented by keeping chickens and cows, trading food with neighbors, and, for Sherelle, occasional junk-food treats from town. The opening scene establishes a way of life centered on subsistence knowledge and skills that testify to generational interconnections among environment and cultural traditions, including food: "They said they loved our Cajun cooking . . . best to talk about business over supper," Rosetta suggests. Presuming the oil men from Houston would be as interconnected with their own homeplace as she, Rosetta, continues: "I went to my freezer and pulled out this rattlesnake Grady killed at some festival in Texas because I thought, 'Let's give 'em a little taste of home'" (3). When Grady returns, he smells what's cooking and exclaims "Mon Dieu, Rosetta! You didn't burn my rattlesnake, maybe!"; Rosetta offers to replace it with duck, but Grady insists "Get me my turtle. . . . I'ma microwave defrost it. I'ma make a sauce piquant" (4). Food – subsistence and traditional or junk food – is a constant presence. Giljour intentionally confuses and blurs the material with the metaphoric through food references, which are peppered throughout character dialogue and at the center of stage action. As characters harvest, prepare, cook, and eat their favorite foods, the shared recipes celebrate their layers cultural heritage. The constant presence of food foregrounds the way this community navigates two worlds – the world of subsistence hunting and fishing, and the world of contemporary convenience, which is slowly eroding that subsistence and the cultural practices it entails.

We meet Inez on a routine fishing trip with Michael who "stared out the window the whole time watching the egrets eating their breakfast in the marsh" and listening to the radio "saying this tropical storm in the Gulf has turned into a hurricane . . . blowing 120 miles an hour . . . expected to hit the Gulf coast in the next three days." They pass their daddy's hunting cabin, and Inez tells Michael, "One day, when we get some money, we gonna fix that place up" (5). Years of dredging by oil companies have shifted wetland water levels, and while the cabin was once on solid ground, it is sinking into the sea. When Inez finally casts her line and Michael is playing in the waves, she relaxes. As she breathes in and out "[t]he air [that] smelled like Coppertone and salt," she affirms the nature–culture soup that is her environment. She tells Michael and the audience that "if that hurricane is bad enough, it'll suck up the sand dune, the road, and the island. Before you know it, I'll be casting my line from my back porch" (6). The erosion of the protective barrier estuaries and islands of Louisiana – already prevalent in the news at the time *Alligator Tales* was produced – becomes personal and palpable. Galjour's performance continues as the wind picks up and the storm begins to touch and rearrange the lives of the characters Galjour has brought to life.

Galjour's performance calls attention to both kinship and commodification and calls out habits of mind that distinguish bright boundaries between species on which human exceptionalism is predicated. Throughout *Alligator Tales*, Galjour's choice of words disrupts and dismantles the linguistic distinctions between human/body and animal/meat, a strategy that elevates the material aspect of both and foregrounds their seamless interface with the watery and polluted world. When Inez catches something big on her line, she reels it in, squealing. Michael helps drag it to solid land, where they realize it is a man – one so overburdened with scuba tank, mask, scientific gear, and jars of specimens that his human form is unrecognizable. He gulps air and wheezes like a fish out of water, and indeed, his name is Marlon Skinner and, like the men in suits who visited Rosetta and Grady, he also works for Shell Oil: "I cut the line and looked at the wound," Inez narrates. "I soaked a towel in water from the waves," but her attempt to treat Marlon's cut ("salt water's the best thing for a cut," she tells him) is vehemently refused: "NO! Not that water!" (7). Yet, as the sole performer, this interchange takes place in one body; as Galjour performs the moment, she is both the hand that treats the wound and the wounded man. Describing Marlon as a kind of creature of the deep, Inez instructs him to lie still: "No flapping," she commands as she draws the fish hook from his head. "There. No meat. What I told you." As Marlon readies to go back in the water to finish collecting his specimens, Inez asks, "[D]id you see any redfish when you were down there?" Marlon cautions her about eating anything from "that water" (8). As the protean storyteller through which characters, voices, and events flow, Galjour also performs the permeable interface among water, food, and body, destabilizing habits of mind that separate human, animal, and environment. For example, 20 minutes of the play pass before the audience realizes that Michael is a dog – Inez's retriever. The moment may catch the listening audience off-guard, as it attests to the lived kinship between human and animal.

   Galjour's performance suggests the idea of a transcorporeal self in playful and disarming ways that disrupt the singular, individualistic, and separate idea of a person, establishing a mode of person that is composed of relations. As Galjour uses her corporeal body to represent human and nonhuman within a dynamically alive and indeterminate world, she conjures a visage of what Stacy Alaimo (2019) calls "transcorporeality," a way of being, of understanding oneself, as not only as multiple but also as ever in flux, a mode of being that emphasizes "the movement across bodies" and "the material interconnections of human corporeality with the more-than-human world." For Alaimo, human corporeality must be understood as "transcorporality, in which the human is always intermeshed with the more-than-human world . . . the substance of the human is ultimately inseparable from 'the environment.'" Galjour performs what Alaimo calls "mobile space that acknowledges the often unpredictable and unwanted actions of human bodies, nonhuman creatures, ecologic systems, chemical agents, and other actors" (2). Human, animal, weather, and swamp come together in her signifying body; she is not an individual but a collective – a neighborhood, an intertidal world in which forms and forces – ecological, economic, and social – intermingle. As Galjour plays all the roles, she conjures an interspecies neighborhood of alligators, fish, turtles, birds, dogs, cows, lightning, wind, rain, and children and neighbors who move in and out of her own corporeal expression. Throughout, Galjour undermines and destabilizes both material and discursive distinctions between nature and culture, human and nonhuman.

   Nothing is as central to human corporeality as food. A journey into the reciprocity between the ecological world, the human body and cultural identity, *Alligator Tales* – if audience members were inclined to take notes – is stuffed with sensual and sensory references Cajun cuisine. Indeed, it is an aspect of the play on which Seattle Repertory's production capitalized by serving up a preshow dinner for eager audience members. But it quickly becomes clear that the traditional knowledge, ecological and cultural, of Grady's turtle sauce piquant, Sherelle's peas and eggs in a roux, Inez's rubbed redfish, and Urus's court-bouillon are signals of a complex cultural heritage emerging from the intersections of colonialism and distinct ecologies. As the material and the metaphoric expression of that life-sustaining exchange, Galjour's play is a litany of seeking, harvesting, gathering, preparing, sharing, ingesting, metabolizing, and, by implication, expelling foods. Food is thus refigured as intimate exchange, the way the land and water pass into bodies and bodies move into other bodies in an ebb and flow of reciprocal sustenance. Consequently, the references to oil in the waters from which these foods are harvested implicitly ask us to ponder the connection between health, culture, environment, and the slow creep of corporate forces in the bayou (18). Does Sherelle's breast cancer, for example, have a connection to the toxins in the water – the toxins that Shell Oil's Marlon does not want poured on his fishhook scrape – or does it have a connection to her love of junk food – the Little Debbies and Hawaiian Punch that she asks Inez to pick up at the grocery store? Or are both – the toxins in the water and the toxic foods – a product of the same forces that are slowly eroding their lives and livelihoods?

As the hurricane approaches, Sherelle and Urus – a farmer who was once hit by lightning and licked back to consciousness by his cow – fall from sensual preparation of peas and eggs in a roux sauce into lovemaking that is as much a kind of intercourse with the forces of the storm as one another. Caressing, they are infused with the same electrical currents that run through the sky, the ground, and every living thing. Unafraid of a force beyond their control or understanding, they ride the storm to new heights of orgasmic pleasure and pregnancy. Human and storm are a cacophonous polyphony of water, fire, earth, animal desire, and love. Meanwhile, before the hurricane hits, Inez and Michael, joined by Marlon, head out for a last fishing trip before the storm makes navigation impossible. As they motor down the canal, slowing "for the natural gas pipeline," the dialogue carries a litany of the life-filled world, and an implicit debate between the two. Marlon then points out, "Inez, you see those slats of wood nailed into the tops of those posts? Chevron made those for all the eagles to build their nests," to which Inez replies, "Marlon, you see how our hunting camp is sinking into the water? Well I believe Chevron and all those oil companies did that by tearing up the marsh to build oil rigs and pipelines" (19).

When they tie off on a tree root to fish, the wind picks up, the "rain start[s] pouring from the sky like salt," and Inez asks, "You ever been in a hurricane Marlon?" And then she tells him a story:

> My great grandmere died in a hurricane . . . she died with all of her children except for my gram. When the hurricane hit the water rose so high she had to tie each of her five children to the branches of the oak tree by their house. But the wind blew so hard and the water rose so high that it swalled up the little children. And all my great grandmere could do was to untie the ones who were left in the hopes they might make it to safety. My gram was the only one who survived. She floated for three days . . . woke up near a cow who let her suck with its calf. . . . When me and my sister Sherelle were little, if we ever wanted to hear the story . . . we had to get her a glass of milk.
>
> *(20)*

It is not a story of fear and fleeing, not of mass exodus, rising waters, and floating cars but of a long history of, yes, sometimes dying, as well as living with and living through, a story of resilience as well as loss. Throughout, Galjour turns the tale away from the drama of catastrophe towards the delicately interwoven lived, and living, experience of her characters.

When the hurricane hits, "[W]e were ready," Rosetta reports, and Sherelle is praying the rosary with "Hostess Cupcakes and Little Debbies" at hand (18). With the help of Seattle Repertory's minimalist use of sound and lighting, we watch Galjour give over to agencies beyond the human – hammering rain, muscular wind charged with lightning, water that rips away the things held most dear. In the storm, Michael has gone missing. Inez and Grady press out into the chaos to find him. The next scene, in the days after the hurricane, Sherelle has prepared Michael's body for burial, and as "Inez wrapped him in a blanket he always loved" she whispers, "Oh Michael. How lucky I was to have had the pleasure and the comfort of [your] good company for twelve years" (26). Full with loss, Inez is "thinking maybe I should leave here" (27), but just then, "I heard a knock at the door. It was Urus. He was holding some kind of squealing animal in a towel . . . when I opened it I saw it was a baby" (28). Galjour's story circles around again and again to a palpable visage of what Donna J. Haraway (2016) calls the "unexpected collaborations and combinations" that propel ongoing and living worlds forward (4).

Act II takes place six years later. Inez's neighbors have sold land to the oil company and used the money to start an alligator farm. With money earned from selling the reptiles for shoes, belts, and pets, Grady and Rosetta enjoy new furniture and a new car. Urus goes to work on the oil rigs, while Sherelle sweeps alligators off the porch: "[T]hey dredge up so much earth from the marsh to build that Hurricane Levee, that by last winter all the alligators were on the move. There were forces all the way up to the canal behind our house" (5). Increased dredging by oil companies has caused the alligator population to migrate closer to human homes; Inez's beloved redfish are few, and the marsh grass is thicker. Inez has adopted the hurricane orphan – Beau – who is six years old but has not learned to speak. Inez shows Beau that she can catch redfish by hand; opening her eyes underwater and following the tail and trail of the fish, she is caught in a gill net herself, submerged, and for a moment, her body is one with the fish in struggle as well as sustenance.

Meanwhile, life carries on, even as the world around them changes in irreparable ways. Sherelle's cancer, in tenuous remission throughout the play, returns when she decides to complete a pregnancy. Inez protests, but Sherelle insists she has planned for all outcomes, including death. In the final scenes of the play, Inez discovers that Sherelle has agreed to sell their father's fishing camp – the land that has already sunk into the water – to the oil company. But even here, as Inez protests, arguing it was the home they grew up in, the place she learned to fish, the site of her father's death (bitten by water moccasins hiding in a cupboard after a hurricane), the typical dialectic generated by a performance in which multiple actors play characters on various sides of an issue or debate, Galjour is both sisters – the one who falls and the one who survives, the one who gives in and the one who resists, the one who rationalizes loss and the one whose rage fights back.

As Sherelle is about to give birth, the second hurricane of the play arrives, consuming everything in "a wall of water rolling toward our home," Inez narrates. Among the sound of "voices . . . being swallowed by the water" and the fish, water moccasins, baby alligators, and pieces of homes, Inez sees one of Urus's cows float by. She grabs onto its neck, and from there climbs into Urus's boat "rocking in the waves" (25). Like Noah, Urus has gathered his loved ones in his boat, including Beau, who pulls his mother in and speaks for the first time: "a baby" (26), he says, and he points to Sherelle nursing her newborn but weakening from the cancer. Later, Beau asks, "[W]here do *I* come from?" (29). Inez hesitates but finally answers, "[Y]ou came from a hurricane" (30). Beau's thriving, like his cousin's birth, not only represents hope as a stance in the face of a changing and damaged world but also a new kind of creation story, one in which the multiple agencies of the wider world are acknowledged, kinship even with consuming chaos, a new possibility of human beingness, born of wind, rain, and fierce resilience.

Like the Louisiana wetlands, a threshold between freshwater and saltwater, between earth and sea, marking a border between worlds, between possibilities of being, Galjour's corporeal performance is just such an invitation to rethink human not as person but as process, a site of ecological exchange.[21] For Louisiana's coastal communities, which include Cajun, Creole, and indigenous communities, the vitality of the estuaries is both life and culture.[22] A prescient play about a future in which communities on front lines of climate change struggle to cope, adapt, and survive, *Alligator Tales* suggests dramaturgical strategies that anticipate Haraway's description of the tasks that accompany profound ecological change. In Haraway's words, *Alligator Tales* goes some distance to

> make kin in lines of inventive connection as a practice of learning to live and die well with each other in a thick present . . . to make trouble, to stir up potent response to devastating events, as well as to settle trouble waters and rebuild quiet places.

As a demonstration of ecodramaturgy going forward, Galjour's performance and text both remind us that we are "mortal critters entwined in myriad unfinished configurations of places, times, matters, and meanings" (2016, 1). In Galjour's play, both place and person, she becomes a site of kinship and exchange that transmits an ecological vision: our bodies and our identities are permeable, awash with the tides; neighborhoods and families include people, plants animals, water, and land; we have all been conceived in chaos; we are what we eat; the flesh of the world around us – including the plastics and pollutants in that world – becomes part of our flesh.

In "Performance, Utopia, and the 'Utopian Performative,'" Jill Dolan (2001) describes theater's primary function as not to formulate the new organizational structures (laws, policies, and programs) but rather to provide a forum in which people might engage the magical "what if" in order to *feel into* experience not their own. Theater allows us to feel into worlds, realities, and events that expand the very sense and sensibility of being human. Feeling into the past, as *The*

Kentucky Cycle aimed to do, theater can increase awareness of the ongoing-ness of exploitation manifesting in patterns of injustice and environmental degradation in everyday life. Feeling into the present, as *Heroes and Saints* did, theater is a place to assert presence, to reclaim histories, re-set values, to engage in the action of building a just and common future. And, like *Alligator Tales*, theater can be a place to stretch the boundaries of who we are in ways that develop a sense of self that does not stop at the edge of our skins, but reaches out in an ever-expanding ecological self through circles of inclusion, reciprocity, and responsibility.[23]

Land as other self

Each of the plays-in-performance discussed earlier (and many others) engage the social and political aspects of environmental issues as they carve out a rhetorical space for theater to tell stories of the reciprocity between people and the lands they call home. Each resists the humanist habit of using nature as a metaphor, or as a mere backdrop for human action, and instead centers ecological interdependence, telling stories of human culture's intertwined relationships with the larger ecological communities of which we are part and on which we depend.

In November 1999, theater performed in the streets of Seattle formed the backbone of direct-action protests against the World Trade Organization (WTO) meeting. Using methods reminiscent of Bread and Puppet, San Francisco Mime Troupe, and El Teatro Campesino, costumed protesters, giant puppets, musicians, and all manner of theatrical activities effectively stopped the WTO meetings and shut down the downtown business districts of the city. A resident and theater practitioner in Seattle at the time, it seemed to me as if Augustin Daly's vision of American theater as the site where national stories vie for power was playing out on the streets. Police, theatrically costumed in intimidating riot gear, faced off against protesters in bandanas and hooded rain jackets. The local news stations broadcast the moment to moment encounters, while both sides attempted to control an unfolding narrative about who was to blame for the violence that shut the downtown business district of Seattle down for several Christmas shopping days. However, the larger story, and the one that brought more than 30,000 people – labor unions, environmental organizations, and indigenous peoples from across the hemisphere – to march through the streets of Seattle concerned the effect of increasing economic globalization on lands and peoples of the planet.[24] Environmentalists, indigenous peoples, workers, and communities of color found common cause against the tyranny of profit-making multinational corporations that run roughshod over lives and land. Going forward, many dramatists would take up the translocal, transnational, and transtemporal impacts of the global economic forces in stories that called on audience to acknowledge the multiple and multiplying ways each and every one of us are implicated in patterns of ecological and cultural destruction. By the end of the twentieth century, a groundswell of theater-makers would be actively wrestling with environmental justice issues, bringing critical viewpoints from cultural studies to bear on environmental concerns. Their struggle to call attention to the material impact of ecocide challenged theater's long tradition of reading/playing nature as metaphor. Interweaving the ecological, economic, social, and spiritual impact of environmental loss, dramatists began to carve out a radical center that neither ignores nor demonizes human agency.

Notes

1 The NASA website provides a volume of information for the curious reader, such as "The History of the Ozone Hole," which summarizes research leading up to the 1980s conclusion about the depletion of ozone over Antarctica. Between 1986 and 1987, several papers likewise suggested possible mechanisms for the ozone hole, including chemical, dynamical (meteorological), and solar cycle influences.
2 In 1995, scientists Frank Sherwood Rowland, Mario Molina, and Paul J. Crutzen won the Nobel Prize in Chemistry for their research leading to the understanding that CFCs rise through the atmosphere to the stratosphere; once there, radiation breaks down CFCs, resulting in the release of ozone-damaging chlorine. See "Ozone" (1995) for more on this discovery; see also William K. Stevens's (1995) *New York Times* article.
3 The Montreal Protocol on Substances that Deplete the Ozone Layer took effect on January 1, 1989, after two years of negotiations.
4 A few examples of such environmental or site-specific theater include María Irene Fornes's *Fefu and Her Friends* (2017), Stefan Brecht's *Peter Schumann's Bread and Puppet Theater* (1988); and Richard Schechner's *Environmental Theater: An Expanded New Edition including 'Six Axioms for Environmental Theater'* (1994).
5 For more on the burgeoning environmental justice movement, see "The Principles of Environmental Justice" in Giovana deChiro's (1996) "Nature as Community: The Convergence of Environment and Social Justice." See also Robert Gottlieb's (1993) *Forcing the Spring: The Transformation of the American Environmental Movement*, which offers a history of contemporary environmentalism.
6 The "Scar Face Trilogy" includes *Intrigue at Ah-Pah*, produced in 1975; *The Road Not Taken* (1984); and *Fear of Falling* (1991). These satirical works were developed by the Dell'Arte Company centered on the character of Scar Face created by Joan Shirle (Jacobson 1992, 2).
7 Like many grassroots theater-makers, as artistic director of *Theatre in the Wild* in Seattle, I began a series of residencies in the schools, in which I wrote plays and developed an arts-based science curriculum with and for fourth and fifth graders about nonpoint source pollution and watershed awareness.
8 Life on the Water Artistic Director Bill Talen later became known for his one-man anticonsumption solo performance, called "Reverend Billy." For more on this topic, see Jane Hindley's (2010) "Breaking the Consumerist Trance: The Reverend Billy and the Church of Stop Shopping" and William Talen's (2003) *What Should I Do If Reverend Billy Is in My Store?*
9 These questions were gleaned from personal conversations at the *Theatre in an Ecological Age* conference in Seattle, Washington, in November 1991.
10 The program notes for the Intiman Theatre's production of *The Kentucky Cycle* included a genealogy chart, which is now appears in the published version of the play (see Schenkkan 1993, xi).
11 These ideas are similarly expressed in the "Author's Note" at the end of *The Kentucky Cycle* (1993). For more discussion of *The Kentucky Cycle* and ecodramaturgy, see Cless's (1996) "Ecology vs. Economy in Robert Schenkkan's *The Kentucky Cycle*."

12 In *Landscape and Memory*, Simon Schama (1996) writes about the reciprocal inscription of landscape on human memory and human culture on the face of the land. For more on this topic, see also Lucy R. Lippard's (1990) *Mixed Blessings: New Art in a Multicultural America*.
13 This milieu of denial was reflected in reviewers' own rhetoric. I examined the critical reception of *The Kentucky Cycle*, which constituted a "hailstorm of frontierism" (164) in "Frontiers. . . ." (May 1999b)."
14 Schenkkan's *Kentucky Cycle* was part of a wider movement in the arts to examine frontier ideology and imagery. In *The Frontier in American Culture*, James Grossman (1995) explains the media response in the early 1990s to revisionist histories of the American West (the New Western History), along with the critical response to a 1991 exhibition at the National Museum of American Art that called out the ideology embedded in a wider constellation of images about the West and the United States. A controversial 1991 exhibition at the National Museum of American Art, titled "The West as America," enraged some visitors, press, and at least two members of the U.S. Senate. The exhibit suggested that the landscapes of the America West painted by, for example, Frederic Remington, were ideological narratives that reflected the attitudes of the painters and their times especially with regard to race, class, gender, and war. This was not a message, apparently, that the American public or its press was ready to hear at the time (1–5).
15 Many additional examples of the rising movement of community-engaged theater-making are discussed by Jan Cohen-Cruz (2005) in *Local Acts: Community-based Performance in the United States*. Downing Cless (1996) focuses on environmental justice themes in community-engaged theater in "Eco-Theatre USA." Wes Sanders and Debra Wise (2017) detail Underground Railway's community-engaged ecotheater in *Underground Railway Theater: Engine of Delight and Social Change*.
16 For more on the history, mission, and projects of TENAZ, see the following resources: Jorge Huerta, *Chicano Drama* (2000), 2–5; Jorge A. Huerta, "Evolution of TENAZ" (1975), 3–6; Jorge A. Huerta, "Concerning Teatro Chicano" (1973); Joanne Pottlitzer, *Hispanic Theatre in the United States and Puerto Rico* (1988); M. Teresa Marrero, "From El Teatro Campesino to the Gay 1990s" (2002); and Olga Sanchez Saltveit, "TENAZ" (2019).
17 Moraga cites Luis Valdez's first full-length play, *The Shrunken Head of Pancho Villa*, as part of her inspiration for the Cerezita character (89). For more on the impact and problematic of Valdez's shrunken head figure, see also the section titled "Cucarachas on the Walls and Vendidos in the Closet: *The Shrunken Head of Pancho Villa*" in Jorge Huerta's *Chicano Theater* (2000, 49–59).
18 For an in-depth discussion and etymology of the term *Xicana*, and its present-day implications and meanings, see Moraga and Celia Herrera Rodriguez's (2009) *A Xicana Codex of Changing Consciousness*.
19 See Anzaldúa's (1987) discussion of the way Aztec and Mayan female deities have informed Mestiza identity, in Chapters 3 and 4 especially; see also Griselda Alvarez Sesma's (2008) "A Short History of Tonantzín, Our Lady of Guadalupe," and Irene Lara's (2008) "Tonanlupanisma: Re-Membering Tonantzin-Guadalupe in Chicana Visual Art" for more about the spiritual and iconographic melding of these two deities in the context of Latinidad.
20 *Alligator Tales* was originally commissioned and premiered by Berkeley Repertory Theatre. For more information, contact the Joyce Ketay Agency.
21 See, for example, Christopher Hallowell's (2001) *Holding Back the Sea,* which provided a compelling and detailed account of the importance of the Louisiana wetlands and the impact of late twentieth century efforts to control the Mississippi River. For more on this topic, see also Craig E. Colten's (2000) *Transforming New Orleans and Its Environs: Centuries of Change*.
22 Both Cajun and Creole describe amalgamated and distinctive cultures in the southern Louisiana area, yet popular meanings vary depending on context. In general, Cajun

refers to descendants of French Canadians who migrated to the southern Louisiana territory while that large section of the continent was under French colonial control. In contrast, Creoles distinguish from Anglo/Euro immigrants who migrated to the region after the Louisiana Purchase of 1803. Creole culture in southern Louisiana includes influences from the Chitimacha and Houma indigenous tribes, as well as French- and Spanish-speaking West Africans ("Indigenous" 2018).

23 The idea of an "ecological self" – a sense of personhood as multiple and interconnected to the more-than-human world, self as part of an ecological whole, rather than as individual – emerged from the work of deep ecologist Arne Naess and was expounded on by several environmental philosophers, including Thomas Berry and Joanna Macy. See, for example, Berry's (1988) *The Dream of the Earth*; Macy's (1991) *World as Lover, World as Self*, also reprinted (2007) as *World as Lover, World as Self: Courage for Global Justice and Ecological Renewal*; and Freya Mathews' (1991) *The Ecological Self*.

24 Many resources on the Seattle WTO protests exist, among them Alexander Cockburn and Jeffrey St. Clair's (2000) *Five Days that Shook the World: Seattle and Beyond*.

References

Alaimo, Stacy. 2019. *Bodily Natures: Science, Environment and the Material Self*. Bloomington, IN: Indiana University Press.

Alvarez Sesma, Griselda. 2008. "A Short History of Tonantzín, Our Lady of Guadalupe: A Bridge of Light Between Cultures." *Indian Country News*.

Angel, Bradley. 1991. *The Toxic Threat to Indian Lands: A Greenpeace Report*. San Francisco, CA: Greenpeace.

Anzaldúa, Gloria. (1987) 2012. *Borderlands/La Frontera: The New Mestiza*, 4th ed. San Francisco, CA: Aunt Lute Books.

Arons, Wendy, and Theresa May. 2013. "Ecodramaturgy and/of Contemporary Women Playwrights." In *Contemporary Women Playwrights: Into the 21st Century*, edited by Penny Farfan and Lesley Ferris, 181–196. New York: Palgrave MacMillan.

Arrizón, Alicia. 2000. "Mythical Performativity: Relocating Aztlán in Chicana Feminist Cultural Productions." *Theatre Journal* 52 (1): 23–49. https://doi.org/10.1353/tj.2000.0001

Berry, Thomas. 1988. *The Dream of the Earth*. San Francisco, CA: Sierra Club Books.

Billings, Dwight, Gurney Norman, and Katherine Ledford, eds. 2001. *Back Talk from Appalachia: Confronting Stereotypes*. Lexington, KY: University Press of Kentucky.

Brecht, Stefan. 1988. *Peter Schumann's Bread and Puppet Theater*. New York and Methuen, MA: Routledge.

Brook, Daniel. 1998. "Environmental Genocide: Native Americans and Toxic Waste." *American Journal of Economics and Sociology* 57 (1): 105–113. https://doi.org/10.1111/j.1536-7150.1998.tb03260.x

Brustein, Robert. 1993. "Hillbilly Blues." *New Republic* 209 (18): 30.

Chaudhuri, Una. 1994. "'There Must Be A Lot of Fish in that Lake': Toward an Ecological Theater." *Theater* 25 (1): 23–31. https://doi-org.libproxy.uoregon.edu/10.1215/01610775-25-1-23

Chaudhuri, Una. 1995. *Staging Place: The Geography of Modern Drama*. Ann Arbor, MI: University of Michigan Press.

Chávez, César. 1989. "Statement to Pacific Lutheran University." Public address at Pacific Lutheran University, Tacoma, Washington, March.

Chiro, Giovana di. 1996. "Nature as Community: The Convergence of Environment and Social Justice." In *Uncommon Ground: Rethinking the Human Place in Nature*, edited by William Cronon, 298–320. New York: W.W. Norton & Company.
Cless, Downing. 1996. "Eco-Theatre U.S.A.: The Grassroots Are Greener." *TDR* 40 (2): 79–102. https://doi.org/10.2307/1146531
Cless, Downing. 1996. "Ecology vs. Economy in Robert Schenkkan's *The Kentucky Cycle*." *Journal of American Drama and Theatre* 8 (2): 59–72.
Cockburn, Alexander, and Jeffrey St. Clair. 2000. *Five Days that Shook the World: Seattle and Beyond*. London and New York: Verso.
Cohen-Cruz, Jan. 2005. *Local Acts : Community-based Performance in the United States*. New Brunswick, NJ: Rutgers University Press.
Colten, Craig E. 2000. *Transforming New Orleans and Its Environs: Centuries of Change*. Pittsburgh, PA: University of Pittsburgh Press.
Das, R., A. Steege, S. Baron, J. Beckman, and R. Harrison. 2001. "Pesticide-related Illness among Migrant Farm Workers in the United States." *International Journal of Occupational and Environmental Health* 7 (4): 303–312.
Dictionary of Louisiana French. 2009. University Press of Mississippi.
Dolan, Jill. 2001. "Performance, Utopia, and the 'Utopian Performative'." *Theatre Journal* 53 (3): 455–479. https://doi.org/10.1353/tj.2001.0068.
Fee, Elizabeth, and Daniel M. Fox. 1992. *AIDS: The Making of a Chronic Disease*. Berkeley, CA: University of California Press.
Floyd-Thomas, Stacey M., and Anthony B. Pinn. 2010. *Liberation Theology in the United States*. New York: New York University Press.
Fornes, Maria Irene. 2017. *Fefu and Her Friends*. New York: PAJ Publications.
Galjour, Anne. 1997. *Alligator Tales*. Unpublished (courtesy of the Joyce Ketay Agency, New York, NY).
Gottlieb, Robert. 1995. *Forcing the Spring: The Transformation of the American Environmental Movement*. Washington, DC: Island Press.
Greenberg, Linda Margarita. 2009. "Learning from the Dead: Wounds, Women, and Activism in Cherríe Moraga's Heroes and Saints." *MELUS* 34 (1): 178. https://doi.org/10.1353/mel.0.0015
Gribbin, John. 1988. *The Hole in the Sky: Man's Threat to the Ozone Layer*. Toronto and New York: Bantam Books.
Grossman, James R., ed. 1995. "Introduction." In *The Frontier in American Culture*, 1–5. Berkeley, CA: University of California Press.
Gutiérrez, Gustavo. 1973. A *Theology of Liberation: History, Politics, and Salvation*. Maryknoll, NY: Orbis Books.
Hallowell, Christopher. 2001. *Holding Back the Sea*. New York: Harper Collins.
Hansberry, Lorraine. 1995. *A Raisin in the Sun*. New York: Modern Library.
Hansen, James. 2009. *Storms of My Grandchildren*. New York: Bloomsbury.
Haraway, Donna J. 2016. *Staying with the Trouble: Making Kin in the Chthulucene*. Durham, NC and London: Duke University Press.
Henry, III, William A. 1993. "America's Dark History." *Time*, November 22.
Hindley, Jane. 2010. "Breaking the Consumerist Trance: The Reverend Billy and the Church of Stop Shopping." *Capitalism, Nature, Socialism* 21 (4): 118–126.
Horkheimer, Max, and Theodor Adorno. 1996. "The Concept of the Enlightenment." *The Continental Philosophy Reader*, edited by Richard Kearny and Mara Rainwater, 199–214. London: Routledge.
Horn, Miriam. 1993. "The Malignancies of History." *US News and World Report* 115 (11): 74.

Huerta, Jorge A. 1973. "Concerning Teatro Chicano." *Latin American Theatre Review* 6 (2): 13–20.
Huerta, Jorge A. 1975. "Evolution of TENAZ." *Pláticas del Sexto Festival Nacional de los Teatros Chicanos*, July 14–19, 3–6.
Huerta, Jorge A. 1982. *Chicano Theater: Themes and Forms*. Ypsilanti, MI: Bilingual Press = Editorial Bilingüe.
Huerta, Jorge A. 2000. *Chicano Drama: Performance, Society and Myth*. Cambridge: Cambridge University Press.
"Indigenous Tribes of New Orleans & Louisiana." 2018. American Library Association, April 13.
Jacobson, Lynn. 1992. "Green Theatre: Confessions of an Eco-reporter." *American Theatre* 8 (1): 17–25, 55.
Kroll, Jack. 1993. *Newsweek*, November 29.
Kushner, Tony. 1993–1994. *Angels in America: A Gay Fantasia on National Themes*. New York: Theatre Communications Group.
Lara, Irene. 2008. "Tonanlupanisma: Re-Membering Tonantzin-Guadalupe in Chicana Visual Art." *Aztlan: A Journal of Chicano Studies* 33 (2): 61–90.
Limerick, Patricia Nelson. 1995. "The Adventures of the Frontier in the Twentieth Century." In *The Frontier in American Culture*, edited by David Grossman, 67–102. Berkeley, CA: University of California Press.
Lippard, Lucy R. 1990. *Mixed Blessings: New Art in a Multicultural America*. New York: W.W. Norton & Company.
Macy, Joanna. 1991. *World as Lover, World as Self*. Berkeley, CA: Parallax Press.
Macy, Joanna. 2007. *World as Lover, World as Self: Courage for Global Justice and Ecological Renewal*. Berkeley, CA: Parallax Press.
Marrero, Teresa M. 2002. "From El Teatro Campesino to the Gay 1990s: Transformations and Fragments in the Evolution in Chicano/a to Latino/a Theater and Performance Art." In *The State of Latino Theater in the United States Hybridity, Transculturation, and Identity*, edited by Luis A. Ramos-García, 39–66. New York: Routledge.
Mason, Bobbie Ann. 1993. "Recycling Kentucky." *New Yorker*, November 1, 58–59.
Mathews, Freya. 1991. *The Ecological Self*. Savage, MD: Barnes & Noble Books.
May, Theresa. 1999a. "Bahktin on Site: Chronotopes in Theatre in the Wild's *Dragon Island*." *On-Stage Studies* 22: 19–38.
May, Theresa. 1999b. "Frontiers: Environmental History, Ecocriticism and *The Kentucky Cycle*." *Journal of Dramatic Theory and Criticism* 14 (1): 159–178.
May, Theresa. 2005. "Greening the Theatre: Taking Ecocriticism from Page to Stage." *Journal of Interdisciplinary Studies* 7: 84–103.
Merriman, Peter. 2010. "Architecture/Dance: Choreographing and Inhabiting Spaces with Anna and Lawrence Halprin." *Cultural Geographies* 17 (4): 427–449. https://doi.org/10.1177/1474474010376011
Moore, Jason W. 2016. "The Rise of Cheap Nature." In *Anthropocene or Capitalocene?: Nature, History, and the Crisis of Capitalism*, edited by Christian Parenti and Jason W. Moore, 78–115. Oakland, CA: PM Press.
Moraga, Cherríe. 1994. *Heroes and Saints*. In *Heroes and Saints & Other Plays*. Albuquerque, NM: West End Press.
Moraga, Cherríe, and Celia Herrera Rodriguez. 2009. *A Xicana Codex of Changing Consciousness: Writings, 2000–2010*. Durham, NC: Duke University Press.
Moraga, Cherríe, and Gloria Anzaldúa. 1983. *This Bridge Called My Back: Writings by Radical Women of Color*. New York: Kitchen Table Press.
Munk, Erika. 1994. "A Beginning and End." *Theater* 25 (1): 5.

"The Ozone Layer – The Achilles Heel of the Biosphere." 1995. *Royal Swedish Academy of Sciences*, October 11.
"Planet of the Year: Endangered Earth." 1989. *Time*, January 2.
"Planet of the Year: What the U.S. Should Do." 1989. *Time*, January 2.
Pottlitzer, Joanne. 1988. *Hispanic Theatre in the United States and Puerto Rico: A Report to the Ford Foundation*. New York: Ford Foundation.
Pulido, Laura. 1996. "The Pesticide Campaign of the UFW Organizing Committee, 1965–71." In *Environmentalism and Economic Justice: Two Chicano Struggles in the Southwest*, 57–125. Tucson, AZ: University of Arizona Press.
Rich, Frank. 1993. "200 Years of National Sorrows in 9 Chapters." *New York Times*, November 15.
Rivera, José. 1994. *Marisol*. New York: Dramatists Play Service.
Rosenthal, Rachel. 2001. "Gaia, Mon Amor." In *Rachel's Brain and Other Stories*, edited by Una Chaudhuri, 139–160. London and New York: Continuum.
Sanchez Saltveit, Olga. 2019. "TENAZ." In "The Latinx Theatre Commons: Feminist Decolonization in the Early Years of a Movement to Update the Narrative of the American Theatre" 28–75. PhD diss., University of Oregon.
Sanders, Wes, and Debra Wise. 2017. *Underground Railway Theater, Engine of Delight and Social Change*. Boston, MA: URT History Project [Smashwords edition].
Schama, Simon. 1996. *Landscape and Memory*. New York: Vintage/Random House.
Schechner, Richard. 1994. *Environmental Theater: An Expanded New Edition including 'Six Axioms for Environmental Theater'*. New York: Applause.
Schenkkan, Robert. 1991. "Theatre, Myth and Social Change." Keynote address at the *Theatre In an Ecological Age* Conference, Seattle, WA, November 21.
Schenkkan, Robert. 1993. *The Kentucky Cycle*. New York: Penguin Books.
Shepard, Sam. 1979. *Buried Child: A Play in Three Acts*. New York: Dramatists Play Service.
Smith, Adrian A., and Alexandria Stiver. 2015. "Power and Control at the Production-Consumption Nexus: Migrant Women Farmworkers and Pesticides." In *Our Chemical Selves: Gender, Toxics, and Environmental Health*, edited by Dayna Nadine Scott, 364–386. Vancouver: UBC Press.
Standing, Sarah. 2008. "Human/Nature: Eco-Theatre Politics and Performance." PhD diss., CUNY.
Standing, Sarah. 2012. "Earth First!'s 'Crack the Dam' and the Aesthetics of Ecoactivist Performance." In *Readings in Performance and Ecology*, edited by Wendy Arons and Theresa May, 147–157. New York, NY: Palgrave Macmillan.
Starna, William A. 2013. *From Homeland to New Land: A History of the Mahican Indians, 1600–1830, Iroquoians and Their World*. Lincoln, NE: University of Nebraska Press.
Stevens, William K. 1995. "3 Win Nobel Prize for Work on Threat to Ozone." *New York Times*, October 12.
Talen, William. 2003. *What Should I Do If Reverend Billy Is In My Store?* New York: New Press, 2003.
Tesich, Steve. 1991. *The Speed of Darkness*. Garden City, NY: Fireside Theatre.
Tuhiwai Smith, Linda. 1999. *Decolonizing Methodologies: Research and Indigenous Peoples*. London and Dunedin, New Zealand: Zed Books; University of Otago Press.
Turner, Frederick Jackson. 1921. *The Frontier in American History*. New York: Henry Holt & Company.
United Farmworkers of America. 1986. *The Wrath of Grapes*. Documentary film. Online.
Weisskopf, Michael. 1992. "U.S. to End CFC Production 4 Years Earlier Than Planned." *Washington Post*, February 12.

White, Richard. 1995. "Frederick Jackson Turner and Buffalo Bill." In *The Frontier in American Culture*, edited by James R. Grossman, 7–65. Berkeley, CA: University of California Press.
Wise, Debram and West Saunders. 1985. *Sanctuary: The Spirit of Harriet Tubman*. Unpublished.
Worth, Libby, and Helen Poynor. 2004. *Anna Halprin*. London and New York: Routledge.
Worthen, William B. 1995. *Modern Drama: Plays, Criticism, Theory*. Fort Worth, TX: Harcourt Brace College Publishers.
Worthen, William B. 2011. "Tony Kushner. *Angels in America, Part I: Millennium Approaches*." *The Wadsworth Anthology of Drama*, edited by William B. Worthen, 1211–1212. Boston, MA: Wadsworth.

7

KINSHIP, COMMUNITY, AND CLIMATE CHANGE

Marie Clements's *Burning Vision* and
Chantal Bilodeau's *Sila*

Since its formation under the United Nations in 1988, the Intergovernmental Panel on Climate Change (IPCC) has collected and reviewed the most respected scientific research from around the world regarding changes in Earth's climate. The IPCC's periodic reports and assessments have provided governments with conclusive analysis, impacts, and recommendations regarding the overwhelming evidence that anthropogenic climate change poses grave risks to our biosphere. Yet scientific facts, it seems, are not enough to inspire the kind of systemic change required to slow the most catastrophic impacts of global warming (see, e.g., Plumer [2017]; and Pachauri, Meyer, and Core Writing Team [2015]). In 2005, Bill McKibben, founder of 350.org – an organization dedicated to generating the political and public will to reduce carbon emissions, mused that "what the warming world needs now is art, sweet art." The arts, he argues, must translate the facts of climate change into felt experience. "[T]hough we know about it, we don't *know* about it. It hasn't registered in our gut; it isn't part of our culture. Where are the books? The poems? The plays? The goddamn operas?" (McKibben 2005) asked.

As the realities of climate change will surely transform the daily configurations of our lives, they also make demands on our collective imagination, reshaping understandings of what it means to be human in the span of geologic time. Stratigraphers – scientists who determine demarcations between earth epochs – suggest that human-caused climate change has precipitated a geologic change of a magnitude not seen since the start of the Holocene (see, e.g., Davies [2016], 1–104). Yet, even in the face of what Elizabeth Kolbert (2014) calls the sixth extinction, climate denial persists. Variously expressed in developed nations as apathy, defiance, entitlement, fear, or powerlessness, climate change denial is as real and potentially as dangerous as the superstorms and mega-fires that have already consumed lives. Sociologist Kari Marie Norgaard (2011) observes that

DOI: 10.4324/9781003028888-8

this collective denial is socially constructed: "[S]ociety teaches us what to pay attention to and what to ignore" (5), she argues, and when new information is inconsistent with long-ingrained "cognitive traditions" (6), we fail to "integrate this knowledge into everyday life or to transform it into social action" (11). Even in the face of overwhelming evidence, the large cultural matrixes of shared understandings, rooted in shared histories, beliefs, and memories, are the building blocks of social norms and cultural values that work to hold denial in place (Norgaard 2011, 3–12; and Swidler 1986, 273–276). These stories, in turn, fuel social action or rationalize the lack of it, for they govern the very cognitive space in which to imagine a future. Some of those stories – of human dominance of the natural world and the entitlement of some humans to property, profit, and prosperity – have been discussed in the preceding chapters.

As a site of culture-making, in which stories are already manifested *as* action, albeit dramatic action, theater is uniquely positioned to bear witness to this ecological moment, amplify the voices of those most impacted by climate change, and to enact what Norgaard (2016) calls a "revolution of our shared imagination." We must, Norgaard argues, "imagine the reality of what is happening to the natural world . . . imagine how those ecological changes are translating into social, political, and economic outcomes, and . . . imagine how to change course" (B4). Like the levies and retaining walls that are crumbling in the face of rising seas, the myths that have shored up business as usual are giving way to a rising tide of new stories.

Theater in the age of climate change

This historical moment requires theater to rise to the ecological occasion, envisioning nothing short of a reimagined human animal in kinship with our world. As the ecological effects of human agency – once so central to the very definition of drama – circle back to dwarf human history, climate change theater has opened up new dramaturgical questions that challenge theater's traditional focus on human-scale narratives. What are the stories that will inspire a new understanding of, and sense of responsibility for, humanity as a geologic force? How do we newly imagine who we are in the scheme of generational and ecological relations and the ethical obligations they imply? Who do we want to be, and how we want to show up, in relation to the changes taking place? How might theater not only raise public awareness and shore up political will but also provide spaces where we exercise our collective imagination to grapple with our part in the current crisis? How might theater call forth a future informed by empathy, community, and justice?

In this final chapter, I consider the recent uprising of plays and performance events that seek to tell the stories of climate change by provoking new felt empathies across cultures and species. In so doing, these plays also engage questions about how democracy and justice are to prevail in the challenging times ahead. Varied in form and focus, climate change theater serves the crucial and collective imaginative work of feeling into the realities of climate change while radically reimagining ourselves and our world. Critically implicated in the perils of climate change, British dramatists have responded with a bulwark of plays that focus on the vulnerability of their island nation. Steve Waters's 2009 *Contingency Plan*, composed of two plays – *On the Beach* and *Resilience* – draws on memories of the devastating flood of 1953 in an examination of the conflicting personal and the political risks of climate change as individuals, families, regional governments, and nations prepare. Chris Rapley and Duncan Macmillan bend dramatic form into a theatricalized science lecture in *2071*, produced by the Royal Court Theatre in 2014. Moria Buffini and an ensemble of four artists interweave diverse narrative threads in *Greenland*, produced by the National Theatre in 2011 – to name only a few.

Climate change has been slower to ignite the imaginations of North American Anglophone theater artists, but the growing array of works are a testament to a rising artistic will. Many of these projects and productions have been part of initiatives that make use of social media, civic partnerships, and international coalition building. In 2014, Chantal Bilodeau's *Sila* premiered in an Underground Railway Theater (URT) production at Central Square Theater in Boston.[1] Directed by Megan Sandberg-Zakian, URT's production was accompanied by two weeks of "Central Conversations," intended to provoke dialogue and invite examination of public policies, cultural assumptions, and shared experience. Events included lectures, panels, and community discussions with scientists, sociologists, and economists from local universities and civic leaders and activists from community organizations. Also, in 2014, Theatre for the New City hosted a "Festival of Consciences" – a series of public engagement events around climate change activism held in association with the production of Karen Malpede/Theatre Three's *Extreme Whether*. Like Water's *Contingency Plan*, Malpede's play is the story of how climate change can divide families.[2] After its premiere at New York's Theatre for the New City, *Extreme Whether* was produced in Paris as part of ARTCOP21, a global festival of arts and culture focused on climate change and an event ancillary to the historic United Nations' 21st Conference of the Parties under the UN's Framework Convention on Climate Change. Climate Change Theatre Action (CCTA) organized a series of short play readings, performances, and theatrical interventions (also in the runup to COP21).[3] Meanwhile, Climate Lens of New York formed as a network of artists who pursue ongoing collaborative projects focused on what Una Chaudhuri (2017) calls "climate chaos" (1). Building momentum through social media platforms, these and other forums are increasingly mobilizing theater artists to tell the stories of climate change.[4]

The impacts of global climate change are profoundly local and often personal. Theater can powerfully connect the dots between climate change, racism, sexism, and economic injustice through stories that resist environmental and cultural imperialism by amplifying the voices of those most affected by environmental risk. In 2016, Native Voices at the Autry in Los Angeles, California, produced Mary Kathryn Nagle's *Fairly Traceable* (see, e.g., McBride [2017]). Both a lawyer and playwright, Nagle is concerned with the important role of environmental law as a force for defining new rhetorical space in climate change discourse for indigenous sovereignty. Nagle's story of two Native environmental law students at Tulane University in New Orleans brings a spotlight to indigenous communities whose homelands and waterways encompass the southern tip of what is now known as Louisiana. Climate-exacerbated weather events such as hurricanes cause many tribes to continue to suffer the loss of their homelands. As she brings together the extreme weather events of the tornado in Joplin, Missouri, and Hurricanes Katrina and Rita, Nagle also demonstrates theater's innate capacity to bend time and space into the hyper-present of the stage, making imaginative connections between temporally and spatially disconnected effects of climate change. In this way, theater serves as a kind collective imaginative *research* into the lived experience of climate change.

Shonni Enelow's 2014 *Carla and Lewis*, for example, uses the space of the stage and the expressive tools of actor's bodies and voices to envision how the impacts of climate change in the nation of Bangladesh bleeds into the lives of New Yorkers, fusing and folding two places into one (Chaudhuri and Enelow 2014). Colleen Murphy's 2017 *Breathing Hole*, commissioned by the Stratford Festival of Ontario, also moves backward and forward in time in order to encompass the magnitude of the change that Inuit peoples and the animal communities on which they depend have endured in recent decades (see, e.g., Nestruck [2017]). Ellen Lewis's (2017) *Magellanica*, which premiered in 2017 at Actor's Rep of Portland, Oregon, tells the story of eight scientists *wintering* over for eight months to conduct research at the international research station in Antarctica in 1986–1987. The play imagines the moment when scientists conclusively understood why the ozone layer of Earth's atmosphere was thinning and how to stop it. Lewis' narrative indicts the persistent voices of climate change denial as it reminds us that ozone researchers confirmed the fact and cause of planetary warming three decades ago. Whether this emergent array of climate change theater would satisfy Bill McKibben (2005), it is clear that theater artists are no longer oblivious to their role in what many experts insist must be the greatest cultural transformation since the industrial revolution.

In the discussion that follows, I examine two plays as demonstrations of how theater artists might not only call attention to the climate crisis, but also nourish resilience. Both plays demonstrate the generative, ceremonial, assertive capacity of theater to carve out presence from the specter of absence, hope from the burden of despair, aliveness from the realities of annihilation, and love from too-frequent excuses for hatred. Commissioned by Rumble Productions and first produced in association with Urban Ink, *Burning Vision* by Métis dramatist Marie Clements (2003) weaves stories of irreversible exposures and losses as it blurs temporalities to reexamine the history of the making of the first atomic bomb from uranium mined on Dene homelands in northern

Canada. While *Burning Vision* is not a play expressly about climate change, it probes a moment in history when the rate of climate change began an exponential increase (see, e.g., Davies [2016], 102–104). The second play discussed, *Sila* by Chantal Bilodeau (2015b), also speaks from a specific regional and indigenous location into a global conversation about climate justice and indigenous sovereignty within a global context. Written a decade after *Burning Vision*, and the first of Bilodeau's *Arctic Cycle* (eight plays set in various nation-states that border the Arctic Circle), *Sila* interweaves the stories of an Inuit family, climate scientists, regional government officials, and polar bears in the face of a changing Arctic.

Kyle Powys Whyte (2018) points out that the apocalyptic, dystopian vision, which often arises in the popular imagination in relation to climate change, is one that indigenous peoples of the Americas have already lived through:

> Sometimes I see settler environmental movements as seeking to avoid some dystopian environmental future or planetary apocalypse. These visions are replete with species extinctions, irreversible loss of ecosystems, and severe rationing. . . . Yet for many Indigenous people in North America, we are already living in what our ancestors would have understood as dystopian or post-apocalyptic times.
>
> *(47–48)*

Theater has not only an opportunity but also an obligation to help sustain this *living through*. The combined ecodramaturgy of *Burning Vision* and *Sila* (as two examples of climate change theater) serve as models for dramatists who are asking questions about how theater might serve justice, democracy, and ecological healing in the face of the realities of climate change and its impacts on human and nonhumans communities. Their multifold and multivocal stories foreground the way culture, community, and place are inextricably entwined, where the slow violence of global environmental degradation is felt through everyday experience. Both put the past in conversation with the present, the local in conversation with the global, asserting indigenous knowledge within the context of transnational conversations about sovereignty, environment, and climate change. Both ground their transnational stories in the context of specific indigenous lands and cultures, centering ecological interdependencies, resilience, and relationship in ways that move through and beyond apocalyptic visions and despair.[5]

Colonialism, climate change, and the bomb

Scientists suggest that we live on the cusp of a new epoch in which human-caused changes to earth systems have outpaced all naturally occurring geologic, biologic, and atmospheric factors (see, e.g., Davies [2016]; and Bonneuil and Fressoz [2016]). Debate continues about whether the Anthropocene began with the age of colonization, the rise of extractive capitalism and the Industrial Revolution or, more recently, just after World War II when the planet saw an exponential increase in population.[6] Coupled with a rise in fossil fuel use, consumer consumption, urbanization, and nuclear radiation, scientists call this sharp rise in carbon dioxide in the planet's atmosphere during the baby-boomer era is known as "the Great Acceleration" (Davies, 2). In *The Birth of the Anthropocene*, Jeremy Davies observes that the deployment of the atomic bomb would be an apt marker of this acceleration. The start of the new epoch, he argues, could be "set with unimprovable specificity on July 16, 1945, 'at 05:29:21 Mountain War Time'. . . . This is the moment of the Manhattan Project's first nuclear weapon test, Trinity: white light in the predawn New Mexico desert" (102–104). The date, however, matters less than the specter of annihilation that both the bomb and climate change have unleashed in the collective imaginary.[7]

In *Burning Vision*, Marie Clements (2003) tracks the making of the first atomic bomb from the initial discovery of uranium ore on Dene land in the Northern Territories to the detonations of the bombs over Hiroshima and Nagasaki. "I had taken a trip to the Great Bear Lake region with my mother," Clements recalls. One of the largest and deepest freshwater lakes in the world, Great Bear Lake is the center of life for traditional Dene subsistence hunting communities (see, e.g., Kenny-Gilday [2000], 107–118). "I wanted to tell this story of my family's genetic connection to the history of the land up there, and to the running of uranium" (Clements 2003). In *Burning Vision*, Clements maps the hand-to-hand route of the "black rock" from which both radium and uranium are harvested and plutonium is made: from the miners who unearthed it; to the Dene ore carriers who loaded it; the boatmen, stevedores, and "sandwich girls" who worked along its watery transport across Great Bear Lake, the black rock travels down the Mackenzie River, across Slave Lake to Fort McMurray as circles of relation take shape on stage. The rock was bound for Ontario refineries and, ultimately, the labs and test sites of the Manhattan Project, where the first atomic bomb was developed.[8] In a personal interview with me (November 12, 2009) about her research to write the play, Clements reflected that "what was extraordinary to me was that one person's decision not only impacts that person and their community, but has an effect beyond, in this case, an effect that encompasses the whole world." The play's interweaving of seemingly distant lives shows the way – subtle, like dust itself – in which humans across the planet are connected through the vulnerabilities of their skins. Clements's counter-geography challenges how we remember and whom we remember, and gives flesh to the far-reaching and intimate impacts of human action, making transtemporal, transnational, previously invisible relationships explicit.

The play's title refers to a prophecy by a Dene medicine man in the 1880s, who had a dream in which he foresaw the impact of the atomic bomb on his people and on another people in a faraway place. According to Dene oral history, this powerful dream caused him to sing through the night and call to his community to listen to what he had seen (see, e.g., Blondin [1990]; Kenny-Gilday [2000], 109–110;

and Van Wyck [2012], 174–185). Cindy Kenny-Gilday notes that for the Dene and other people of the Northwest Territories "dreams have as much impact on daily life as the cycle of caribou migration." The narrative of the medicine man's dream told of a "deadly material would come from around the rock outcrop of what was to become Port Radium" and that this rock would be "put into a big stick" and taken by a "metal bird." As Kenny-Gilday recounts, the medicine man went on to tell that it would be "dropped on people far away and it burned them all. These people looked just like us" (109). The dream was verified when a delegation of Dene traveled to Hiroshima in 1998 on the anniversary of the detonation of the first atomic bomb (109–110).

Burning Vision affirms the significance and authority of dreams for the Dene and oral tradition for indigenous people more generally. Produced in 2002 at the Firehall Arts Centre in Vancouver, British Columbia, and then in subsequent productions, including at the National Arts Centre in Ottawa, *Burning Vision* pays homage to the Dene delegation's visit to Japan and underscores the authority of indigenous oral tradition. In 1997, the Canadian Supreme Court recognized aboriginal oral history as legitimate evidence of historical fact, similar to written accounts (see, e.g., Kenny-Gilday [2000] 107; and Van Wyck [2012], 174–185). This meant that facts long known by aboriginal people, and communicated over generations, could finally serve as legal evidence of human rights and treaty violations. The decision promised recognition of claims by Dene of the Great Bear Lake region, who had lived with the legacies of radiation exposure for decades. Weaving stories of Dene ore carriers, their widows, and others whose labor was used to make weapons of annihilation, Clements (2003) also calls attention to the "bomb" that is settler colonialism – a force that continues to send out shockwaves of destruction and suffering.

In 1921, according to Dene oral history, a white trapper who lived on Great Bear Lake, was given a black rock by a Dene man named Beyonnie (Kenny-Gilday 2000, 108). The white man traveled south and returned the next season with prospectors. Their "discovery" made Canada a leading producer of uranium in the 1930s and 1940s. At the center of the boomtown that would become known as Port Radium, Eldorado mine exposed both white miners and Dene ore carriers to radiation poisoning and contaminated the lands and waterways.[9] Kenny-Gilday writes of similar stories all along the transportation route: debris and mud that killed fish; dust from the mine that made the snow melt faster; sickened "young Dene men hired to carry one-hundred-pound bags of uranium ore on and off boats, barges and trucks" (108). Contamination destroyed subsistence fishing and hunting and tore at the social fabric of Dene culture. In the 1970s, studies conclusively showed high rates of cancer deaths in Port Radium miners. However, despite community members' stories of the devastating effects of radiation, no studies on Dene workers were ever conducted. The premature deaths of Dene ore carriers were so high that entire villages of widows were left behind. It would not be until the 1990s that studies

Kinship, community, and climate change   **245**

finally documented the 1.7 million tons of radioactive tailings that ended up saturating the land around Great Bear Lake and connected the environmental contamination to health impacts in Dene communities (Kenny-Gilday 2000, 114–117; and Van Wyck 2012, 174–185). After the 1997 decision, Dene stories of the effects of radium mining on their land, which had been ignored for decades, gained legal authority, and Dene people were increasingly able to see how their own loss was connected to loss on a global scale.

Burning Vision is not a linear story, however, but a circular one, and it must be understood as a score for performance rather than a dramatic text to be read. The play's circles of human and environmental relation and impact move out like rings formed by dropping a stone in the water, like the concussive rings expanding from the detonation of the bomb. Time seems to stand still on stage as the arc of the plays begins and ends with a countdown, followed by the "sound of a long, far-reaching explosion that explodes over a long, far-reaching time," and then a cascading flash of detonation (Clements 2003, 20). The story thus transpires in the split second between that first flash of light and its reign/rain of sudden death, providing an expanding, spatial, and relational contemplation of the varied impacts of that moment.

The play is structured in four movements (rather than acts or scenes). The first movement begins with the sounds of the atomic blast, quickly followed by the voices of two white prospectors, the Labine brothers, who, in 1921, claimed to "discover" the black rock. Uranium is personified in the play as Little Boy, "a beautiful Native boy. Eight to ten years old. . . . [T]he darkest uranium found at the center of the earth" (10). The juxtaposition of these dual inciting moments asks the audience to attend to the way settler colonialism finds a material and metaphoric equivalent in the atomic bomb. Expanding circles of action move out from this moment as Little Boy runs away from them into the darkness of the stage. Like valences around the central moments of discovery and detonation, more stories emerge, coming (literally, on stage) into light: a Dene Widow mourns her deceased ore carrier husband by keeping a vigil fire; a young Métis woman earns her living making bread to sell in the boomtown that grows up around the mine. An Icelandic captain shepherds the ore down the waterways of northern Canada, on its journey to Los Alamos, New Mexico. Movement by movement, these individuals collide with one another like particles in a nuclear explosion. There, Little Boy meets a hybrid character named Fat Man, both a test dummy and a character who personifies the ideology of domination that drove the development of the bomb. Named for the two bombs that the United States dropped on Japan, Fat Man and Little Boy live in a mock home in Nevada's Jornada del Muerto, where the atomic bomb was first tested. As circles upon circles of impact widen, temporal realities fold in on one another. Characters continue to emerge, imply seemingly impossible relations: a miner who works deep in the tunnels under Great Bear Lake meets a blonde radium dial painter working in a factory, where, in the 1920, female workers were exposed to radiation poisoning as they painted dials on military clocks. She is drawn deep into the tunnels of the mine by her desire to see where the substance came from that has caused her disfiguring cancer. Each movement returns to the detonation and the See-er's prophecy. Dene and Japanese people are hurled into relation by an event that shattered time and space. Koji, a fisherman in Hiroshima, has just caught a trout on the other side of the world. At the time of detonation, he falls through the center of the earth and into Great Bear Lake. There two Dene Stevedores working aboard the *Radium Prince* drag him from the lake like a giant fish.

As the Dene See-er's voice propels the four movements forward, temporal and cognitive boundaries blur and geographic distances collapse. Acts of remembering across time take shape as relation. Like a wheel of consequences that lead us back, again and again, to examine and reexamine the extractive act that took the black rock from Dene homelands and sent shockwaves out across the world. As each movement opens up a different aspect of how Beyonnie's prophetic dream came true, casting a widening net of relations and moral obligations (see, e.g., Abel [2005]). In this way, *Burning Vision* functions within what Laguna Pueblo poet and theorist Paula Gunn Allen (1992) calls "ceremonial time." This "Achronology," Allen explains, represents the "tribal concept of time [that is] timelessness" in which both time and space are multidimensional. In this indigenous perspective, which is consistent with theories of relativity in contemporary physics, the self is conceived "as a moving event within a moving universe" (69–70). As time expands and contracts, bending on and within itself into a hyper present – the simultaneous time/space in the now/here of the stage – 18 characters from various space/times emerge, collide one into the other like electron valences in an atom. As the characters come into relation with one another, the historical events that have impacted them open up for examination, like petals of a rose.

In 2010, I directed *Burning Vision* at the University of Oregon. While not a professional production, a description of the staging may help readers to envision its horizontal storytelling.[10] We wanted a scenography that would honor Clements's invocation of ceremonial time/space, serve the circular movements, and invoke the play's implicit aim to show solidarity between the people of Japan and the Dene villagers of Great Bear Lake. Designer Jonathon Taylor constructed four entrances to a circular stage space, echoing the four movements, the four seasons, the four geographic locations, and the four elements of the play's imaginary landscape. One of these entrances was a Torii Gate, from the Shinto tradition, which signifies a transition out of everyday life into a sacred space. The other entrances were a wooden structure that was at once the inside of mine, a fishing dock at water's edge, and the deck of a transport ship; another, the bow of a cherry tree; another, the tall tower of the Trinity test site. Actors entered and exited through these transitional "gates" into a circular playing space organized around the Widow's fire. When not playing a character in a scene, actors sat on the floor around the circle, bearing witness with the audience to the unfolding stories. The actors playing the radio voices (Canadian broadcast announcers, Dene community radio voices speaking Slavey, and the voice of the Dene See-er) performed from a walkway outside and above the stage, as a space typically used only by the crew to set lighting and electrical instruments. Using this area as the *sphere of orality* allowed the players to be seen, yet separated them into a liminal space – the world of airwaves and oral histories. This staging meant the carriers of information and stories were seen by the audience – a reminder that oral histories and virtual spaces also contain bodies and tellers whose stories touch lived experience of listeners.

Tuned to a time immemorial when traditional Dene communities follow the migration of caribou around Great Bear Lake in the Northern Territories, Clements asserts a set of values consistent with the moral arc of seven generations. She returns to the prophecy of the Dene See-er at the start of each movement, accompanied by the sound of caribou hooves performing their seasonal migration around the lake. When so many men were tapped to work in the uranium transport system, these traditional lifeways were also interrupted. Radiation, as well as debris and disturbance of the land by mining operations, also resulted in declines to animal populations. The reoccurring sound of the caribou carries the story forward and back, at the same time, to a time when the land sustained the people.

The first movement quickly weaves Dene traditional lands at Great Bear Lake with the shoreline of the fishing village of Hiroshima to the Nevada test site in New Mexico, binding these places – which represent not only nations but also people, families, and lifeways. In the darkness of the theater, the Brothers LaBine collide with walls and objects, displaying both their ignorance and fear as they discuss what to trade for the rock they have come to claim and exploit.

> What's an Indian gonna do with money? We'll give him some lard and baking powder and he can bake some bread. Sure! What the hell! What the hell is an Indian going to do with a rock anyways, at least he can eat the bread.
>
> *(14)*

Kenny-Gilday (2000) writes that Dene oral tradition long told of the dangers of the black rock through stories that advised people to keep a distance from the area now known as Port Radium (see also Blondin [1990]; and Van Wyck [2012]). This Dene worldview, in which all elements of the land have aliveness, is reflected most clearly in the character of Little Boy. Indigenous and autonomous, he asks the audience to imagine into the point of view of this vibrant matter: "Every child is scared of the dark, not because it is dark but because they know sooner, or later, they will be discovered" (20). The implicit critique of the doctrine of discovery is spoken from the point of view of that which is discovered, that which is touched by the act of laying claim.[11] "It is only a matter of time," Little Boy whispers, "before someone discovers you and claims you for themselves. . . . Not knowing you've known yourself for thousands of years." (20–21). Like Native children who were removed from their families and placed in residential schools, Nelson Gray (2012) points out that Little Boy is also "a potential survivor of such abuse, whose very expression of what he fears contains the seeds of resistance and hard-won self-respect" (30). Once loose on the earth, Little Boy cannot return. He is chased, captured, and then escapes and runs away, always seeking to go home. Little Boy is also "the voice of the earth itself and what we all might make of it, depending, that is on how we treat it and we treat another" (30). The theme of *making* raises questions about the relational obligation implied in both the act of making and the transaction that follows. What will we make from the substances of the Earth, things that will sustain or destroy life?

Meanwhile, the character of Rose, a young Métis woman, enters the circle, carrying a sack of flour over her shoulder. As she walks, a thin stream of flour leaks out, inscribing a circle on the stage in which audience members are implicitly included. Throughout the performance, Rose makes bread, recounting her mother's recipe – knowledge that has been handed down. Her actions become ceremonial as she returns to the dough, kneads again, lets it rise, punches it down, and folds it over into itself. As the central material metaphor of the play, Rose's bread-making helps the audience to make sense of how the play unfolds: worlds fold in on one another as time and space bend and twist in a transformation of substance. As Rose mixes the ingredients for her bread throughout the four movements, ("substances meeting like magic") characters fall into sudden relationship (39). Both bread and ore are the earth's body-becoming-human-body; the flesh of our bodies interweaves with the plants and animals we take for sustenance, both transmutable and permeable. The play kneads multiple realities together as a reminder that making anything carries a moral obligation.

Rose likens herself to a "perfect loaf of bread" that "is plump with a rounded body and straight sides. I have a tender, golden brown crust which can be crisp, or delicate. This grain is fine and even, with slightly elongated cells; the flesh of this bread is multi-grained" (58). Each of us is just such a grainy substance; we make and unmake ourselves, Rose suggests, by the way we engage the elements of the earth. Rose's bread-making calls attention to the web of ecological and personal relationships that are a product of our makings. As Rose draws attention to things made by human hands – bread, bombs, enemies, love, family, beaded jackets – she suggests that people must take collective responsibility for what they have made. Labor, after all, is a site of human kinship with the things of the earth, an expression of our intimacy with the environment.[12] We touch others and are touched by them in every moment of each day through the products and goods that we exchange.¹³ Work implies both a transaction and a transformation, Rose's kneading implies, one by which we continuously remake the world and ourselves.

By the third movement, the sacks of flour become indistinguishable from the sacks of radioactive ore. The wind mixes the white flour leaking from Rose's sack with the black dust that infects the environment. "The wind's blowing it everywhere," Rose observes. "The kids are playin' in sandboxes of it, the caribou are eating it off the plants, and we're drinkin' the water where they bury it. . . . I guess there's no harm if a bit gets in my dough" (103). In performance, this infiltration of the dust is reinforced by the physical closeness of the audience to the playing space. The audience can see, feel, and perhaps taste the flour dust spreading from the stage. They might viscerally imagine how the "dust" of the uranium mine became so much a part of the environment that people were breathing it and consuming it in water and subsistence foods, that caribou were eating it as they grazed on vegetation: Earth's body and becoming human bodies.

As she sits by her fire, the Widow of a Dene ore carrier talks to her husband's boot – his figurative and material point of connection to the land. She cannot seem to let the boot go, even though traditional Dene practice is to burn the earthly possessions of those who die so that they may cross over. Neither can she let go of his clothes, especially a beaded jacket that she made for him. Her refusal is a kind of resistance. She will not be silent; she will not let him go silently:

> It is always the little things of his that take my breath away. The real things like a strand of his hair lying on the collar of a caribou hide jacket he loved . . . the real things like the handle of his hunting knife worn down from his beautiful hands that loved me. The real things.
>
> *(87–88)*

The Widow's refusal to release the material remains of his life and hers constitutes a refusal to let his death remain invisible as it calls attention to events that were ignored by the Canadian government for decades. The Widow affirms the vibrancy of matter and material culture when she speaks to his boots and the beaded jacket she made for him. She insists he not be forgotten and honors his life and living. Her unwillingness to let go of her ore-carrier husband is not denial in the face of grief but rather an insistence that attention be paid to the still-ongoing effect of radiation in the lives and on the land of Dene villagers. In performance, her vigil *becomes an invitation* (to the audience as well as the players) to bear witness to a past that has not been forgotten by those who lived it, and that demands to be remembered by those who benefited from it.

Clements's storytelling moves freely and fluidly between times and places to reclaim and assert Dene rights to traditional lands. By firelight, the Widow conjures her ore-carrier husband in an act of mourning that expresses the permeable interweaving of bodies and land and asserts her long-standing kinship with traditional Dene lands: "I miss the smell of sweat on his clothes after a long day hunting. I miss how the land stayed in the fabric even when he got inside the cabin" (44–45). She sees him in the flames and pulls him to her as if in a dream. Calling on their historic kinship with the earth and resisting the doomsday change that her waking hours struggle to comprehend, she coaxes him to return: "There are plenty of trout and caribou to last us till we die," she whispers to his ghost (70). Reciprocity with the land, which has sustained the Dene since time immemorial, is vulnerable and at risk as a result of uranium extraction. But here, the Widow's conjuring is proactive, standing in solidarity with Dene activists who work to have these environmental injustices redressed. At the time of the Dene delegation's trip to Hiroshima in 1998, the Canadian government had not admitted knowledge of the health impacts of uranium mining in Dene communities. Her presence at the center of the play and center stage asks witnesses not only to remember but also to consider what reparations are due to families and communities whose suffering went unacknowledged and unstudied for decades (see, e.g., Kenny-Gilday [2000]; and Van Wyck [2012]).

Finding heart in the world

In the fluid, unstable, interdependent world of Clements's play, where time stands still and space is permeable, people fall into one another's worlds as easily as cherry blossoms fall from the branch, as easily as flour is carried on the wind. When Koji is caught in the detonation that begins the first movement, he cries out, "Pika!" – a word meaning "the light of two suns" (36). In the third movement, as soundscape and lighting effects carry the memory of the fireball that swept out from ground zero in Hiroshima across the stage, Koji does not die. Instead he journeys through the center of the earth, surfacing in Great Bear Lake. Here, Clements re-renders the Sahtu' Dene legend of the Waterheart, or Tudzé, in which a medicine man journeys and finds a giant heart beating at the bottom of Great Bear Lake.[14] As Koji follows the trajectory of the Dene story of Tudzé, Clements seems to suggest that both Dene and Japanese might find heart – courage, forgiveness, love, solidarity – not only in shared loss but also shared survivance.

Koji falls between worlds, slipping through the tunnels carved out by legends that connect past and present. Thinking he is a giant trout, two Stevedores on the *Radium Prince* haul him out of the water. On board the *Radium Prince*, Koji receives dry clothing and bread from the ship's newest crewmember, Rose, the bread-maker. Like the unexpected discovery of the great heart that gives life to the world, or, as Rose says, like "two substances meeting like magic," Koji and Rose fall into love (39). Even in the face of destruction, the Dene See-er suggests, the opportunity to be in relationship exists:

Can you read the air? The face of the water? Can you look through time
and see the future? Can you hear through the walls of the world? Maybe
we are all talking at the same time because we are answering each other
over time and space.

(75)

Koji's journey is possible only in legend or in theater: "There was this light in the
sky when I was talking to this trout and then I was flung and I landed on a branch
and then I tried to grab a loaf of bread then and I was swimming" (89). This
crashing together of worlds – the Dene's and the Japanese's – not only makes the
breakage palpable in both communities but also asserts a possible future where
that which is *not* broken carries on.

Tough like hope

First Nations playwright Monique Mojica (2009) affirms that

> by performing possible worlds into being . . . by embodying that wholeness
> on the stage, we can transform the stories that we tell ourselves and project
> into the world that which is not broken, that which can be sustained, not
> only for Aboriginal people, but for all people of this small, green planet.
>
> *(2)*

Mojica suggests radical generosity, one that refuses apocalyptic futures, asserting
instead living through and loving into the future. Even as she recognizes and
recounts the past through an indigenous point of view, Clements carves an opening into that history where light can shine through to nourish new life – a way
of living forward that responds to the sudden revelation of relatedness contained
in the ancestral vision.

By the fourth movement, Rose is pregnant with Koji's child, and friendship has grown between Rose and the Widow. The final scene wraps back to
the moment of impact, superimposing the two catastrophes so the audience
now understands that the bomb "fell" not only on Koji and his grandmother
but also fell over time on the lands of the Dene. Like the Widow's ore-carrier
husband, Rose loses her life to the radioactive dust that contaminates her
environment and the very bread she made. But the child she conceived with
Koji, the Japanese fisherman who fell through the world, lives on, and the
Widow embraces him as her own. "You look like her. You look like him.
You are my special grandson. My small man now. My small man that survived. Tough like hope," the Widow tells him (212). In the Widow's refusal
in the face of apocalyptic reasons for despair, Clements resists the narratives
of annihilation so common to stories not only of nuclear holocaust but also
of futurist fictions about climate change. She answers the question, "What

carries on?" with a clear message of determination not only to survive but also to thrive.

As ceremonial theater, *Burning Vision* turns the connections forged by the making and use of the atomic bomb into a vision of an intimate world in which people across time and across geography come into relation. We are not one flesh, but we are *of* one flesh – the fibers and sinews of the ecological world that connect us in material and spiritual ways. What we do or fail to do has consequences for people, other creatures, and the land in which we all live – this is how we fall into one another's worlds.

Kinship, community, and climate justice

In 2005, Sheila Watt-Cloutier, chair of the Inuit Circumpolar Commission, sued the United States for its precipitous role in global climate change. She argued that actions (and inaction) of the largest producer of carbon emissions – the cause of climate change – constitutes a violation of the human rights of indigenous peoples of the Arctic under the 1948 American Declaration of the Rights and Duties of Man (Inter-American Commission on Human Rights).[15] In 2007, Watt-Cloutier gave testimony before the Inter-American Commission on Human Rights, articulating how climate change harms traditional peoples of the Arctic, as well as indigenous people and vulnerable populations throughout the hemisphere and the world (Watt-Cloutier 2015, 224–258).[16] She explained that climate change has already effected the lifeways of indigenous peoples of the Arctic hemisphere in multiple ways. Damage to land, loss of permafrost, sea-level rise, the loss of sea ice, and other changes to the environment have, in turn, caused a loss of traditional lifeways, values, and culture, all of which negatively impact the physical and mental health of Inuit communities. Her case, in partnership with Earthjustice and 62 Inuit hunters and community members, was based on the 2004 Arctic Climate Impact Assessment and the oral testimony, memories, and stories of Inuit community members – sources respected as authentic and legitimate historical evidence under Canadian law since 1997. In 2007, even as her petition was denied by Inter-American Commission, Watt-Cloutier was a nominee for the Nobel Peace Prize. Her activism shifted the conversation such that today it is not possible to talk about mitigating the impacts of climate change without also talking about climate justice.

In *Sila*, Chantal Bilodeau (2015b) takes up questions of climate justice in relation to geopolitics, local politics, and ecological reciprocity on Baffin Island, in the Inuit territory of Nunavut, Canada.[17] Set in and around the village of Iqaluit, a community on the front lines of climate change, *Sila* interweaves three storylines of a climate scientist, an Inuit family, and a mother polar bear and cub. Together these threads interrogate the tensions between indigenous and nation-state sovereignty, and between diverse ways of knowing. At the center of the intersecting stories, Leanna, an Inuit climate justice activist inspired by

Watt-Cloutier, struggles to balance the two worlds that need her – her daughter and grandson and her geopolitical work. Leanna's daughter, Veronica, a schoolteacher and poet, is modeled on contemporary Inuit vocal artist Tanya Tagaq and spoken-word poet Taqralik Partridge, whose poems are used with permission in the play.[18] Leanna and Veronica embody a commitment to family, community, land, and culture through contrasting choices – one working tirelessly in the international arenas; the other taking a stand locally through education, cultural sovereignty, and art-as-resistance. While Leanna struggles with international politics, Veronica's son Samuel is in crisis. Bilodeau draws on Watt-Cloutier's testimony regarding the effect of climate change on the fabric of lifeways in Inuit communities for the central crisis in the play. The loss of Inuit cultural practices has caused increases in rates of youth addiction, depression, and suicide, and in the play Veronica's son Samuel is at risk (see, e.g., Vinyeta, Powys Whyte, and Lynn [2015]; and Nagel [2016]).

Inuit tradition is centrally represented in *Sila* by the character of Tulugaq, an Inuit elder, hunter, and stone carver. His knowledge of the land and skills as a guide are sought by Jean, a New York–based Quebecois climate scientist who has come to Iqaluit to conduct research. The relationship between Jean and Tulugaq changes over the course of the play, but in the opening scenes, Jean demonstrates what Veronica calls a "typical *qallunaaq*" (39) – another white researcher who objectifies indigenous people as objects of study.[19] Jean's relationship with Tulugaq dismantles the primacy of scientific knowledge as the dominant discourse, asserting that the knowledge of lived experience and traditional ecological knowledge are also fundamental to meeting the challenges of climate change.

Sila raises questions of nation-state claims to sovereignty through the characters of two Canadian Coast Guard officers: Thomas, an Anglophone commanding officer, and Raphaël, a young Québécois second officer, who manage the feeding frenzy by nation-states that would lay claim to the oil under the melting Arctic ice where neither Canadian nor Nunavut sovereignty extend. The coast guard station becomes the site of struggle that lays bare the global implications of melting Arctic ice as a German exploratory ship becomes stranded in a storm. The play suggests that sovereignty in/of a place reflects the real, lived investment of a people (or species) on land and is earned by roots in, love of, and commitment to it. Geopolitics comes into sharp contrast with place-based sovereignty as Thomas and Raphaël struggle to save crew members aboard the German ship.

The third storyline follows a Mama Polar Bear and Daughter, *Paniapik*, as they try to survive in a world where the winter ice on which they depend is melting too soon and too fast. However, these polar bears are not the megafauna so ubiquitous in climate change iconography. Instead these bears are the ecologically and culturally situated autonomous beings, which with whom Inuit communities have long shared a mutually life-sustaining relationship. This relationship is established in the first moment of the play as Tulugaq stands on stage, listening to the sound of "wind [that] morphs into a sort of breathing. . . . The breathing turns into Inuit throat singing" (11). Tanya Tagaq (2009) explains that this form of singing is performed by pairs of Inuit women through "vocalizations on both the inhale and the exhale." In practice, performers face one another, voices fueled by a shared circular breath.[20] In the play, Mama teaches Paniapik that "*sila*" refers to the material-ecological life-sustaining connection between and among Inuit communities, polar bears, and the land. Through this indigenous framework, the play repositions the polar bears as knowledgeable, feeling beings who live in a world that also happens to be inhabited by humans. As Mama teaches her cub the ways of their kin, and as she knows that there will come a time when her cub must survive in the world without her, she speaks the knowledge that is indigenous to her and to the people with whom she shares the continuous vital world:

> All life is breath. From the original and eternal breath from which Creation is drawn to the world itself, Sila wraps itself all around us . . . Sila holds the stars above our heads, carries the clouds across the land, ripples the ocean with waves . . . and moves in and out of our lungs. *(She breathes)* See? That's Sila. And with each breath, Sila reminds us that we are never alone. Each and every one of us – from the lichen on the tundra to the beluga whale in the oceans to the snow goose in the sky – is connected to every other living creature.
>
> *(35)*

Sila is not the mere notion of oneness, but an understanding that arises from the lived experience of Inuit people whose distinct ways of knowing have enabled them to thrive within the specific ecological relationships of the Arctic. It is the fabric of this sustaining relationship over time and the represents bodies of knowledge, lifeways, and livelihoods at risk from the impacts of colonialization and climate change.

From the opening scene, *Sila* centers recognition of indigenous knowledge and sovereignty within a geopolitical arena. In a passage reminiscent of Watt-Cloutier, Leanna asserts the authority of indigenous traditional ecological knowledge to the Circumpolar Commission about her land, culture, and people:

> I come from a place of barren landscapes and infinite skies. I come from a place of rugged mountain, imperial glacier, and tundra covered permafrost. I come from a place where north is where you stand and south, everywhere else. . . . I come from a place where you can walk onto the ocean, and if you're lucky, beyond the horizon itself. I come from a people who have kept accounts of the early days when the world was rich and

> urgent and new. . . . When spirits roamed the land like polar bears and muskoxen and caribou.
>
> *(12)*[21]

Staged with Leanna speaking directly to the audience, Leanna interrupts mainstream (northern)hemispheric bias, orients us to an ecologically distinct world, and implicates us in the circumstances she describes through a strategic use of "we" that links the loss of lifeways to changes in the land:

> This place I come from we call Nunavut. It means "Our Land" in Inuktitut. It's where we, Inuit, have thrived for more than four thousand years. It's where we strive to realize our full potential. It's where we nurture our knowledge of who we are. But Nunavut, our land, is only as rich as it is cold. And today, most of it is melting.
>
> *(2)*

As Leanna describes the land as the foundation of Inuit identity, she asserts Inuit cultural and geographic sovereignty, echoing Watt-Cloutier's argument that the willful destruction of place and culture constitutes a violation of sovereignty and human rights.

Leanna's argument, and the struggle of Inuit communities within a geopolitical context, becomes more apparent in the next scene. Thomas and Jean walk across an expanse of ice, stopping to watch a polar bear and her cubs in the distance. They discuss the mounting pressures by nation-states and oil companies to conduct Arctic drilling. Thomas uses the bears to make a political point as he pressures Jean to conduct an environmental assessment for a new seaport that would assert Canadian sovereignty over Arctic waters:

> You know, biologists use to think polar bears followed the movement of ice around the Pole. They'd have the babies in Canada, raise them Russia, and breed again in Svalbard or Greenland the following year. But turns out, polar bears are extremely faithful to where they come from.
>
> *(15)*

Jean, who now lives in New York and has different research plans – specifically, to study "one of the last remaining sheets of multiyear ice predicted to break away from the coast this summer," reads between the lines and flashes back: "Parce que tu penses que c'est une affaire de territoire?" (13). Thomas, who is not bilingual, still understands the retort: "Yes. It's about national security, control, diplomatic relations, and most of all, money. . . . Somebody's gonna drill, Jean. If it's not us, it'll be the Americans, the Chinese, the Arabs, whoever the fuck, but somebody's gonna drill!" (16). Thomas goes on to make the case for Canadian sovereignty that stands in sharp and ironic contrast to Leana's argument for Inuit territorial sovereignty over the lands and waters of Nunavut.

The question of *ecological* sovereignty compounds that of national and indigenous sovereignties through Mama and Daughter polar bears, who are fully developed characters with lives and lineage. The polar bears address one another in the familiar, as *anaanaa* and *paniapik* (mama and daughter), signaling that these are not just anybody's polar bears. Bilodeau has invoked storytelling as an indigenous framework for understanding the workings of the ecological world and invites a suspension of disbelief on the part of the non-Native audience. Mama and Daughter share homelands with the Inuit, and this cultural ecology gives rise to shared cosmology. "It is my duty to make you a good hunter. . . . So you may live a long life and be a great *nanuq*," Mama tells her. "Like the *nanuq* who climbed into the sky? . . . [T]ell me the story again!" the cub cries (18). The story Mama tells establishes a cosmology of interdependence and reciprocity, of an era "when strange power were shared among animals, people, and the land, when all creatures spoke the same tongue and traded skins with ease" (19). *Sila*'s bears thus resist objectification because Bilodeau situates them with an indigenous framework of Inuit cultural knowledge and philosophical tradition provides a kinship place for the ice bear as co-habitant of this Arctic environment.

The cosmology of *sila* is not a universe of mute and inert matter but one that is brimming with intelligence, in which all elements are in constant dialogue. Survival is insured through the transfer of hunting skills from one generation to the next. In an early scene, Mama teaches Paniapik to "read" the surface of the water, a survival skill that underscores the bear's knowledge gained through lived experience, and demonstrated through her intentional transference of that knowledge to her cub. Mama and Daughter wait at the edge of a seal breathing hole in the ice:

MAMA: Stay low, *paniapik*. . . . The seal sees you. He sees the shadow of your paws moving across the ice. The seal hears you. He hears the symphony of ice crystals shifting under your weight. You must learn to be attentive and silent.
They wait.
DAUGHTER: (*whispering*) Anaanaa, how will I know when the seal comes?
MAMA: The bubbles in the water will tell you that he is here. Stay low, *paniapik*. Downwind of the aglu and low.

(17)

The scene establishes that polar bears have knowledge specific to their lived experience that is interconnected with that of the indigenous people with whom they share place and sustenance. Mama explains to *Paniapik* that the seal, too, has knowledge, perception, and wit. The seal can also read. It listens to the language as articulated by ice crystals as they bend and whisper the predator's presence. The world in which both bear and seal live is alive with a symphony of meaning.

The problem of staging polar bears

Many dramatists writing on ecological themes have struggled with the task of representing nonhuman others. Una Chaudhuri (2014) has described this effort as "theatre of species," which, in the face of environmental risk and disaster "reminds us that we humans are one species among many, among multitudes, all equally contingent and threatened." The theater of species, she writes, "brings the resources of performance to bear on what is arguably the most urgent task facing our species: to understand, so as to transform, our modes of habitation in a world we share intimately with millions of other species" (50). Bilodeau's Mama and Daughter polar bears fall within this project of representation for the purpose of critiquing, rather than asserting, human-centric values. Yet it does so within a wider cultural milieu in which the very image of the polar bear is an oft-used rhetorical device, calling up almost instantly the very meanings that theater of species and ecodramaturgy work to resist. For example, polar bears featured in climate change discourse pull at sentimentality, the Coca-Cola bear is a commercial strategy, and in popular culture, the stereotype of "Nanuq of the North" has functioned as a stand-in for racist depictions of Arctic peoples.

Representing polar bears on stage carries ethical responsibilities. Alaimo (2014) observes that in light of the very real suffering – both human and nonhuman – exacted by environmental degradation, we must proceed with caution in representation. Meaning-making has become a "swirling landscape of uncertainly where practices and actions that were once not even remotely ethical or political matters suddenly become so" (16–17). In 2014, I directed a production of *Sila* at the University of Oregon, where we struggled to resist the tendency of audience members to read the polar bears as cute but helpless victims. Instead, we sought to invite the audience to understand the specific cultural and ecological context in which Bilodeau had constructed these polar bears and to resist their oversaturated symbolic meaning.

Alaimo (2014) proposes that as embodied animals, humans are necessarily part of an ecological entanglement that includes not only our biological interdependence with all that surrounds us and includes all spheres of culture. Humans "cannot be disentangled from networks that are simultaneously environmental, economic, political, cultural, scientific, technological, and substantial." These entanglements constitute what she calls "transcorporality," which "connect human bodies, animal bodies, ecosystems, technologies" (16–17; see also Alaimo [2010]). In our investigation of how to represent the polar bears of *Sila*, we wanted to foreground not the "object" of the bear, but rather the interweaving of human and animal lives that sustains both. We approached the representation of Mama and Daughter *as a process of inquiry* through physical engagement, rather than as the task of making and (re)presenting a *thing*. In *Zoographies: The Question of the Animal*, Matthew Calarco (2008) suggests that such an "encounter with what we *call* animals" produces a kind of "clearing of the space for the *event* of what we call animals" (4, ellipses in original). Taken together, Alaimo and Calarco's ideas ultimately prompted us to consider the characters of Mama and Daughter as an open question, as an invitation to explore what we might share, and confront what we cannot know.

Can embodied performance intervene in species privilege and instead perform an inquiry into the possibility of kinship? We hoped that representing nonhuman animals through human bodies could be a collective contemplation on both the socially constructed animal and our ecological animal kin and cohabitant. After all, we have much in common: a requirement for food and shelter, rest, attraction and affection, familial connections, and social associations; as animals, we experience pain and contentment. Shared flesh, breath, and blood is a way of knowing our animal character(s) beyond, and in ways inaccessible by, discursive language. My cast and production team worked toward a representation of the bears that invited an active contemplation of kinship. It required imagining the flesh of our common flesh and the concurrence of our suffering.

Philosophical arguments alone, Calarco (2008) maintains, will not "suffice to transform our thinking about what we call animals." He suggests that "any genuine encounter with what we call animals will occur only from within the space of surrender" (4). In the case of my production of *Sila*, our work began with our own bodies, for even as actors may stand in for imagined others, the actor's body is an ecological system in present time: bodies burn fuel, exchange oxygen and carbon molecules, and otherwise carry on their aliveness. The community of bodies – actors and audience – also form an ecological system that affirms the interweaving of body and environment merely by breathing. Six actors used a contact improvisation process to breathe and move together as one body/community. Actors began with touch, then breath, and then began moving, breathing together, always moving, always touching, bearing one another's weight when necessary. The interwoven lives of bears and humans could be signaled through the actors' embodied connection to one another. Collectively and nonverbally actors *felt* into imagined experiences of the polar bear characters, experiences that they, too, have had: hunger, fear, love, generosity, celebration, grief, rage, loneliness. The actors moved as if one organism, allowing these imaginative suggestions to filter into and inform their interwoven movements. What began in rehearsal as an exploration in rehearsal continued in performance as the ensemble of moving human bodies played the Mama bear from scene to scene, with the single addition of a puppet bear head, designed to reflect Inuit stone carvings suggested by the Tulugaq carving.[22] The result was an imagined polar bear that was also a community of bodies. Indeed, Inuit use every part of the bear when hunted, and one bear sustains many people. In this way, we not only felt into what Calarco calls "the space of the animal," but we also metaphorically represented the central value of the bear to a village (4; see also Wolfe [2003]).

Thus conjured as a process of inquiry, the Mama bear was more than human-playing-animal. Rather, she was an active, embodied moment-to-moment act of wonder, as the actors' energies engaged in a conversation with the idea of bear. Applying theater as a way of knowing in which a dynamic, emergent, embodied experience allow us to encounter not only our concepts of the polar bear but also the unknown Other, who exists beyond those constructions. Through their kinesthetic engagement as a community of bodies, performers not only opened a space of contemplation regarding the kinship between bear and human. In a practice of active meditation, they committed their minds and muscles to imagining into all they know and did not know about polar bears.

Language and storied knowledge

Three human languages are spoken in *Sila* – English, French, and Inuktitut – and while Bilodeau has written primarily for an Anglophone audience, she has deployed language in an incisive critique of Euro/Canadian/American colonialism and its culpability for the impacts of climate on Arctic lives and lifeways. Some characters speak a combination of English, French, and Inuktitut, thereby calling attention to not only the ways language is part of the legacies and ongoing structures of colonialism but also how language is connected to land, bodies, and breath. The use of multiple languages in *Sila* argues for a multivocal understanding of, and response to, climate change, one that seeks to include not only the language of Western science but also the policy priorities of nation-states and multinational corporations. In performance, context allows non-speakers of any of the three languages to understand the sense of what has been said.

The characters of Leanna, Veronica, and Tulugaq code switch between Inuktitut and English, sometimes struggling because ideas do not always translate, sometimes finding comfort in the words and expression that cannot be translated from the indigenous language into the language of the settler. Veronica's spoken-word poetry powerfully reclaims indigenous words (such as *kayak*) that have been appropriated into colonial languages spoken by people like Jean. Similarly, Raphaël and Jean, both bilingual, move fluidly between French and English, sometimes code-switching mid-sentence, leaving Thomas to struggle in his monolingual understanding. The play is written primarily in English, and context allows nonspeakers of any of the three languages to understand the sense of what has been said.

Early in the play when the two men meet, the exchanges merely consist of friendly chatter about living in New York and the fact that Raphaël's wife is pregnant. The scene, like the Inuktitut spoken by the polar bears, also opens up an experience of partiality that becomes a useful metaphor for granting that maybe no one language, no one point of view, no one discipline (and no one species) has all the answers to the impending changes that are coming. But even as Jean blithely leaves Thomas out of portions of his conversation with Raphaël, he is angry when Thomas informs him that as a researcher in Nunavut, he is required to collaborate with the Inuit community. Gone are the days when a Western scientist can conduct research on indigenous lands without consultation: "I hope you know there's a community requirement," Thomas warns. "Every research project has to involve the community somehow. New rule from the Nunavut Research Institute," to which Jean replies: "I'm a SCIENTIST. Not a SOCIAL WORKER!" (14). As the play unfolds, Jean confronts several additional opportunities to learn that his knowledge, his perspective, and his personae are not paramount but part of a larger whole.

Jean's first lesson occurs in scene nine, in which he arrives to give a talk to Veronica's class at a local high school. While they wait for the students to arrive, Veronica probes: "[I]s this your first time on Baffin?" Jean confesses he's been doing research and coming to Baffin Island for some time. "In fifteen years you must have learned some Inuktitut." He makes excuses; there is no time, he says. "It's just how science works." Veronica holds him accountable – after all, it's her classroom: "what about *nanuq*?', or "*Qajaq*." Jean gets it, "Oh! Kayak right?" "*Iglu* . . ." Veronica's point is the Inuit culture has already contributed much to Jean's, but he is been unaware. "I guess I know more than I thought," he confesses. "Colonialism has a sneaky way of leaving its traces. *Qallunaat* got the land, but Inuit managed to infiltrate the language," she retorts and then hands him a flyer inviting him to hear her spoken-word poetry (40). In the next breath, she also invites him to shift his worldview:

> You know, Jean, things have changed over the last fifteen years. If you want to work in Nunavut, it's not enough to talk AT us anymore. You have to talk WITH us. That's just *pitsiaqattautiniq*. . . . You're a smart man, figure it out.
>
> *(40–41)*

Here Bilodeau turns the colonial gaze embodied in Jean back on itself, demanding research methods that require the researcher himself to grow and to change.

In *Decolonizing Methodologies*, Linda Tuhiwai Smith (1999) writes that Western research (like Jean's) reflects "the complex ways in which the pursuit of knowledge is deeply embedded in the multiple layers of imperial and colonial practices" (2). The idea that research benefits mankind – or as Jean says to Thomas, "What ever happened to science for its own sake? Science to understand the world?" (33) – reveals how researchers, as Tuhiwai Smith observes, "assume that they as individuals embody this idea and are natural representatives of it when they work with other communities" (2). Tuhiwai Smith continues by affirming that

"Indigenous people across the world have other stories to tell which not only question the assumed nature of those ideas and the history of Western research through the eyes of the colonized" (3). The systematic data gathering that Jean has initially planned is both a product of and a continuation of the process of colonialization. Tuhiwai Smith points out that such research is often "wielded as an instrument" of power over the communities that are the subject of the studies. She also demands more from those who critique colonial research, for to deconstruct the damage it has done to communities, she writes, "provides words, perhaps, an insight that explains certain experiences – but it does not prevent someone from dying" (3). Instead, she calls for research and *creative work* envisioned to serve the forward-looking needs and goals of indigenous communities. The critical questions that researchers and artists must ask are, "Whose research is it? Who owns it? Whose interests does it serve? Who will benefit from it? Who has designed its questions and framed its scope? Who will carry it out? Who will write it up? How will its results be disseminated?" (10).[23] In the case of *Sila*, these types of questions have not occurred to Bilodeau's Jean, an arrogant, a self-professed expert who is used to being listened to as an authority in his field. Throughout the play, Jean is confronted directly with these questions through the Inuit characters of Tulugaq and Veronica. Slowly and not without protest, Jean begins to engage in decolonizing his own mind. The journey immerses him, body and soul, in Inuit ways of knowing and costs him his prestige and power.

In Scene 12, which takes place in a hotel lobby, Jean meets Tulugaq, the hunter, community elder, and friend of Leanna who also works as a professional guide and tracker for non-Inuit scientific and hunting parties. After hearing about Jean's data collection project and equipment needs, Tulugaq turns to leave and says nothing. Jean follows him and once outside Tulugaq points up to the *Aqsarniit*, the Aurora Borealis. Tulugaq whistles, "They come closer. See?" Both men whistle, watching the northern lights move and dance in the sky.

> That is *Inuit qaujimajatuqangit*. Inuit traditional knowledge. Old learning about living in peace with people, animals, nature. Arctic is not just numbers. Arctic is stories. . . . *Inuit qaujimajatuqangit* is alive. Observation, experience. Always changing. Numbers are not enough. We need stories.
>
> *(52)*

In the language of place, kinship, and family, traditional stories have made sense of the elements of water, wind, ice, snow, and sun since time immemorial. Stories, Tulugaq aptly observes, are their own language and carry knowledge vital to survival.

Land-speaking

As understanding derived from adaptive practices handed down through generations, communicated through stories, ceremony, and cultural practices, traditional ecological knowledge is crucial to meeting the challenges of climate change (see, e.g., Cajete [2000], 269). Tulugaq, for example, listens to the sound of ice underfoot to determine if it's thick enough to traverse. Later in the play Jean ignores Tulugaq's warning and falls through into the icy Arctic sea. The experience destabilizes the primacy of Jean's faith in scientific knowledge alone and sets him on a path informed by Inuit traditional ecological knowledge.

In this near-death experience, Jean encounters the spirit world that Tulugaq understands as simultaneous with the material world. Stage directions charge artists to realize Jean's encounter on stage as "[d]eep blue darkness. Sunlight streams through a hole in the ice above, illuminating the underwater world . . . JEAN, seemingly unconscious . . . strands brush his face" (81). There he meets a "larger than life NULIAJUK," Inuit goddess of the sea. She is angry. Terrified and drowning, Jean begs for his life to be spared. He must listen to Nuliajuk and acknowledge her anger: "Like my father. Don't protect me. At a time. When. I need it most." Her accusations are fierce, and Jean submits, pleading, "I will protect you, I promise. I'll do anything" (86). Two scenes later, Jean sits wet and shivering with Tulugaq, who has saved Jean's life by fishing him out of the water. Tulugaq tells him the Inuit legend of Nuliajuk. A young girl, Nuliajuk married a kind young man, who was really an evil bird spirit. Nuliajuk's father came to rescue her, but when the bird attacked him, Nuliajuk's father threw her overboard into the icy sea. He cut off her fingers when she tried to climb back in. Ever grieving and angry at her father, Nuliajuk lives under the water where sea creatures, sprung from her fingers, are caught in her matted hair. Tulugaq charges Jean to respond – a response that will require Jean to step out of his worldview and assumptions of scientific rationality, and into another way of knowing and being-in-relation. "You have to comb her hair. . . . Nuliajuk is very angry. If you don't comb her hair, she keeps all the sea animals away from hunters," Tulugaq tells him. When Jean protests, Tulugaq refuses to continue working with him. "She asked you," Tulugaq reminds him (95). Framed as an invitation, Jean's encounter with Nuliajuk begins Jean's process of transformation during which he must acknowledge his paternalistic cultural viewpoint.

Stories are one of several modes of communication, or *ways of speaking*, that Bilodeau utilizes in *Sila*, effectively driving home the necessity of giving voice to all forms of knowledge. Ways of speaking include Inuit stories and legends, which carry and communicate indigenous traditional ecological knowledge passed down through generations that teach the ways of hunting, subsistence, thriving in the cold of the Arctic. Other ways of speaking include the voice of the land itself – in the voice of the wind, and the voice of ice as it moans underfoot telling the human or bear how thick, how safe; the spoken-word poetry of Veronica; the policy-speak and presentational formal speeches of Leanna as she makes her case before the commission; Jean's scientific-speak, and the language of geopolitics; the Canadian Coast Guard's Morse Code; and, finally, the language of grief.

Sila makes use of a vast vocabulary of expression and so calls into question the very the primacy of human language, suggesting instead that language is part of our fluid enmeshment in the world, no more distinctly human than bodies themselves. Like Veronica's throat singing, language, *Sila* suggests, arises from breath and exists on a continuum of breath–utterance–song–language that connects rather than divides the human and animal worlds. In *The Logos of the Living World*, Louise Westling (2014) argues that "[o]ur very bodies attest to the impossibility of a world without animals and to our own biological enmeshment within and inclusion on a continuum of other creatures" (72). Taking the audience into the play through a circle of shared breath, the sound of the wind together with the sound of the actors breathing are the first voices in performance.

Veronica's spoken-word poetry is in the tradition of Inuit throat singing and is a traditional form of song in which breath is directed both the inhale and the exhale to produce a range of sounds and notes quite distinct from Euro-American musical and vocal traditions. Throat singing is typically performed by women in pairs who face one another, inhaling and exhaling as one, and sharing breath and song. As a potential soundscape for the performance, throat singing is an embodied demonstration of shared breath, shared song, that suggests a new discursive space in which the power of voice is granted to peoples, animals, and the land itself. Indeed, in performance, as actors and audience breathe together, sharing the same air, an exchange of molecules is a testament to the absolute intimacy of being embodied in and of the world. The polyphony of languages – each intelligible and crucial in specific ways but none in universal ways – asserts not only knowledge but also voice across culture, species, and material form.

This same breath gives animals voice. Arising from shared bodies-that-breathe, breath becomes utterance becomes song becomes speech. These sign-systems and utterances that can be understood as meta-languages, languages of body, decoded across difference by millions of species every day, as animals interpret clues to survival and sustenance. Okanagan poet Jeanette Armstrong (1998) calls this "land speaking," writing that the "land as language surrounds us completely, just like the physical reality of it surrounds us. Within that vast speaking, both externally and internally, we as human beings are an inextricable part – though a minute part – of the land language" (178). Armstrong describes how *Sila* establishes a world in which speaking and language itself are fluid and mutable aspects of aliveness: "All my elders say that it is land that holds all knowledge of life and death and is a constant teacher. It is said in Okanagan that the land constantly speaks. It is constantly communicating" (176).[24] As the elder, Mama teaches Daughter to read this language of the land: the bubbles on the water, creaking of ice crystals, and direction of the wind are words – utterances of breath that is *sila*.

In Western storytelling traditions, talking animals are most often understood as allegorical humans, stand-ins for human experience. Indeed, the Western intellectual traditions have long prescribed speech only to humans; animals have been discursively silent (see, e.g., Wolfe [2003]; and Chaudhuri [2014]). While polar bears in popular culture might be "read" as merely the favored megafauna of climate change iconography or as human caricatures of humans, *Sila* resists this interpretation by representing its polar bears as integral to a specific culture as well as specific ecological niche. Indeed, the play demonstrates the ways in which ecologies and culture are intertwined and indivisible. Mama and Daughter polar bear are meant as representatives of their real-world counterparts and are part of a larger set of community relations and kinship between human and environment.[25] They are integral to the Inuit world of the play in which indigenous traditional knowledge of the ecological fluidity between animal and human worlds begins with the vitality of the land itself, which maintains continuity between human affairs and all other events. *Sila* asks its audience to consider themselves creatures who are part of a complex and dynamic web of relationships, reciprocal in their interdependence, a transcorporeal world in which we are enmeshed, entangled with all matter, and implicated in the ethical complications that such a perspective carries.[26]

The breakage is increasingly real to people of the Arctic, resulting in profoundly devastating, personal, grievable losses – losses that Watt-Cloutier (2007) argues are legally grievable violations of human rights. Speaking to the Human Rights Commission, Watt-Cloutier connects the dots between increasing social problems and the loss of ice. Ice is fundamental to Inuit hunting traditions, to communication and connection between villages, to identity, and to the fabric of the environment in which Inuit lifeways are integral:

> The ice is not only our 'roads' but also our 'supermarket.' Deteriorating ice conditions imperil Inuit in many ways. Ice pans used for hunting at the floe edge are more likely to detach from the land fast ice and take hunters away. As the ice is melting from below, hunters can no longer be certain of its thickness and how safe it is to travel upon. Many hunters have been killed or seriously injured . . . thinner ice also means much shorter hunting seasons. . . . In turn, some ice dependent species such as ringed seals, walrus and polar bears are experience impacts and the Arctic Climate Impact Assessment projects that these species will likely be pushed to extinction by the end of this century.
>
> *("Testimony" 2007)*

Returning to the play, in Act 2 the cost of ice loss becomes clear. Polar bears have evolved to be good swimmers, yet as the ice melts earlier, they must swim farther distances to access the ringed seal, their primary food.[27] Sometimes, the exertion causes overheating, and bears die not from drowning but from heat exhaustion. When mother bears weaken because they cannot access enough food to support lactation, cubs may die. No matter how efficiently Mama trains Paniapik to hunt, her skills of patience, endurance, observation, and precision cannot possibly keep pace with the rapid changes to the environment with which their kin have coevolved. Even as Mama praises Paniapik for taking her first successful seal, and the cub celebrates, "I am a good hunter!", the ice shelf has detached and floated farther than Mama realized. "Who broke the ice, Anaanaa? Who broke it? Who broke it? . . . Did the human break the ice?" Mama explains through a story: "The ice broke because Nuliajuk is angry. And Nuliajuk is angry because the humans have angered her. . . . We need to swim back to shore before the wind pushes us farther out. Can you do that?" (54–55). Not only polar bears but also people are caught in a too rapidly changing world.

Inuit hunters, Watt-Cloutier (2007) reminds us, not only sustain families and communities but also learn valuable life skills through the cultural practice and protocols of hunting. This embodied traditional ecological knowledge, she argues, forms the foundation of healthy identities; it develops the discernment, leadership, and generosity that ensure community survival:

> [H]unting culture that I come from is not only about the pursuit of animals and the technical aspect of a hunt. Hunting is, in reality, a powerful process where we prepared our young for the challenges and opportunities not only for survival on the land and ice, but for life itself. The character skills learned on the hunt of patience, boldness, tenacity, focus, courage and sound judgement and wisdom are very transferable to the modern world that has come so quickly to the Arctic world. We are seeing this powerful training ground on the land and ice being destroyed.
>
> *("Testimony" 2007)*

The loss of cultural lifeways, such as hunting, has resulted in increased depression, suicide, and drug use, especially among young people.

These are impacts are made palpable for the audience in the frayed relationship between Leanna and Veronica. Woven between scenes of the swimming polar bears, mother and daughter argue, and Veronica pleads, "Samuel got caught stealing gas yesterday" (57). Leanna hardly hears her daughter's immediate concern because she is preoccupied with a phone call regarding her global activism. Veronica screams at her:

> Mom, MY SON, YOUR GRANDSON, may be inhaling gasoline. Can you take that in? . . . You're going on and one about a stupid call with some stupid radio station while I'm here, doing everything I can to hold this family together. Your GRANDSON, Mom!

Moments later, the phone rings: "Yes, this is Samuel's mother . . . where is he? . . . where is he?" Here, the audience understands from Veronica's visceral response ("The blood drains from her face") that Samuel is dead (59). Meanwhile, Mama's panting is heard, and in the next scene, Mama and Daughter struggle to swim, to survive, in the open sea. Fighting exhaustion, Mama swims on, but *Paniapik*, also exhausted and weak, slips off her back and is taken by Nuliajuk. "Mama lets out a series of long desperate wails," the stage directions read, as *sila* becomes the voice of mourning (61).

The dialogue does not completely reveal if Samuel's death was an accident or suicide, but like the Daughter bear's drowning, both losses are the result of the disruptive changes in land and sea. Both mothers grieve a world that has cracked open; both deaths are caused by climate change. Bear and human, ice and land, are part of one body. Samuel's mother Veronica, a spoken-word poet, is mute with grief. In scene 18, Tulugaq comes to comfort Veronica; he enters her dreamworld of grief where Inuktitut and English letters swirl in chaos around her. She is adrift in a sea of language, searching for a reason to keep breathing. Meanwhile, Mama wanders the ice, searching and calling out to her daughter, her tongue licking the wound of her loss in a litany of grief that transcends species identities and language, giving voice to the wail of all mothers, indeed all parents who have lost a child to unfathomable changes that crack and tear at the structures that sustained their lives.[28]

Climate change theater as ceremonial practice

Like *Burning Vision*, *Sila* suggests that ceremonial remembering and grieving in relation to loss of land as well as loved ones may be a necessary response to climate change. In "Climate Change as the Work of Mourning," Ashlee Cunsolo Willox (2012) argues that

> grief and mourning have the unique potential to expand and transform the discursive spaces around climate change to include not only the lives of people who are grieving because of the changes, but also to value what is being altered, degraded, and harmed as something mournable.
>
> *(141)*

Leanna's community knows this loss well, as she tells us: "I come from a world where life and death walk hand in hand like giggling teenagers" (12). At the top of Act II, Leanna proposes that what keeps us in denial about climate change is not that we are ill prepared to make the kinds of changes to our lives that will be required, but that "we are totally unprepared for loss" (63). The "we" in this case includes audience and their counterparts in the world beyond the theater. *Sila* invites its audiences to feel into that loss. When a species, a culture, a language, or an ecosystem is lost, Willox suggests, part of who we are as individuals and communities is diminished. Each loss damages the body of knowledge that assures all people have a place in the fabric of things.

Coming full-circle back to the production of *Sila* I have described, like Jean's deep-sea journey, the endeavor changed us. The cultural demands of the play required partnerships with indigenous communities in our region. Yupik students, faculty, and staff became involved as cultural consultants and cast members, making rehearsals a space of connection and exchange around the lived experience of climate change. *Sila* reoriented our sense of home and kinship. Its nonhuman characters required us to empathize beyond the human. And its

North American indigenous characters challenged the white privilege of those of us from Euro-settler ancestry. As we worked toward production, Native community members watched rehearsals and offered comments; the role of Tulugaq (Kuvageegai in an earlier version of the script) was played by a Native elder from our community. He shared his personal knowledge of climate change impacts on regional indigenous communities.

The Inuit culture, cosmology, and language at the heart of the play provoked a radical understanding of the land as a living breathing system of relatedness. During the first cast meeting for the *Sila* production that I directed in 2014, Yupik Native law student Meghan Siġvanna Topkok, who played the role of Daughter bear, explained to the cast why kinship with animals is central to indigenous Arctic identities: "At one time in history we were the animals and they were us. People and animals were one being. It's not a metaphor for climate change" (2014). The polar bears were rehearsed not as objects/puppets, or even as individual characters, but as possibilities of becoming. "Becoming polar bear" but never "acting" bear, sharing breath and continuous movement, an ensemble shaped and reshaped their bodies, feeling into the question of kinship. Inspired by Inuit depictions of animals and humans as interwoven images of multiple forms, our Mama bear was always shape-shifting: an intermittent apparition.

Making use of theatrical practice as a mode of inquiry, actors also practiced crucial climate change skills and personal transformations, including an awareness of themselves as permeable, interconnected, mutable, and multiple. A year later, students surveyed affirmed that something fundamental in their worldview had shifted: "It was because the land was always alive – it was us – we were breathing together that I can't look at anything the same way: that rock, that river is alive," one cast member wrote to me.

Theater, like the world, arises from breath; the actors breathe in and out to animate the story. The audience breathes in and out, exchanging not merely artistic conjuring but also material breath – oxygen and carbon dioxide. This alone makes theater both a political and an ecological act. The utterance of another catches in our ears as the beats of a small drum and is then transformed into electrical impulses: the breath of utterance is inhaled into our bodies where it becomes blood that carries the molecular gifts of *sila* to our muscles and bones. The exchange is not metaphoric; it is as material as our bodies. In *Sila*, the bodies of the actors are like a drum that beats out its meaning and messages, changing the shared air, pressing on the cells of the audience, and touching them with the very stuff of *sila*.

Staying with the trouble

The environmental and climate crisis is a crisis of relationship and a crisis of story. Colonial assumptions that humans are separate from the earth's collective of lands and living others and as such can be mined for profit have brought all peoples to the brink of a shared catastrophe. The same evolutionary thinking that was fundamental to Anglo-Saxonism of the nineteenth century and justified North American colonialism is at work in linear, time-based imaginings of climate change. Donna Haraway (2016) argues that cataclysmic visions conjured by discussions of tipping points, peak oil, mass extinctions, and the collapse of civilization as we know it are fraught with a dangerous temptation to forget that individuals, families, and communities of earthlings have already and will continue to live through and beyond.

Meanwhile, indigenous people across North America have already lived through such cataclysm to survive and thrive. Powys Whyte (2018) argues that figuring out "what exactly needs to be done will involve the kind of creativity that Indigenous peoples have [already] used to survive some of the most oppressive forms of capitalist, industrial, and colonial domination" (47–48; see also Vinyeta, Powys Whyte and Lynn [2015]). Tuhiwai Smith (1999) recognizes that scholarly and creative deconstruction of hegemonic systems (like those that precipitated climate change) provides "insight that explains certain experiences" but does not "prevent someone from dying (3). While her admonitions are aimed at researchers in general, culture-makers are in a unique position to enact decolonizing by unsettling of the stories, systems, and structures that have brought the world to the present crisis. Decolonizing, Tuhiwai Smith asserts, consists of (re)claiming (stories, lives, land), celebrating (culture, women, survivance), indigenizing, or "centering of the landscapes, images, languages, themes, metaphors and stories in the indigenous world" (142–162).

In the same vein, Haraway (2016) stresses that culture-makers should consider what we wish to *call forth* through our anthroposcenic visions. In particular, she encourages artists to nurture the "ongoing and living worlds" through a practice of "tentacular thinking," which reaches out in multiple directions, through varied worlds, to make connections that are not only nonlinear and nonhierarchical but also sometimes defiantly antilogical and necessarily unsettling (33). Tuhiwai Smith (1999) similarly suggests that decolonizing is a proactive, generative, life-giving process and one in which stories and storying are central. Both *Sila* and *Burning Vision* are two of a raft of plays and performance projects that utilize theater's potential as a place to envision and call forth a future informed by empathy, community, and justice. Both works affirm the forward momentum in what Gerald Vizenor (1999) calls "survivance," which encompasses not only survival but also thriving culturally, spiritually, ecologically: the continuing and expanding presence of indigenous lifeways (vii).[29]

In *Burning Vision*, Clements (2003) honors those who lived before, *through*, and after unsettling notions that colonialism is in the past. In the ceremonial practice of her play, Clements reclaims a world that is intact, whole, and culturally rich before, during, and after, and in which people carry on their cultural traditions even as they cope with the ongoing impacts of colonialism. *Burning Vision* carries out what Tuhiwai Smith (1999) calls "indigenizing projects" (142) that not only work to dismantle the ideologies and systems of plunder that make all humans test dummies but also assert and affirm thriving reciprocity and relatedness among indigenous cultures and their living ecologies. *Burning Vision* stages improbable intimacies and incongruous solidarities in a story of survivance and sovereignty that challenges mainstream histories as it forges solidarity between Dene and Japanese victims. As a story that must be both enacted and witnessed, Clements's circular telling draws past, present, and future into a portrait of a hyperpresent in which the lived experience of indigenous people and Japanese people are brought into relation. In this way, Clements fixes theatrical attention on the *relations* that are implicated in the light of the earth-shattering human-made catastrophe of the bomb/colonialism. She conjures multiple historical moments in the hyper-present of the stage, where they cannot be denied or ignored and where a ceremonial remembering calls forth a kind of hope-in-action that carries lifeways forward.

Similarly, Bilodeau's *Sila* (2015b) envisions the shared vulnerabilities among people and the land in a way that functions as an invitation to come into relation, to respond even to catastrophic events with sober open-heartedness and self-reflection. The way each playwright uses the live, embodied, immediate, and communal qualities of theater to open questions that reach across borders of nation-states, continents, and species demonstrates how theater might envision and nurture ongoing living worlds in which we all might thrive.

The work of climate change is arduous and multifold, asking at once for social and political activism, personal exploration, and a reorganization of the imaginative understanding of what it means to be human. This diverse upsurge of artistic will on the part of climate change cultural workers shares an unwavering faith in the civic function of arts practice, acknowledging that the work of humanity in the Anthropocene is not to master geoengineering but rather to imagine into the possibilities of being human in a way that exercises and expands our capacity for radical empathy – across differences of person, body, community, nation, geography, and species. Theater provides this imaginative space where people – as audiences and artists – can encounter one another in the joint work of understanding, processing, and responding to climate change. Composed of embodied and communal storytelling that provokes felt connections of responsibility and reciprocity, theater as a way-of-knowing is a critically useful decolonizing methodology, one that can help not only to envision but also to enact compassion, justice, and democracy going forward.

Notes

1 *Sila*'s production at Central Square Theater in Boston was accompanied by a slate of ancillary events that brought scientists, civic leaders, and climate activists together, putting the play at the center of a community conversation about climate change. Bilodeau ("In Search" 2015a) describes a new aesthetic in her playwriting process that includes multiple protagonists, nonhuman characters, and putting community and the land at the center of the plot.
2 Amy Brady (2018) of the *Village Voice* notes the similarity of the two plays in her review, writing that Malpede's play is "[s]et not at, say, a climate convention, but on a plot of

remote family land, the drama feels removed – literally and figuratively – from actual policy debate." Brady's review is also pertinent to the role of the arts in the climate change debate.
3 Concurrent with international COP21 in Paris, Climate Change Theatre Action (CCTA) called for short plays or performance events initiated by artists and communities around the world and then published and promoted those events on social media (Bilodeau 2016). CCTA continues as an increasingly international consortium of playwrights and participating community organizations; in 2018, the group published a collection of these short plays in a volume titled *Where Is the Hope?* (Bilodeau 2018).
4 *Howlround*'s climate change series provides firsthand reports on the many expressions of climate-related performances. For more on this topic, see Bilodeau ("In Search" 2015) and May (2016).
5 Similar themes of transnational effects of environmental injustice are evident in *Ruined*, by Lynn Nottage (2009), and *Bhopal*, by Rahul Varma (2004). However, these plays do not as readily demonstrate a new dramaturgy – one that insists on bending the logic of time and place in the theater to draw attention to the interconnectivity of peoples across the global – nor do they track the historic patterns of ideologies that find their way into projects such as the atomic bomb.
6 For slightly differing narratives of the first use of the term *Anthropocene*, see Davies (2016), especially 42–45, Haraway (2016), 44–47, and Kolbert (2014), 107–110.
7 The discussion of Clements's play here reworks two earlier arguments about the production of Clements's *Burning Vision*, which I directed in 2010 (May 2010, 5–11; 2017, 1–18).
8 Gordon Edwards provides a chronology of the mining of uranium at Great Bear Lake for use in the U.S. atomic program in "How Uranium from Great Bear Lake Ended Up in A-Bombs: A Chronology" (2019).
9 Eldorado, and the obvious allusion to the Spanish conquistadors' search for gold in North America, is the actual name of the Canadian mining corporation founded by the LaBine brothers. See, for example, Robert Bothwell (1984), in which the author valorizes the mine in light of World War II but ignores its impacts on the environment and Dene lives.
10 Others have directed *Burning Vision* in an educational setting, and those production histories are also relevant here. See, for example, Annie Smith (2010).
11 The phrase "doctrine of discovery" refers to the rationale by which European explorers laid claim to indigenous lands. It is still used today to justify ongoing land takings. See, for example, Roxanne Dunbar-Ortiz (2014), 197–217.
12 Environmental historian Richard White (1994) writes about this significant interconnection of nature and work in "Are You an Environmentalist? Or Do You Work for a Living?"
13 For more about the ways that objects (can) imply connection and complicity, see Arons and May's (2013) discussion of Lynn Nottage's (2009) *Ruined*, in which the mining of coltan and other minerals used in contemporary technology implicate all of us in the violence against women in the context of the Congolese civil war.
14 Several short documentaries concerned with this legend are archived on the Canadian Broadcast Company website, which provides this description of the legend of the Waterheart of Great Bear Lake, according to the Sahtu' Dene of Déline: "[The Waterheart is a Dene legend] about a medicine man who found a giant heart beating at the bottom of Great Bear Lake. The lake is among the largest freshwater lakes in the world. After a trout steals the medicine man's hook he takes on the spirit of a loche, the largest fish of the lake, and dives deep into the lake's abyss to retrieve his hook. In his journey he finds much more. He finds a living, breathing heart, called the Tudzé in the Slavey language. This medicine man finds that the fragile Tudzé is what gives life to the everyday physical world of trees, fish, water, and human beings. The heart was also surrounded by every species of fish found in Great Bear Lake, guardians of the powerful Tudzé. The

Waterheart documentary explores the metaphors of this legend. The heart is the culture of the Sahtu' Dene people. They live in Déline, a small cluster of houses on the tip of the eastern arm of the lake. The waterheart and the culture represent everything the Sahtu' Dene live for."

15 The full text of the "1948 American Declaration of the Rights and Duties of Man" can be found at the Inter-American Commission on Human Rights from the Organization of American States.

16 The full text of Watt-Cloutier's testimony before the Inter-American Human Rights Commission can be found on the Earthjustice website.

17 Nunavut was formed as a new territory of Canada in 1999, when it gained autonomy from the Northwest Territories. The change has brought increased regional independence and indigenous cultural and linguistic autonomy (Kikkert 2019).

18 Taqralik Partridge is a celebrated Inuk spoken-word artist and throat singer. See, for example, Taqralik Partridge and Martin Keavy (2016). In 2014, Tanya Tagaq, Inuk vocal artist and throat singer, won Canada's Polaris Prize, one of the nation's highest literary honors. See, for example, Allan Kozinn (2014).

19 In her analysis of the long history of Western/white patriarchal research practices, Linda Tuhiwai Smith (1999) writes that "the term 'research' is inextricably linked to European imperialism and colonialism. The word itself . . . is probably one of the dirtiest words in the indigenous world's vocabulary" (1). She attests to its usefulness to those who wield it as an instrument of control and domination as "research became institutionalized in the colonies . . . from the imperial centres of Europe" (8). She calls instead for an indigenizing research methodology composed of "more that rhetorical acts of defiance," methods that have as their strategic purpose "cultural survival, self-determination, healing, restoration and social justice" (142).

20 Tanya Tagaq's lecture demonstration of Inuit throat singing can be found on YouTube.

21 In Leanna's speech in Scene 2, Bilodeau draws on the words of Watt-Cloutier. Because Watt-Cloutier has primarily shared her knowledge orally through her activism, I encourage readers to seek out her many talks available on YouTube, including her Keynote Address at Climate 2050, "Climate Change and Human Rights."

22 The origins and meanings of transformation pieces in Inuit art are complex; this art often entails artwork depicting an animal becoming human (or vice versa). For more on the interconnections between Inuit art-making and spiritual practices, see Frédéric Laugrand and J. G. Oosten (2008), the last chapter of which includes commentary from contemporary Inuit artists.

23 Linda Tuhiwai Smith is writing specifically about scholarly research and not expressly about artistic practice. I have included artists here, however, because her questions can (and I believe should) be applied to creative work as well, particularly productions like *Sila*, which endeavors to represent indigenous characters and cultural knowledge. While not a community-based play in the strict sense of the term, Bilodeau consulted closely with Inuit communities on Baffin Island in writing this play, and both the premiere production at Central Square Theater and the production I directed involved Native cultural consultants throughout the production process, and cast Native actors in Native roles to the extent possible.

24 For a more comprehensive discussion of "land speaking" as one of the modes linking Native theater and storytelling traditions, see also Christy Stanlake (2009), 23 and 39–42.

25 While *Sila* is not a Native play (the playwright is French Canadian), Bilodeau has drawn on strategies employed by many Native playwrights, which Stanlake (2009) describes in more detail. Scholarship in Native and First Nations drama and performance shares much common ground with ecodramaturgy, which in turn has benefited from the methods and theoretical modes of Native and First Nations theater as examples of effective eco-centered theater making. See, for example, Gray (2013) and Gray and Sheila Rabillard (2010).

26 Whether we call it the web of life, Gaia, or what Tim Morton (2010) calls "the mesh," *Sila* suggests that we are definitely enmeshed, entangled, interdependent, and implicated.
27 Molnár, Derocher, Thiemann, and Lewis (2010) argue that "sea ice is a complex and dynamic habitat that is rapidly changing. Failure to incorporate climate change effects into populations projects can result in flawed conservation assessments and management decisions" (1612). See also Richard Ellis's *On Thin Ice* (2009).
28 The poetry that Bilodeau ascribes to Mama in the yet unpublished version of the script, which I produced in 2014, included a litany of grief spoken by the mother polar bear. In the published version of *Sila*, Bilodeau (2015) has revised that segment, removing the poetry and indicating that Mama wanders the ice in search of the place where she gave birth to her cub (71–72).
29 A term coined by Anishinaabe scholar Gerald Vizenor (1999), *survivance* involves an "active sense of presence" and continuance. Beyond mere survival, survivance connotes thriving in the face of and a refusal of victimhood and tragedy.

References

Abel, Kerry. 2005. *Drum Songs: Glimpses of Dene History*. Montreal: McGill Queen University Press.
Alaimo, Stacy. 2010. *Bodily Natures: Science, Environment and the Material Self*. Bloomington, IN: Indiana University Press.
Alaimo, Stacy. 2014. "Thinking as the Stuff of the World." *O-Zone: A Journal of Object-Oriented Studies* 1: 13–21.
Allen, Paula Gunn. 1992. *The Sacred Hoop: Recovering the Feminine in American Indian Traditions*. Boston, MA: Beacon Press.
Allen, Paula Gunn. 2000. "The Ceremonial Motion of Indian Time: Long Ago, So Far." In *American Indian Theater in Performance: A Reader*, edited by Hanay Geiogamah and Jaye T. Darby, 69–75. Los Angeles, CA: UCLA American Indian Studies Center.
Armstrong, Jeanette. 1998. "Land Speaking." In *Speaking for the Generations*, edited by Simon Ortiz, 174–194. Tucson, AZ: University of Arizona Press.
Arons, Wendy, and Theresa May. 2013. "Ecodramaturgy in Contemporary Women's Playwriting." In *Contemporary Women Playwrights*, edited by Penny Farfan and Lesley Ferris, 181–196. New York: Palgrave MacMillan.
Bilodeau, Chantal. 2015a. "In Search of a New Aesthetic." *Howlround*, April 19.
Bilodeau, Chantal. 2015b. *Sila: The Artic Cycle*. Vancouver, BC: Talonbooks.
Bilodeau, Chantal, ed. 2015. "Theatre in the Age of Climate Change." *Howlround*, April 19.
Bilodeau, Chantal. 2016. "As the Climate Change Threat Grows, So Does the Theatrical Response." *American Theatre*, March 30.
Bilodeau, Chantal, ed. 2018. *Where is the Hope? An Anthology of Short Climate Change Plays*. Toronto: Center for Sustainable Practice in the Arts.
Blondin, George. 1990. *When the World Was New*. Yellowknife: Outcrop, The Northern Publishers.
Bonneuil, Christopher, and Jean-Baptiste Fressoz. 2016. *The Shock of the Anthropocene: The Earth, History, and Us*. London and Brooklyn, NY: Verso.
Bothwell, Robert. 1984. *Eldorado, Canada's National Uranium Company*. Toronto and Buffalo: University of Toronto Press.
Brady, Amy. 2018. "'Extreme Whether' Falls Short at Turning Climate-Change Awareness Into Compelling Drama." *Village Voice*, March 7.
Buffini, Moira. 2011. *Greenland*. London: Faber and Faber.

Cajete, Gregory. 2000. *Native Science: Nature Laws of Interdependence*. Santa Fe, NM: Clear Light Publishers.
Calarco, Matthew. 2008. *Zoographies: The Question of the Animal*. New York: Columbia University Press.
Capra, Fritjof. 1996. *The Web of Life: A New Scientific Understanding of Living Systems*. New York: Anchor Books.
Chaudhuri, Una. 2014. "The Silence of the Polar Bears." In *Readings in Performance and Ecology*, edited by Wendy Arons and Theresa May, 45–58. New York: Palgrave.
Chaudhuri, Una. 2017. "Climate Lens: Birth of a Post-Nation." *Howlround*, April 17.
Chaudhuri, Una, and Shonni Enelow, eds. 2014. *Research Theatre, Climate Change and the Ecocide Project*. New York: Palgrave.
Clements, Marie. 2003. *Burning Vision*. Vancouver: BSL Talon Books.
Davies, Jeremy. 2016. *The Birth of the Anthropocene*. Oakland, CA: University California Press.
Dunbar-Ortiz, Roxanne. 2014. "The Doctrine of Discovery." In *A Peoples History of the United States*, 197–217. Boston, MA: Beacon Press.
Edwards, Gordon. 2019. "How Uranium from Great Bear Lake Ended Up in A-Bombs: A Chronology." *Canadian Coalition for Nuclear Responsibility*.
Ellis, Richard. 2009. *On Thin Ice*. New York: Alfred A. Knopf.
Enelow, Shonni. 2014. "*Carla and Lewis*." In *Research Theatre, Climate Change and the Ecocide Project*, edited by Una Chaudhuri and Shonni Enelow, 87–116. New York: Palgrave Macmillan.
Environmental Assessment Panel. 1978. *Report of the Environmental Assessment Panel on the Port Granby Uranium Refinery Proposal, Eldorado Nuclear Ltd. = Rapport De La Commission Environnementale Sur Le Projet De Raffinerie D'uranium De L'Eldorado Nucléaire Ltée, À Port Granby*. Federal Environmental Assessment and Review Process. Ottawa, Ontario: Fisheries and Environment Canada.
"Gasp at Emotional Depth of 'The Breathing Hole' at Stratford Festival." 2017. *The Star*, August 21.
Gray, Nelson. 2012. "Bringing Blood to Ghost: English Canadian Drama and the Ecopolitics of Place." In *Readings in Performance and Ecology*, edited by Wendy Arons and Theresa May, 23–32. New York: Palgrave.
Gray, Nelson. 2013. "The Dwelling Perspective in English Canadian Drama." In *Greening the Maple: Canadian Ecocriticism in Context*, edited by Ella Soper-Jones and Nicholas Bradley, 511–526. Calgary, Alberta: University of Calgary Press.
Gray, Nelson, and Sheila Rabillard. 2010. "Theatre in and Age of Ecocrisis." *Canadian Theatre Review* 144: 3–4. https://doi.org/10.1353/ctr.2010.0016.
Haraway, Donna. 2016. *Staying with the Trouble: Making Kin in the Chthulucene*. Durham, NC and London: Duke University Press.
Kenny-Gilday, Cindy. 2000. "A Village of Widows." In *Peace, Justice and Freedom: Human Rights Challenges for the new Millennium*, edited by Gurcharan Singh Bhati, 107–118. Edmonton: University of Alberta Press.
Kikkert, Peter. 2019. "Nunavut." In *The Canadian Encyclopedia*, last modified January 22. www.thecanadianencyclopedia.ca/en/article/nunavut
Kolbert, Elizabeth. 2014. *The Sixth Extinction: An Unnatural History*. New York: Henry Holt and Company.
Kozinn, Allan. 2014. "Tanya Tagaq Wins Canada's Polaris Prize." *New York Times*, September 23.
Laugrand, Frédéric, and J.G. Oosten. 2008. *The Sea Woman: Sedna in Inuit Shamanism and Art in the Eastern Arctic*. Fairbanks, AK: University of Alaska Press.
Lewis, Ellen M. 2017. *Magellencia*. Portland, OR: Actors Repertory Theatre. Unpublished.

Lovelock, J.E. 2000. *Gaia: The Practical Science of Planetary Medicine*, Revised ed. Stroud: Gaia.
Macmillan, Duncan, and Chris Rapley. 2017. *2071*. London: Royal Court Theatre.
Malpede, Karen. 2014. *Extreme Whether*. New York: Theatre Three Collaborative. Unpublished.
May, Theresa. 2010. "Kneading Marie Clements' *Burning Vision*." *Canadian Theatre Review* 144: 5–11. https://doi.org/10.1353/ctr.2010.0002.
May, Theresa. 2016. "Radical Empathy, Embodied Pedagogy and Climate Change Theatre." *Howlround*, April 20.
May, Theresa. 2017. "*Tú eres mi otro yo* ~ Staying with the Trouble: Ecodramaturgy and the Anthropocene." *Journal of American Drama* 29 (2): 1–18.
McBride, Dara. 2017. "The Climate of Change in 'Fairly Traceable'." *American Theatre*, March 14.
McKibben, Bill. 2005. "What the Warming World Needs Now is Art, Sweet Art." *Grist*, April 22.
Mojica, Monique, and Ric Knowles. 2009. "Creation Story Begins Again: Performing Transformation, Bridging Cosmologies." In *Performing Worlds Into Being: Native American Women's Theater*, edited by Ann Elizabeth Armstrong, Kelli Lyon, and William A. Wortman, 2–28. Oxford, OH: Miami University Press.
Molnár, Péter K., Andrew E. Derocher, Gregory W. Thiemann, and Mark A. Lewis. 2010. "Predicting Survival, Reproduction and Abundance of Polar Bears under Climate Change." *Biological Conservation* 143 (7): 1612–1622. https://doi.org/10.1016/j.biocon.2010.04.004
Morton, Timothy. 2010. "Guest Column: Queer Ecology." *PMLA* 125 (2): 273–282. www.jstor.org/stable/25704424
Murphy, Colleen. 2019. *Breathing Hole*. Ontario, CA: Stratford Festival.
Nagel, Joane. 2016. *Gender and Climate Change: Impacts, Science, Policy*. New York and London: Routledge.
Nagle, Mary Kathryn. 2014. *Fairly Traceable*. Los Angeles, CA: Native Voices at the Autry. Unpublished.
Nestruck, Kelly. 2017. "Existence, Pursued by a Polar Bear in Stratford Festival's 'The Breathing Hole'." *Globe and Mail*, August 22.
Norgaard, Kari Marie. 2011. *Living in Denial: Climate Change, Emotions, and Everyday Life*. Cambridge, MA: MIT Press.
Norgaard, Kari Marie. 2016. "Climate Change Is a Social Issue." *The Chronicle of Higher Education* 62 (19): B4.
Nottage, Lynn. 2009. *Ruined*. New York: Theatre Communications Group.
Pachauri, Rajendra, Leo Meyer, and the Core Writing Team. 2015. "Climate Change 2014 Synthesis Report." Intergovernmental Panel on Climate Change (IIPCC).
Partridge, Taqralik, and Martin Keavy. 2016. "'What Inuit Will Think': Keavy Martin and Taqralik Partridge Talk Inuit Literature." *The Oxford Handbook of Canadian Literature: Oxford Handbooks of Literature*, edited by Cynthia Conchita Sugars, 191–208. New York: Oxford University Press.
Plumer, Brad. 2017. "Stay In or Leave the Paris Climate Deal? Lessons From Kyoto." *New York Times*, May 9.
Powys Whyte, Kyle. 2018. "Let's Be Honest, White Allies." *Yes* 85: 47–48.
Smith, Annie. 2010. "Atomies of Desire: Directing *Burning Vision* in Northern Alberta." *Canadian Theatre Review* 144 (1): 54–59. https://doi.org/10.1353/ctr.2010.0004
Stanlake, Christy. 2009. *Native American Drama: A Critical Perspective*. Cambridge and New York: Cambridge University Press.
Swidler, Ann. 1986. "Culture in Action." *American Sociological Review* 51: 273–276.

Tagaq, Tanya. 2009. "The Sounds of Throat Singing." YouTube, 3:07, July 7. 2009
Topkok, Meghan Sigvanna. 2014. Cast meeting Notes for *Sila*. Personal collection. Unpublished.
Tuhiwai Smith, Linda. 1999. *Decolonizing Methodologies: Research and Indigenous Peoples.* Dunedin, New Zealand: University of Otago Press.
Van Wyck, Peter C. 2012. "Northern War Stories: The Dene, the Archive and Canada's Atomic Modernity." In *Bearing Witness: Perspectives on War and Peace from the Arts and Humanities*, edited by Sherrill Grace, Patrick Imbert, and Tiffany Johnstone, 174–185. Montreal: McGill-Queen University Press.
Varma, Rahul. 2004. *Bhopal.* Toronto: Playwrights Press.
Vinyeta, Kirsten, Kyle Powys Whyte, Kathy Lynn, and Pacific Northwest Research Station. 2015. *Climate Change through an Intersectional Lens: Gendered Vulnerability and Resilience in Indigenous Communities in the United States.* General Technical Report PNW; 923. Portland, OR: U.S. Department of Agriculture, Forest Service, Pacific Northwest Research Station.
Vizenor, Gerald. 1999. *Manifest Manners: Narratives on Postindian Survivance.* Lincoln, NE: Nebraska.
Waters, Steve. 2009. *The Contingency Plan: On the Beach* and *Resilience.* London: Bush Theatre.
Watt-Cloutier, Sheila. 2007. Testimony before the Inter-American Human Rights Commission. *Earthjustice*, March 1.
Watt-Cloutier, Sheila. 2015. *The Right to Be Cold.* New York: Random House.
Watt-Cloutier, Sheila. 2018. "Climate Change and Human Rights." Keynote address at *Climate 2050.* Montreal, Quebec, Canada, October 24.
Westling, Louise. 2014. *The Logos of the Living World: Merleau-Ponty, Animals, and Language.* New York: Fordham University Press.
White, Richard. 1994. "Are You an Environmentalist? Or Do You Work for a Living?" In *Uncommon Ground: Rethinking the Human Place in Nature*, edited by William Cronon, 171–185. New York: W.W. Norton & Company.
Willox, Ashlee Cunsolo. 2012. "Climate Change as the Work of Mourning." *Ethics & the Environment* 17 (2): 137–164. https://doi.org/10.2979/ethicsenviro.17.2.137
Wolfe, Cary. 2003. *Animal Rites: American Culture, the Discourse of Species, and Posthumanist Theory.* Chicago, IL: University of Chicago Press.

EPILOGUE

Theater as a site of civic generosity

Of all the human qualities that contribute to environmental destruction – greed, carelessness, self-centeredness – none is as devastating as how quickly we forget. On July 12, 2017, the Larsen C ice shelf broke from the coast of Antarctica. It was a monumental climate-related event that scientists had been anticipating for years (Marchant 2017, 443). Between the writing and the reading of these words, there will be news of even more breakage. Headlines of new disasters layer one on the other like geologic stratification on overdrive. Elders who carry the lived experience of the Dust Bowl are fewer in number each year. Today, the environmental catastrophes of my youth – Chernobyl, Bhopal, Biafra, Love Canal, *Exxon Valdez* – are ghosts of headlines past. So, too, are the 2005 images of victims of Hurricane Katrina clinging to rooftops; the 2010 Deepwater Horizon/BP oil spill, when in excess of 200 million U.S. gallons devastated more than 1,000 square miles of the Gulf Coast, killing 11 and injuring many others; the 2012 superstorm Sandy, which pushed torrents of water through Manhattan subways; the 2014 poisoned water of Flint, Michigan; the destruction in Puerto Rico from Hurricane Maria in 2017; the devastation of Hurricane Dorian in the Bahamas.[1] The 2016–2017 Dakota Access Pipeline protests at Standing Rock (still ongoing) have earned a place in environmental justice and U.S. history. And yet the work of water protectors everywhere continues every day, headlines or not. Just today, to name but one example, the Environmental Protection Agency rolled back wetlands protection. Meanwhile, in Alaska, the Pebble Mine development threatens the largest wild salmon habitat in the world. The loss of rainforests worldwide, like other processes that take place over time, are often dismissed until it is too late. As new climate-related ecological disasters compete for public attention and outcry within or near U.S. borders, communities around the world live with the ongoing effects of those eclipsed events. Globally, suffering and loss related (directly or indirectly) to climate change (including the

DOI: 10.4324/9781003028888-9

mass migration of peoples) already defy the limits of imagination. Where does this destruction and heartache leave the "What if?" of theater considered in earlier chapters?

On February 10, 2017, I sat for five hours in a small, hot theater in Portland, Oregon, watching a play about a group of scientists who spend eight months wintering over in the international research station at South Pole, Antarctica. On stage, it is 1986. The scientists also occupy a small, enclosed space, and – like those of us in the audience – they too are sometimes hot, cold, hungry, irritated, angry, and inspired. The characters of Ellen Lewis's *Magellanica* (2018) include climatologists from the United States and the U.S.S.R., a Chinese American physicist, a British glaciologist, a Bulgarian cartographer, a Norwegian ornithologist, and crewmembers. They have come to collect data for various studies on ozone, glacial movement, topography, astrophysical plasma, ice, and penguin behavior. At first, each character seems isolated within the parameters of their discipline-specific interests and quirky personalities. They share a capsule-like accommodation in an inhospitable environment, and soon learn that they are entirely dependent on one another. There is no going home for these characters – at least not until the sun returns. On stage, they are housed within a thin layer of shelter powered by generators, where all supplies – everything they will eat, wear, use, burn, or break – have come with them on their military flight from New Zealand. Lewis and her production team have conjured the very visage of what Buckminster Fuller (1969) called the Spaceship Earth. When one of the generators stops working, two team members must venture out in subzero weather to fix it. As they risk death, Todor, the 70-something Bulgarian cartographer, speaks directly to the audience:

> We treat this world as if it is endless, but it is not endless. We have mapped its circumference, we have found its top and bottom, we have traveled past the edges of the maps until there were no more edges. This is the last edge, and here we are, standing on it!
>
> *(59)*

This is not a play about climate change, however, or at least not directly.

Magellanica premiered at Artist Repertory in January 2017, 30 years after the National Aeronautics and Space Association (NASA), in concert with scientists from the around the world, confirmed the thinning of ozone[2] – part of the vital protective layer of atmosphere that protects life on Earth. Throughout the play's five acts, which span six hours, we learn a great deal about both history and science. In a series of scientific reports beginning in 1985, NASA scientists, as well as their counterparts around the world, began to report an *ozone hole* over Antarctica. Experts from multiple disciplines had been tracking, measuring, and puzzling over ozone concentrations since 1956. By 1987, chemical analysis and direct observation by high-altitude aircraft confirmed the theory that ozone depletion was being caused by a particular class of gases called chlorofluorocarbons, or CFCs, which break apart the ozone molecule when they reach the stratosphere. Ubiquitous to modern life at the time, CFCs were commonly used in air-conditioning, refrigeration, polystyrene products (such as to-go cups), and propellants of all kinds, including hairspray and asthma inhalers. It was the most global, stunning, and imminently dangerous threat to the planet since the invention of the atomic and hydrogen bombs, and scientists around the world warned their governments to take action.[3] If the ozone continued to deplete at the current rate, the science indicated, life as we know it on Earth would be irrevocably harmed, perhaps extinguished. In 1987, the United States joined Canada and 12 European nations in negotiating the Montreal Protocol on Substances that Deplete the Oxone Layer, after which participating nations agreed to phase out the use of chlorofluorocarbons.[4]

Magellanica (2018) imagines into the convergence of scientific research between 1985 and 1987 when the "hole in the sky" became too apparent to ignore. Lewis has also written a metaphor for our own historic moment. We too stand at a "tipping point," at risk of losing our own fragile but superbly provisioned spaceship earth. *Magellanica* asks its 2017 audiences to conjure, with the actors on stage, a moment in 1986 when scientists reached across the divide of the Cold War, across disciplinary boundaries and across language and culture, saying that "the evidence of what's happening is clear, now" (231). Throughout the long winter, we watch as tempers flare and food and fuel seem precious. Meanwhile, Todor envisions a map in four dimensions. Decades before digital modeling of climate changes or hurricane trajectories, he longs for a map that can track changes over time as well as locations in space, a map that can tell us where we are going as well as where we have been. Comparing his own maps with those made in the seventeenth century, Todor reminds the audience that researchers are often blinded by their own assumptions and that as human beings we sometimes chart our lives by maps that are nothing more than outdated pictures of a world that may have never existed in the first place.

On June 1, 2017, my students and I watched the president of the United States turn logic on its head and withdraw the United States' commitment to the Paris Climate Agreement. While the cognitive dissonance between the story told in Lewis's *Magellanica* and Donald Trump's performance could not have been more mind-boggling, one student characterized Trump's speech as a perfectly timed culmination to the seminar. It suddenly brought home for each student the call-to-action that theater, too, must engage. In the face of conclusive scientific evidence, as well as the lived experiences of people at risk around the world, the president dismissed international partnerships, denied ethical obligations, and rejected sensible strategic planning. It was a radicalizing moment for some who watched Trump put their own futures in jeopardy in a single act of hubris. The president's Rose Garden speech drove home the urgency of the issues we had been talking about all term, including the perpetuation of climate denial through powerful and familiar narratives. In a final paper, one of my students wished that the president could comprehend the "struggles and stories represented in the plays" we had just studied:

> If Trump saw the ways that families in the Louisiana bayou are harmed by rising waters in Mary Kathryn Nagle's *Fairly Traceable*; or how loss of cultural practices resulting from climate change is causing epidemic depression and teen suicide in Chantal Bilodeau's *Sila*; or if he could watch the father in Steven Waters' *On the Beach* risk death in a mega-storm to prove to his son that government must do more to protect ordinary people. . . . If the president could see *any* of the plays we've read, then he might see the flaw in this idea of "fairness" he keeps talking about.

That student's faith in the potential of theater to tell stories that exercise empathy was compelling (if perhaps optimistic, in this case). Indeed, my abiding faith in the power of theater in times of crisis has been the primary reason for this book.

Surely, then, the world needs theater's "What if?" now more than ever. If, as artists, theater is our homeplace, we are presented now – as in at past historical crossroads – with the opportunity to *take a stand where we stand*. From here, we have a unique opportunity to tell new stories in new ways by applying the sometimes searing, sometimes silly, edge of our critical and creative practice to the predicament in which we, and all life on Earth, now face together. From this standpoint – the fluid, mutable, palpable location of theatrical experience – and through the civic practice of conjuring possible futures together, the vast panorama of Earth's ecologies and cultures might be empowered and safeguarded.

At a time when the master narratives of empire have induced a global ecological crisis, with implications for human and animal suffering of catastrophic proportion, the critical role of the performing arts as a site of counter discourse, resistance, and reimagining can hardly be more apparent. The task of dismantling the stories that have sanctioned destruction must proceed apace with the task of generating new stories that help flesh out the possibilities of a just, humane, and sustainable world. Theater is a site of confluence in which ecological relationships might be (re)imagined, explored, articulated, and, if not healed, at least brought to consciousness in ways that might sustain us on the path ahead. In the face of the ecological challenges of the coming century, how might theater serve as an arena for the exploration of reciprocity and relationship wherein fluid notions of self, other, and environment can be transfigured and transformed? How does theater participate and/or intervene in the ideologies that frame our thinking and understanding about environmental conditions? How does it reshape, dismantle, undo, and/or rearrange those ideas such that what was not visible before (perhaps due to conceptual blinders) becomes visible and actionable?

As ARTCOP21 (Chapter 6) and other examples in the preceding chapters have shown, theater as direct action can intervene in business-as-usual by blocking space, interrupting proceedings, making demands, and asserting the voices, presence, and experience of those who may have been ignored in public debate.[5] Theater can also intervene in other ways, such as in how we think, thereby shaping our social and ecological imaginations. When we leave the theater, for example, are things around us more alive, do we listen better, care more, have a deeper or more complex sense of our own ecological identity?

Theater can consciously and directly engage the complex economic, scientific and cultural challenges of climate change and climate justice, (re)imagining what it means to be human in the age of the Anthropocene. Through its unique embodied, immediate, and communal qualities, theater can advocate for environmental justice, develop a sense of connection between human and nonhuman communities, and strategically animate the ecological world so that the very boundaries between nature and culture, self and other, begin to dissolve.

As society becomes more polarized – itself a symptom of the social pressures of global climate change and environmental damage – theater's democratizing potential also becomes more crucial. In this dangerous civic climate, theater can serve as a kind of *bracketed time* in which we set aside both the demands and the differences that tear at democratic values. Simultaneously multivocal and multi-affective, theater provides a means to come together, suspend judgments, and witness the consequences of actions unfolding through the ruse of a drama. Theater thus can serve as a site of – as a practice of – civic generosity where people of all viewpoints and lived experience can be invited back into a conversation through story. In this way, theater can encourage us to exercise the very muscles of tolerance and empathy that seem to be atrophying before our eyes.

As a way of knowing that is at once imaginative, affective, immediate, embodied, and communal, theater can become a place to cope with shifting realities and to envision ways of being that preserve humane democracy. As artists and scholars, we must, Chaudhuri and Enelow (2104) write, apply "the vast resources of live embodied performance at the service of the program of radical re-imagining called for by the perilous predicament we find our species – and others – in today" (2). What that theater looks like, how it feels, and how it interfaces with the community bears witness to the unfolding present and to presence. It makes visible and palpable the interwoven ways in which, as philosopher Donna Haraway (2016) writes, "we require each other in unexpected collaborations and combinations, in hot compost piles. *We become-with each other or not at all*" (4, my emphasis). What Haraway calls staying-with-the-trouble includes understanding that empathy and compassion *are actions*, and offering a vision of how to *inhabit* a living-if-turbulent present.

Thus, ecodramaturgy is both a stance from which artists might proceed as well as from which to conduct dramaturgical analysis and historical study. Going forward, ecodramaturgy must seek to reimagine and revitalize the relationships between and among communities (human and otherwise) and places (material and imagined) even as they continue to be at risk. As a living art form, the product of which is lived, affective experience, theater invites us to live into our historic moment and unfolding crises and changes with open minds and feelers forward. As practice, theater can breathe life into the infinite enmeshment of us-ness, giving form to unexpected intimacies across isolation, bearing witness to ecological vulnerabilities, and in this way continuing to animate resilience during this epochal transition. "[M]any different paths forward are possible," historian Jeremy Davies (2016) writes, reminding us that "the chaotic nature of the crisis means that the flap of any given butterfly's wings might have disproportionate influence on the new world" (200). Each generation of artists must define anew what it means to put their creative shoulders to the wheel of change. Now is the time for butterfly-wing theater: conceived as a state of vigilance, a practice of humility; informed by the work of mourning, and the necessity of rage. Women pregnant with the future child of a future child will see our marks and hear our voices across time, and – like Clements's (2003) Dene See-er – look back at a history that has not yet happened. Whether it is a comic send-up of the why-can't-we-fix-this frustration of test dummies, or an invitation to honor kinship with polar bears, caribou, sturgeon, or radioactive rocks, theater must become a verb rather than a noun. Such theater offers us all an opportunity to enter into the interlocked fate of human and other-than-human worlds through embodied exploration of transcorporealities, where curiosity leads to compassion and perhaps interspecies solidarity. It is also an opportunity to break and break open the tired conceptual frames that have been used to colonize lands and peoples and, through stories, to live into what Tulugaq in Bilodeau's *Sila* (2014) calls "*pitsiaqattautiniq*" (41) – that is to say, a (respectful) relationship with one another and the places we share. Indeed, if theater's relational aliveness – an art form that occurs only as a living, embodied, present, and communal event – suggests anything at all, it is that all our relations – civic, familial, and ecological – require continual renewal.

Notes

1 A number of timelines of significant environmental events of the twentieth and twenty-first centuries exist online. One of the more reliable sources is environmentalhistory.org.
2 Ozone is a molecule composed of three oxygen atoms, one more than its typical, breathable, earthly, and stable state. While ozone can be hazardous in everyday air, high in the stratosphere it is part of the vital protective layer of the atmosphere, without which life on Earth would burn away. The NASA Goddard Space Flight Center provides a summary of research leading up to the mid-1980s conclusion that the depletion of ozone over Antarctica was accelerating. This research includes J.C. Farman, B.G. Gardiner, and J.D. Shanklin's (1985) observations of springtime losses of ozone over Antarctica. In 1986, NASA further confirmed ozone depletion over Antarctica based on the analysis of satellite data. Ozone depletion was confirmed in 1987 by aircraft, linking CFC-derived chlorine to ozone depletion.
3 In 1995, scientists Frank Sherwood Rowland, Mario Molina, and Paul J. Crutzen were awarded the Nobel Prize in Chemistry for their research, which led to the understanding that CFCs rise through the atmosphere to the stratosphere, where, broken down by radiation, ozone-damaging chlorine is released.
4 Despite the protests of multinational corporations like DuPont, one of the largest producers of CFCs, the Montreal Protocol came into effect January 1, 1989, and the United States led the way in phasing out CFCs. See, for example, Michael Weisskopf's (1992) *Washington Post* article.

5 Theater was a central means of direct action in the 1999 World Trade Organization protests, as performances by small bands of activists literally took the streets, shutting down traffic and business as usual. See, for example, Rysavy (1999) and Travers (1999). For a more general discussion of theater-as-direct-action, see Standing (2012) and Reed (2005).

References

Bilodeau, Chantal. 2014. *Sila*. Vancouver, BC: Talonbooks.
Chaudhuri, Una, and Shonni Enelow. 2014. *Research, Theatre, Climate Change and the Ecocide Project*. New York: Palgrave.
Clements, Marie. 2003. *Burning Vision*. Vancouver, BC: Talonbooks.
Davies, Jeremy. 2016. *The Birth of the Anthropocene*. Oakland, CA: University California Press.
Farman, J.C., B.G. Gardiner, and J.D. Shanklin. 1985. "Large Losses of Total *Ozone* in Antarctica Reveal Seasonal ClOx/NOx Interaction." *Nature* 315 (6016): 207–210. https://doi.org/10.1038/315207a0
Fuller, Buckminster R. 1969. *Operating Manual for Spaceship Earth*. Touchstone Book. New York: Simon and Schuster.
Haraway, Donna. 2016. *Staying with the Trouble: Making Kin in the Chthulucene*. Durham, NC: Duke University Press.
Lewis, Ellen M. 2018. *Magellanica*. Unpublished (post-production draft). February 6. Personal collection.
Marchant, Jo. 2017. "Iceberg Unveils Secret Ecosystem: Biologists Rush to Study Life Exposed under Antarctica's Larsen C Ice Shelf Before It Changes." *Nature* 549 (7673): 443. https://doi.org/10.1038/549443a
National Aeronautics and Space Administration Goddard Space Flight Center. 2019. "NASA Ozone Watch: What is the Ozone Hole?"
Reed, T.V. 2005. *The Art of Protest: Culture and Activism from the Civil Rights Movement to the Streets of Seattle*. Minneapolis, MN: University of Minnesota Press.
Rysavy, Tracy. 1999. "From the Trenches: Opening Day of the WTO Ministerial Meeting." *Yes!*, November 30.
Standing, Sarah. 2012. "Earth First!'s 'Crack the Dam' and the Aesthetics of Ecoactivist Performance." In *Readings in Performance and Ecology*, edited by Wendy Arons and Theresa May, 147–155. New York: Palgrave Macmillan.
Travers, Mary. 1999. "Art and Revolution: Playing the Fool to Beat the Man." *The Stranger*, December 2.
Weisskopf, Michael. 1992. "U.S. to End CFC Production 4 Years Earlier than Planned." *Washington Post*, February 12.

INDEX

Act Green xii
Actor's Rep, Oregon 241
Adams, Henry 92, 93
The Adding Machine (Rice) 90
Adorno, Theodor 209
African Americans: lived experience of environmental racism 164; post-Civil War settlement in Oklahoma 135; as sharecroppers 106–7, 171
the age of anxiety 123
Agricultural Adjustment Act (1933), ruled unconstitutional 95–6, 100, 103, 107–8; *see also Triple-A Plowed Under* (Arent) (FTP)
Ah Sin: A Dramatic Work (Harte and Twain) 28
AIDS epidemic 201
air pollution 146–7
Alabama (Thomas) 58
Alaimo, Stacy 226, 257–8
Alarcón, Francisco X. 187
Allen, Paula Gunn 246
Alligator Tales (Galjour) 207; ecological self 230, 233n24; reviews of 229; self-as-place 223; toward a transcorporeal self 223–30
Alurista 178
American, use of term 6–7
American Society for Theatre Research: Performance and Ecology Working Group xiv
American Theatre (journal) xiii
American Communist Party 95, 107, 115n7, 145

American Progress (Gast) 43, 61
American Theater, on eco-canon of plays 205
American theater, through historical environmental lens 1–17; ecodramaturgy as methodology 4–5; frontier ideology and environmental destruction 7–12; potentials of storytelling and social change 12–14; terminology usage 6–7; theater as civic discourse 2–4; theater as civic generosity 276–81; *see also* ecodramaturgy; theater
Angels Fall (L. Wilson) 188
Angels in America (Kushner) 201
Anglo-Saxonism: absence of relatedness of all being from 267–8; acceptance of lawlessness by white men 37–8; celebration of at Chicago's Columbia Exposition 19–20; and Darwin's theory of evolution 25–6; defined 51n9; private property concept 33–4; technology and utopia 69; *see also* frontier mythos; settler colonialism; white supremacy
Animal Ecology (Elton) 88
Anthropocene: birth of 12, 15n12; debates on beginning of 242–3; Moore on capitalist system as cause of 212; potential purpose of 269
Antiquities Act (1906) 62–3
Anzaldúa, Gloria 177, 178, 185–6, 187, 219
Appia, Adolphe 89

Arctic Circle (Bilodeau) 242; *see also Sila* (Bilodeau)
Arctic Climate Impact Assessment (2004) 252
Arizona (Thomas) 58
Armstrong, Jeanette 263
Arons, Wendy 219, xiv
Arrizón, Alicia 177
The Arrow Maker (Austin) 59
ARTCOP21 240, 279–80
assimilation, as violence 25, 37
Austin, Mary 59
Australia, apology to aboriginal peoples 49
Aztlán mythos 176–88, 190, 194n19

Back Talk from Appalachia (Norman) 214
Bacon, J.M. 8, 22
bad-boy-hero persona 19, 20, 40, 41, 132
Barthes, Roland 51n2, 131
Bates, Barclay W. 145
The Beginning and End of Rape (Deer) 39
Belasco, David 59; *see also Girl of the Golden West* (Belasco)
Belasco Theatre, New York 64
Berkeley Repertory Theatre, California 223
Bernabé (Valdez) 191; and Aztlán mythos 182–8; Catholicism themes in 183; feminist viewpoints on 184, 188; as *mitos* 182; plot summary 182–3
Berry, Wendell 126, 127, 148–9, 150, 151
"Beyond Bambi" (May) xiv
Beyond the Horizon (O'Neill) 90
Bierstadt, Albert 63, 66–7
Bilodeau, Chantal 242, 269n1, 270n23, 270n25; *see also Sila* (Bilodeau)
Biltmore Theatre, New York 97
binary framework of land 8
Birth of a Nation (Griffith) 134
The Birth of the Anthropocene (Davies) 243
bison annihilation 29–30, 102
Black Arts Movement 160
Black Elk 44
Black Hills, South Dakota 72, 126
Blackmon, Douglas 106
Black Power Movement 172
Blackstone, Sarah 45
Bliven, Bruce 105–6
Bloody Bloody Andrew Jackson (Friedman and Timbers) 8, 49–51; bad-boy-hero character 19, 20, 41; financial success of 18–19; racial stereotypes in 19, 28, 129, 216; reviews of 19, 49–50
Bonneville Power Administration (BPA) 87, 110

Borderlands/La Frontera: The New Mestiza (Anzaldúa) 177, 185–6, 187
Boulder Canyon Project Act (1928) 92, 109–10, 112
Brantley, Ben 19
Brava! Women for the Arts, San Francisco 218
Bread and Puppet Theater, Vermont 230
Breathing Hole (Murphy) 241
The Bronco Buster (Remington) 133
Brown, Kirby 121
Broyles-González, Yolanda 180, 181
Buffalo Bill's First Scalp for Custer (Cody) 24
Buffini, Moria 240
Buntline, Ted 23
Buried Child (Shepard) 208, 213; destructive legacies of frontierism 188–92; received Pulitzer Prize 188; reviews of revivals 190
Burning Vision (Clements) 241–5; as ceremonial time/theater 246, 252; Dene uranium extraction for atomic bombs 243–4; four movement structure 245–50; as model for serving justice 242; plot summary 11–12; re-rendering of Dene Waterheart legend 250–1, 270n14; reviews of 268–9; title based on Dene prophecy 243–4; tough like hope 251–2
Bush Theatre, England 240
Buss, Kato 51n7

Calarco, Matthew 258–9
California agribusiness 181, 185
Canada: Dene delegation trip to Hiroshima (1998) 244, 250; Supreme Court ruling on aboriginal oral history as evidence of fact 244, 245; uranium extraction for atomic bombs 243–5; *see also Burning Vision* (Clements)
Carla and Lewis (Enelow) 241
Carson, Rachel 9–10, 170, 171, 174
"The Case for Reparations" (Coates) 165–6
cattle industry 131, 133, 154n12
The Celebrated California Bandit (Howe) 64
Center of Puppetry Arts, Atlanta 205
Central Square Theater, Boston 240
ceremonial time (Allen) 246
Chaudhuri, Una: on climate chaos 240; on ecological theater xiii, 4–5, 161, 201, 203–4, 257, 280; *Staging Place* 202, 223; on transvaluation 1
Chávez, César 175, 180–1, 217–18, 222
Cherokee Nation: Oklahoma state dissolution of (1906) 122

Chicago Columbia Exposition (1893) 19–20
Chicana/o *teatros* 160, 194n22; *actos* 181, 182; *mitos* 182, 194n25; objectives of 179; *see also El Movimiento*; El Teatro Campesino (ETC); TENAZ – El Teatro Nacional de Aztlán
Chicano Movement *see El Movimiento*
Chief Joseph 44, 58
Christianity: conquering of wilderness as biblical directive 38, 60, 104; rhetoric of dominion of feminized landscape and technology 92; theatrical portrayal of 34
Civilian Conservation Corps (CCC) 113, 142
Civil Rights Act, Title VI (1964) 175
civil rights movement: as precursor to environmental justice movement 10; role of the arts in 159–61; *see also* social justice
Civil War 27
Clean Air Act (1970) 174
Clean Water Act (1970) 174
Clements, Marie 243; *see also Burning Vision* (Clements)
Cless, Downing 11, xiii, xiv
climate change: greenhouse effect 200–1; Inuit Circumpolar Commission lawsuit against U.S. 252; IPCC reports 201, 238; ozone depletion due to CFCs 200, 231n2, 277–8, 281nn2–3; scientific community warnings of 11; socially constructed denial of 238–9; theater in age of 239–42; unequal risks of 11
"Climate Change as the Work of Mourning" (Willox) 12, 266
Climate Change Theatre Action (CCTA) 240, 270n3
Climate Lens of New York 240
The Closing Circle (Commoner) 161
Coates, Ta-Nehisi 165–6, 167, 168, 170
Coatlicue State (Anzaldúa) 187
Cody, William F. "Buffalo Bill" 42, 45, 79; as adventure capitalist 46–7, 53n24; as Calvary fighter 23–4; as guide for elite hunting parties 23; legacy of 50; reenactments of hunting/fighting experiences 23–4; *see also The Wild West: The Drama of Civilization* (Cody)
Columbia River 110, 112, 113
The Coming of the New Deal (Schlesinger) 98–9
Commoner, Barry 161
communist agitprop theater, in Russia 91

communities of color 159–60; dramatists from 191–2; and environmental injustice 174, 175; scope of social injustice 175; *see also* indigenous peoples of North America
"The Concept of the Enlightenment" (Horkheimer and Adorno) 209
conservation: New Deal conservation-for-use 9, 59–60; and Pinchot's forestry techniques 8, 62–3, 69, 70; and progress as reclamation of biblical garden 8; during Progressive era 58, 59–61, 65–6, 69, 76–7, 81–2
Conservation and The Gospel of Efficiency (Hays) 63
consumer capitalism: negative impacts of 10; portrayed as patriotic 9–10; *see also Death of a Salesman* (Miller)
Contemporary Arts Center, Louisiana 205
Contingency Plan (Waters) 240
Coolidge, Calvin 111
Cooper, James Fenimore 35–6
Corbin, John 76
Craig, Edward Gordon 89
Cronon, William 62
cultural imperialism: naming as tool of 33, 41; and unequal risks of climate change 11
Cummins, Frederic T. 58
Custer, George Armstrong 23, 41, 45–6, 53n24, 72, 132, 154n13

Daly, Augustin 50, 230; *see also Horizon* (Daly)
Darwin, Charles 25–6, 40, 43
Davies, Jeremy 243, 281
Davy Crockett (Murdock) 36
Dawes Allotment Act (1887) 42, 113
Days of Gold (Rohrbough) 70
DDT, banning of 174
Death of a Salesman (Miller) 9, 144–53; call to witness catastrophe 144–8; dreams that kill us 148–53; extractive capitalism vs. habitation 144–50; film production 145; received Pulitzer Prize 123; reviews of 145
decolonization: as generations-long project 173; theater's role in 12, 216; Tuhiwai Smith on 268
Decolonizing Methodologies (Tuhiwai Smith) 216, 260–1
Deer, Sarah 39
Dell'Arte Company xiv, 161, 205
Deloria, Philip J. 44

de Mille, Agnes 121, 129
DeMille, William 59
Dene peoples, Canada *see Burning Vision* (Clements)
Deukmejian, George 217
Dietrich, William 94–5
Dietz, Steven 188
Dippie, Brian 133
doctrine of discovery 150, 155n22, 247–8, 270n11; *see also* frontier mythos; settler colonialism
Dolan, Jill 229–30
dry-land farming 179
Dunbar-Ortiz, Roxanne 24, 50
duplicitous and strategic amnesia (Moraga and Herrera Rodriguez) 222
Dust Bowl (Worster) 96
Dust Bowl era: Black Sunday 103; as ecological disaster 102–3; history of 124–5; *see also Triple-A Plowed Under* (Arent) (FTP)
Dynamo (O'Neill): infatuation with technology themes 9, 90, 92–5, 109, 110; plot summary 93–4
"The Dynamo and the Virgin" (Adams) 92, 93
dynamo mechanism, in electrical current 92

"Earthbound on Solid Ground" (hooks) 171
Earthjustice 252
ecocide *see Alligator Tales* (Galjour); *Heroes and Saints* (Moraga); *The Kentucky Cycle* (Schenkkan)
ecodramaturgy: during 1990s 205–6; Chaudhuri on xiii, 4–5, 161, 201, 203–4, 257, 280; defined i; early history of xi–xv; land as other self in 230–1; as methodology 4–5; as (re)illumination of the ecological 161, 280–1; responsibility for complicit history of theater 50
"Ecodramaturgy of Contemporary Women Playwrights" (Arons and May) 219
ecofeminist theater and performance art 202; *see also Alligator Tales* (Galjour)
ecological disasters: due to automobiles 142–3, 146–7; due to cattle industry 131, 133; due to insecticide use 171; due to westward expansion 29–30, 48; *see also* Dust Bowl era; settler colonialism
ecological self 230, 233n24
ecological theater *see* ecodramaturgy
"Ecological Theater" (Munk) 203

ecological violence of settler colonialism (Bacon) 8
ecosystem, use of term 6
Eisenhower, Dwight 141, 142, 151
Ellis, Jeffery Charles 140–1
El Movimiento: description of 14n9; *El Plan Espiritual de Aztlán* (Alurista) 178; theater as activating force for 10; *see also* Aztlán mythos; El Teatro Campesino (ETC)
El Plan Espiritual de Aztlán (Alurista) 178
El Teatro Campesino (ETC): and ecological understanding 10; on grounds for justice 175–80; legacy of 217, 230; staging Aztlán 180–2; *Vietnam Campesino* 192; *see also Bernabé* (Valdez)
El Teatro Campesino: Theater in the Chicano Movement (Broyles-González) 180
Elton, Charles S. 88, 89, 91
EMOS Ecodrama Playwrights Festival xiv, xvn5
Enelow, Shonni 241, 280
Enemy of the People (Ibsen) 205
Enlightenment, and frontier ideology 208–9
environmental activism 159; first-wave 192n2; second-wave 160–2, 179, 191, 192n2, 204; and social justice 9–10; and white privilege 10
environmental dramaturgy *see* ecodramaturgy
Environmentalism and Economic Justice (Pulido) 175–6
environmental justice 10, 11, 153, 168, 174, 175, 204; *see also* Aztlán mythos
environmental racism 163–4, 174, 175; *see also A Raisin in the Sun* (Hansberry)
environmental theater *see* ecodramaturgy
Environmental Theatre (Schechner) 202
ethnic cleansing *see* genocide and ecocide, in U.S. frontier expansion
extractive capitalism *see* industrial/extractive capitalism
Extreme Weather (Malpede) 240
Exxon Valdez oil spill (1989) xii

Fairly Traceable (Nagle) 240–1
Farmer–Labor Party 9, 95, 97, 100, 108, 109
Farmers' Alliance 128
Farmers Union 105, 128
farmworkers 175–6; *see also* El Teatro Campesino (ETC); *Heroes and Saints* (Moraga)

Federal Environmental Pesticide Control Act (1972) 174
Federal Theatre Project (FTP) 9; Living Newspapers 90–1; *One-Third of a Nation* 90, 95; *Timber* 91, 95; *see also* Power (FTP); *Triple-A Plowed Under* (Arent) (FTP)
Firehall Arts Centre, British Columbia 244
Flanagan, Hallie 91
Forepaugh, Adam 44
Fornes, María Irene 202
Forrest, Edwin 35
43rd Street Theatre, New York 111
14th Amendment 175
Fried, Larry K. xii, xiii, xiv
frontier mythos: and environmental domination 7–12, 19, 21–2, 25–6; Nelson Limerick on 14n5, 21, 215; in present day corporate logos/depictions 50; present day euphemisms from 21; white supremacy in 19; *see also* settler colonialism; Turner, Frederick Jackson (frontier thesis)
Fuller, Buckminster 277

Gaia Mon Amour (Rosenthal) 202
Garden of Eden myth 8, 26, 60–1, 72–3
Garrick Theatre, Chicago 72
Gast, John 43, 61
gendering of nature, as virgin/whore binary 60–1, 64–6, 74–5, 76, 111–12, 183–4, 185
genocide and ecocide, in U.S. frontier expansion 18–19, 24–7, 36–41, 47–8, 102
Geronimo 44, 58
Ghost Dance resistance movement 44
Girl of the Golden West (Belasco) 8, 59, 64–72, 73; masking of mining the west 69–72, 126; preservation/conservation tension 61; as recovery narrative 65–6; reviews of 65, 67–8; stagecraft technology 66–9
The Girl of the Golden West (Puccini) 64
Glacier National Park 62
Globe, Danny 125–6, 127–8
going/playing native (Huhndorf) 40, 75
gold rush 24, 44, 48, 58, 70, 72; *see also Girl of the Golden West* (Belasco)
Goodman Theatre, Chicago 205
Gore, Al 174, 193n15
Gottlieb, Robert 141, 142, 146–7
Grand Coulee Dam 87, 110, 112, 113
The Grapes of Wrath (Steinbeck) 122, 135, 220; received Pulitzer Prize 124

Gray, Nelson 248
The Great Divide (Moody) 8, 59, 72–82, 147; geographic typologies and environmental destruction 75–8; marriage to the land 72–5; plot summary 57; preservation/conservation tension 61; relations and ability to take 78–82; reviews of 72, 76, 79
Great Migration 135, 164
Great Sioux Nation 126
"Green Aztlán" (Ontiveros) 178
Greenberg, Linda Margarita 218, 220
Green Grow the Lilacs (Riggs) 9, 120, 121–2, 124, 129, 130, 132, 139
Greening Up Our Houses (Fried and May) xiii
Greenland (Buffini) 240
green theater/dramaturgy *see* ecodramaturgy
grief work 12
Griffith, D. W. 134
Grotowski, Jerzy xi
Group Theatre 90
Guthrie, Woody 87–8, 92, 110

Halprin, Anna 203
Hammerstein, Oscar 120, 124; *see also Oklahoma!* (Rodgers and Hammerstein)
Hanford Site, Washington 113
Hansberry, Lorraine 172; *see also A Raisin in the Sun* (Hansberry)
Hansen, James 11, 201
Haraway, Donna J. 229, 268, 280
Harden, Blaine 110
Harte, Bret 28
hate crimes 37
Hays, Samuel P. 63
Hearst, William Randolph 72
Helburn, Theresa 124, 143
Henry V (Shakespeare) 3
herbicides 193n15
Heroes and Saints (Moraga) 10–11, 188, 191, 217–23, 230; Catholicism themes in 221–2; farmworker bodies at risk 217–23; plot summary 218–19; reviews of 220; slow violence impacts on land and bodies 206–7
hero-scout persona 47
Herrera Rodriguez, Celia 222
Hetch Hetchy dam 63–4, 81
Hiawatha (Longfellow) 35
Hill, Robert F. 124–5
Hofstadter, Richard 25–6, 51n9
hooks, bell 171
Hoover Dam 112

Horizon (Daly) 8, 23–41, 59, 67, 147; assimilationist views in 25; Darwinian bravado in 26; demonization of Native characters 34–8; plot summary 28–9, 52n14; racist ideologies in 27; railroading land and lives 31–4; reviews of 27, 35, 36, 40; role in propagating frontier mythos 22–3, 24, 25, 48; standing (in) for wilderness 38–41; weaponizing the stage 26–31
Horkheimer, Max 209
House of Atreus (play) 207
Howe, Charles E. B. 64
Huerta, Dolores 175, 180–1, 218, 222
Huerta, Jorge A. 179, 181, 182–3, 220
Huhndorf, Shari M. 40, 75
human rights: environmental degradation as violation of 12; Inuit Circumpolar Commission lawsuit against U.S. for role in climate change 252; and social health issues 12
hydropower *see Power* (FTP)

Ibsen, Henrik 205
Indian, use of term 51n4
Indian Congress (Cummins) 58
Indian Relocation Act (1956) 123
Indian Removal Act (1830) 18, 35, 122, 125
Indian Reorganization Act (1934) 113
indigenizing projects (Tuhiwai Smith) 268–9, 270n19
indigenous, use of term 51n4
indigenous peoples, global risks of climate change for 11
indigenous peoples of North America: Cherokee identity in *Lilacs* 121; climate change harmful to 252; conflation with wilderness 38–41; delegitimization of land/water rights 34, 58; deterritorialization and forced removal of 35, 36–7, 62, 122, 125, 177; and doctrine of discovery 155n22; ecological violence of Dawes Allotment Act on 42; and genocidal frontier expansion 18–19, 24–7, 36–41, 47–8, 102; Inuit throat singing 263; land ethic of 114; national parks as forced deterritorialization of 62; Native performers in plays 44; Native prisoners of war used as exhibitions 44, 58; protection of cultural sites from artifact theft 62; return of tribal lands to 113; subsistence hunting practices 29; Turner's thesis on 20, 33, 40; white racialization of sexual desire 37; *see also* military occupation of indigenous lands; sovereign tribal nations
industrial/extractive capitalism: access to resources as basis of 38; binary separation of scenic beauty from economic utility 160; Black Hills gold 44; and conservation views 8; domination of nature viewpoint of 96–7; labor exploitation by 27, 48; and nature as matrix of economic exchanges 88–9; Nixon on the slow violence of 162; and private property concept 34; during Progressive era 57–8; uranium extraction for atomic bombs in Canada 243–5; and westward expansion 24, 25, 43, 46
In Mizzoura (Thomas) 58
insecticide use: estimates of usage in U.S. 174; and farmworkers 175–6, 217–18; by government 170–1, 175
institutionalized racism: defined 163–4; exclusion from wealth accumulation 168; infant mortality rates 169, 193n9; legacy of 168; redlining in housing markets 165–7; *see also A Raisin in the Sun* (Hansberry)
Inter-American Commission on Human Rights: Watt-Cloutier's appeal to 252, 264, 265
Intercourse Act (1832) 125
International Workers of the World (IWW) 130, 136–8, 154n16
Intiman Theatre, Seattle xii, 206
Inuit Circumpolar Commission 252

Jackson, Andrew 18, 35; *see also Bloody Bloody Andrew Jackson* (play)
Jacobson, Lynn xiii, 205
Jefferson, Thomas 24, 148
Jordan, John M. 69, 92–3, 94

Kazan, Elia 144
Kenny-Gilday, Cindy 244–5, 247
The Kentucky Cycle (Schenkkan) 10–11, 191, 203, 223, 229–30; ethos of habitation 212–13; individual choices as based on conscience 212; plot summary 207–8; received Pulitzer Prize 206; reviews of 214–16; settler colonialism and environmental destruction 207–17
King, Martin Luther, Jr. 175
King, Thomas 22
kinship, in ecological stewardship 2, 13, 114, 179, 185, 188, 225–7

Knights of Liberty 137
Kolbert, Elizabeth 238
Kroll, Jack 215
Kushner, Tony 201

labor strikes 105, 154n16
labor unions 176, 180
land, use of term 6
land as commons 112, 113, 114; *see also Power* (FTP)
land ethic (Leopold) 113–14
landscape, use of term 6
land speaking (Armstrong) 263
the language of myth (Barthes) 51n2
La Raza Cósmica (Vasconcelos) 186–7
lawlessness, by white men 37–8, 40–1; *see also* bad-boy-hero persona
League of Workers' Theatres 90
Leopold, Aldo 113–14
Letters to Harriet (Moody) 73, 78–9
Levine, D. M. 18
Levittown, Pennsylvania 166–7
Lewis, Ellen M. 241, 277, 278
liberation theology 222
Life on the Water Theatre, San Francisco: Earth Drama Lab 205, 223
Lilienthal, David 110, 112
Little, Frank 137
Little Theatre Movement 89
The Logos of the Living World (Westling) 263
Longfellow, Henry Wadsworth 35
Louisiana Purchase (1830) 97
Louisiana Purchase Exhibition, St. Louis (1904) 58
L.O.W. Gaia (Rosenthal) 202

Machinal (Treadwell) 90
Machine Age Ideology (Jordan) 69, 92–3
machismo, use of term 185–6
MacKay, Percy 46, 73
MacKay, Steele 23, 42–3, 44, 46; *see also The Wild West: The Drama of Civilization* (Cody)
Macmillan, Duncan 240
Madison Square Garden, New York 23, 42, 44, 58
Magellanica (Lewis) 241, 277, 278
Malpede, Karen 240
Mamoulian, Rouben 121, 130
Manes, Sheila 125
Manhattan Project 243
Manifest Destiny ideology 19–20, 25, 43, 48, 51n9, 155n22, 183–4; *see also* frontier mythos; settler colonialism

Marisol (Rivera) 203
Marker, Lise-Lone 66–7, 68
Marlboro cigarettes 50
Marshall, Robert 97
Martin, Edward Winslow 32
Marxist theory, on machines as a kind of "second nature" 92, 93
Mason, Bobbie Ann 214
May, Theresa J.: "Beyond Bambi" xiv; as co-founder of EMOS at Humboldt State University xiv; as co-founder of Performance and Ecology Working Group, American Society for Theatre Research xiv; as director of *Burning Vision* at University of Oregon 246–7; as director of *Sila* at University of Oregon 257–9, 266–7, 270n23, 271n28; as founder of Theatre in the Wild xii–xiii; *Greening Up Our Houses* xiii; as participant in *Mountain Project* at Teatr Laboratorium xi; *Readings in Performance and Ecology* xiv; *Salmon Is Everything* xiv; course on climate change theater 278–9
McCabe, Edward Preston 135
McCarthy, Joseph 145
McGee, W. J. 69, 70
McKibben, Bill 238
men conflated with agency 60–1
Merchant, Carolyn 60–1, 64–5, 76, 111
Mercury Theatre 90
Merrill Lynch bull 50
Merrimack Repertory, Massachusetts 205
Mesa-Baines, Amalia 177
Mestiza/o peoples *see* Aztlán mythos
Metamora, or the Last of the Wampanoags (Stone) 35, 36
Mexican–American War 176, 177
Mexico, indigenous history of 176
Mielziner, Joe 146
military occupation of indigenous lands: and Custer's defeat 41; federal extermination policies 24–5; theatrical legitimization of 8, 23, 25, 27, 28–31, 33, 39–40
Miller, Arthur 144, 145, 146; *see also Death of a Salesman* (Miller)
mining industry 48, 69, 71–2, 77–8; *see also Girl of the Golden West* (Belasco)
Mojica, Monique 251
Montreal Protocol on Substances that Deplete the Ozone Layer 200, 278
Moody, William Vaughn 72, 76, 78–9; *see also The Great Divide* (Moody)
Moore, Jason W. 212

Moraga, Cherríe 219, 223; *see also Heroes and Saints* (Moraga)
Moran, Thomas 66–7
Morgan, Matthew 43
Morosco Theatre, New York 123
Mother Earth (Shearer and Shearer) 161, 191
The Mound Builders (L. Wilson) 188
Muir, John 62, 63–4, 67, 88, 90
Mumford, Lewis 97
Munk, Erika 4–5, 203, 204
Murdock, Frank 36
Murphy, Colleen 241
My Civilization (Center of Puppetry Arts) 205
Myers, Bill 166–7
Myers, Daisy 166–7
Mythologies (Barthes) 131
myths: Barthes on 51n2; Schenkkan on 209–10; Slotkin on 131, 143, 151

Nagle, Mary Kathryn 240–1
National Aeronautics and Space Administration (NASA) 277; Goddard Institute for Space Studies 200–1
nationalism and vigilantes 40
Native, use of term 51n4
Native drama, use of term 26
Native theater 160, 270n25
Native Voices at the Autry, Los Angeles 240–1
natural world, use of term 6
nature: as demonic/sublime 72; as matrix of economic exchanges 88–9, 91; as organic machine 8–9, 88, 90, 92, 112; as series of dynamic energy exchanges 89; *see also* gendering of nature, as virgin/whore binary
Nature Is Leaving Us (Goodman Theatre) 205
Nelson Limerick, Patricia 14n5, 21, 49, 215
Nemiroff, Robert 162–3
New Deal 59–60, 122; *see also* Federal Theatre Project (FTP)
New Deal Soil Conservation Act (1937) 9
New Ecologists 88, 92, 96, 99–100
news media, role in propagating frontier mythos 37
New Stagecraft 89–90
New York Theatre Guild 9; *see also Oklahoma!* (Rodgers and Hammerstein)
New Zealand, incorporation of Maori culture into conservation strategy 49
Nixon, Rob 162

noble savage concept 27, 35–6, 59
Norgaard, Kari Marie 238–9
Norman, Gurney 214
nuclear threat 9–10, 159, 243

Oil, Wheat and Wobblies (Sellars) 136
oil industry pollution *see Alligator Tales* (Galjour)
Oklahoma: agribusiness and agrarian unrest 127–8, 130, 135–7; Black Wall Street Massacre 135; Green Corn Rebellion (1917) 136; history of 124–6; Oklahoma Land Run (1889) 126; post-Civil War migration of African Americans to 135
Oklahoma! (Rodgers and Hammerstein) 9, 120–44, 152–3; comparisons to precursor *Lilacs* 121–2, 124, 129, 130, 132, 139; cowboy as American identity 130–3; economic privilege and entitlement to land 128–30; as exclusionary white America 122; manufacturing desire for consumer goods 140–3; masking of agribusiness 124–8; nationalism within storyline 122–3; as propaganda for hegemonic order 143–4; race, radicalism, and dismissal of Jud's death 134–40; reviews of 131, 143; revivals of 153n2, 154n15
Oklahoma Organic Act (1890) 122
Oklahoma Progressive Party 128
Oklahoma Socialist Party 136
Olympic Theatre, New York 23
O'Neill, Eugene 90; *see also Dynamo* (O'Neill)
One-Third of a Nation (FTP) 90, 95
Ontiveros, Randy J. 178
outlaw persona *see* bad-boy-hero persona

Pan American Exposition 58
Park Theatre, New York 35
Partridge, Taqralik 253
Payne, David L. 125–6
Peabody, Josephine Preston 73
Peña, Devon G. 179
Pensamiento Serpentino (Valdez) 187
People of Color Environmental Leadership Summit (1991) 10, 174, 216–17
"Performance, Utopia, and the 'Utopian Performative'" (Dolan) 229–30
Perseverance Theatre, Alaska xii
Pinchot, Gifford 8, 62–4, 69, 70, 76–7, 81, 82, 82n7, 111

Piscator, Edwin 89
Planetary Dance (Halprin) 203
"Planet of the Year: Endangered Earth" (Time) 201
The Politics of Upheaval (Schlesinger) 103
postwar consumer culture *see Death of a Salesman* (Miller); *Oklahoma!* (Rodgers and Hammerstein)
Power (FTP): advocated public access to hydropower 91, 110–14; and invention of dynamo mechanism 92
Powys Whyte, Kyle 114, 242, 268
predatory housing mortgages 165–6
Princess Theatre, New York 72
Principles of Environmental Justice 168, 174
private property concept 33–4
pro-electricity movement *see Dynamo* (O'Neill); *Power* (FTP)
Progressive conservation *see* conservation
Provincetown Players 90
Public Theater of New York: Native Advisory Committee 18; *see also Bloody Bloody Andrew Jackson* (play)
public utilities *see Power* (FTP)
Puccini, Giacomo 64
Pulido, Laura 175–6, 179, 184–5

Quinn, Arthur Hobson 28, 35, 76

racial stereotypes: of new enemies of the nation-state 50–1; in plays and literature 19, 28, 37
railroad industry: and agribusiness 127; and bison hunting 29–30; corruption within 32; and westward expansion 24, 27, 31–4
A Raisin in the Sun (Hansberry) 10, 160, 179, 191, 192, 192n5, 206; (re)claiming home 162–5; bodies at risk 168–71, 193n13; environmental racism 163–4; homeplace, as environmental right 165–8; plot summary 164; received New York Drama Critics Award 192n5; redefinition of environment 161; reviews of 163, 192n6; revivals of 192n5; tending roots of belonging 171–5
Raker Act (1914) 64
Ramírez, Elizabeth C. 183, 184
rape: of natural resources 80–1; as weapon of colonization 39
Rape of the Sabine Women myth 73, 80–1, 83n12; *see also The Great Divide* (Moody)

Rapley, Chris 240
Readings in Performance and Ecology (Arons and May) xiv
Reclamation Act (1902) 60
redlining 165–6
The Red Right Hand (Cody) 24, 51n7
Reinhardt, Max 89
Reinheimer, Hermann 88
"Reinventing Eden: Western Culture as a Recovery Narrative" (Merchant) 60–1
Remington, Frederic 133
Reno, Milo, *New Republic* interview 104–5
research, use of term 253, 260–1, 270n19, 270n23
Rice, Elmer 90
Riggs, Lynn 9; *see also Green Grow the Lilacs* (Riggs)
"The Rise of Cheap Nature" (Moore) 212
Rivera, José 203
A River Lost (Harden) 110
Rivers of Empire (Worster) 92
The Road to Aztlán (Mesa-Baines) 177
Robinson, Edwin Arlington 73
Rodgers, Richard 120, 131, 143; *see also Oklahoma!* (Rodgers and Hammerstein)
Rohrbough, Malcolm J. 70, 71
"Roll, Columbia, Roll" (Guthrie) 92
Roosevelt, Franklin D. 9, 64, 82, 87, 95, 110; *see also* New Deal
Roosevelt, Theodore 21, 62, 81, 82, 113
Rosenthal, Rachel 202
The Rose of the Rancho (Belasco) 59
Royal Court Theatre, England 240
Royale, Edwin Milton 59
Rumble Productions 241

The Sabine Woman (Moody) *see The Great Divide* (Moody)
Salmon Is Everything (May) xiv
Salsbury, Nate 42
Sandberg-Zakian, Megan 240
San Francisco Mime Troupe 160–1, 230
satire, stereotypes within 50
savage, use of term 38, 40, 130, 134
"Scar Tissue" trilogy (Dell'Arte Company) 205
Schechner, Richard 202
Schenkkan, Robert 191, 209–10, xii
Schlesinger, Arthur, Jr. 95, 98–9, 103, 104
Schuman, Peter 202
Seattle Repertory Theatre, Washington 223, 226, 228
self-determination 168, 178
Sellars, Nigel Anthony 127, 135, 136–8

settler colonialism: and ecological ethos of destruction xiii–xiv, 8, 22, 41, 102–4, 207–17; racist legacy for minorities 170; reparations for 49; theatrical legitimization of 27, 36; *see also* gendering of nature, as virgin/whore binary; Manifest Destiny ideology
sexual desire, racialization of 37
sharecroppers and sharecropping 27, 105, 106–7, 128, 130, 135, 171
Shepard, Sam 188; *see also Buried Child* (Shepard)
"The Significance of the Frontier in American History" (Turner) 20–1, 33
Sila (Bilodeau) 12, 240, 252–69, 271n28, 281; as ceremonial practice 266–7, 269; climate justice and indigenous sovereignty 242; climate justice in Inuit Territory on Baffin Island, Canada as setting for 252; indigenous knowledge and sovereignty 254–7; as model for serving justice 242; reviews of 263; staging of polar bears 257–9; use of multiple languages and storied knowledge 259–66
Silent Spring (Carson) 9–10, 171, 174, 193n15
Sitting Bull 44
sixth extinction 238
slavery 102, 106, 167
Slavery by Another Name (Blackmon) 106
Slotkin, Richard 37, 45–6, 50, 131, 134, 143, 151
Smith, Molly xii–xiii
social Darwinism 26
socialist movements, of 1930s: domination of nature viewpoint of 96–7, 100
Socialist Party 137
social justice 159–60; absence from early environmental movements 9–10, 160; inclusion in second-wave environmentalism 160–2, 179; *see also* communities of color
social protest plays 89–90
Soil Conservation Act (1935) 96, 108, 113
Soil Conservation Service 113
sovereign tribal nations 41, 42–3; *see also* indigenous peoples of North America; *specific tribal nations*
the space of the animal (Calarco) 258–9
The Speed of Darkness (Tesich) 203
Spencer, Herbert 26
Squaw Man (Royale) 59
Staging Place (Chaudhuri) 202, 223
Stanislavski, Konstantin 89

"Starvation in Arkansas" (*The New Republic*) 107
staying-with-the-trouble (Haraway) 280
Stein, Howard F. 124–5
Steinbeck, John 122; *see also The Grapes of Wrath* (Steinbeck)
St. James Theatre, New York 120
Stoking the Fire: Nationhood in Cherokee Writing 1907–1970 (Brown) 121
Stone, Augustus 35
stories (King) 22
Stowe, Harriet Beecher 174
Street Scene (Rice) 90
Strongheart (DeMille) 59
Sunrise in the Hetch Hetchy Valley (Bierstadt) 63
superpower, origin of term 110
survivance (Vizenor) 268, 271n29
Synge, J. M. 90

Tagaq, Tanya 253, 254
Tansley, Arthur G. 89, 91
Taylor, Jonathon 246
Taylor Grazing Act (1934) 113
technological utopia *see Dynamo* (O'Neill); *Power* (FTP); *Triple-A Plowed Under* (Arent) (FTP)
tenant farmers and farming 95, 100, 101, 106–7, 124, 126, 127, 135–6, 137, 143
TENAZ – El Teatro Nacional de Aztlán 160, 217
Tennessee Valley Authority (TVA) 110, 112–13
tentacular thinking (Haraway) 268
Tesich, Steve 203
theater: Chaudhuri on antiecological bias of 161, 203; as civic discourse 2–4, 223–4, 230–1, 269; as civic generosity 276–81; as force for justice/social change during 1960s-1970s 160–1; as political and ecological act 267; potentials of storytelling 13–14; role in propagating frontier mythos 22–4; as site of generosity 276–81; and social change xi–xii; *see also* ecodramaturgy; *specific plays*
Theater (journal) 4–5
Theatre in an Ecological Age conference (1991) xii–xiii
Theatre in the Wild (TITW) xii
Theatre Topics (journal): "Beyond Bambi" xiv
Theatre for the New City, New York 205, 240
Theatre Guild, New York 90, 120, 121

Thienemann, August 88
This Bridge Called My Back: Writings by Women of Color (Moraga and Anzaldúa) 219
Thomas, Augustus 58
Thompson, Gary 126
350.org 238
Timber (FTP) 91, 95
timber industry 48
Topkok, Meghan Siġvanna 267
Torrence, Ridgely 73
transcorporality (Alaimo) 226, 257–8
transnational environmental justice activism 11
Treadwell, Sophie 90
Treaty of Guadalupe Hidalgo 176–7
Triple-A Plowed Under (Arent) (FTP) 9, 90, 95–109, 111, 114; cost of dust 101–5; dialogue about Agricultural Adjustment Act 101–2, 104; Drought (Scene 16) 100, 102; Dust Bowl era and discontent 95–7; farmer-worker solidarity 97–101; labor strikes and protests 105–8; legacy of 109; visages of the *un*natrual 108–9
Trump, Donald 278–9
Tugwell, Rexford 98–9, 103
Tuhiwai Smith, Linda 216, 260–1, 268–9, 270n23
Turner, Frederick Jackson (frontier thesis): Belasco's narrative consistent with 65; free land concept 33, 38–9, 47, 149, 209–10; habitation follows conquest concept 148; ideology in *Death of a Salesman* 147–9, 151–2; ideology in *Oklahoma!* 124, 128, 130, 131, 132, 138–9; legacy of 50; on pioneer savagery 20–1, 40; on re-wilding 34; savage turned businessman concept 79, 80
Twain, Mark 28
2071 (Rapley and Macmillan) 240

Uncle Tom's Cabin (Stowe) 174
Underground Railway Theater (URT), Boston 205, 216–17, 222, 240
United Farmworkers Movement (UFW) 175, 176, 180–1, 217–18
United Nations Intergovernmental Panel on Climate Change (IPCC) 12, 201, 238
United States/U.S.: absence of reparations for indigenous genocide 49; electrical energy generation 95; genocide and ecocide in frontier expansion 18–19, 24–7, 36–41, 47–8, 102; Inuit Circumpolar Commission climate change lawsuit against 252; Louisiana Purchase (1830) 24; subsidization of railroad construction 32; tribal termination policies 9, 24–5, 122, 123; use of term 7; withdrawal from Paris Climate Agreement 278; *see also* industrial/extractive capitalism; military occupation of indigenous lands; settler colonialism; white supremacy; *specific agencies, leaders and legislation*
University of Oregon: EMOS performance and ecology symposium xiv
The Unsettling of America (Berry) 126
unsustainable farming practices *see Triple-A Plowed Under* (Arent) (FTP)
uranium and radiation poisoning *see Burning Vision* (Clements)
Urban Ink 241
U.S. Bureau of Reclamation (BOR) 60
U.S. Calvary 23
U.S. Congress: designation of Earth Day 159; dissolution of treaty system (1871) 41; House Un-American Activities Committee 145; land grants to railroads 31–2; resistance to T. Roosevelt's conservation initiatives 63; *see also specific legislation*
U.S. Environmental Protection Agency 276; formation of 174
U.S. Farm Security Administration (FSA) 103, 122, 124
U.S. Federal Housing Administration 166
U.S. Forest Service 63–4, 113
U.S. Indian Bureau 24–5, 33, 37, 39, 44
U.S. National Park Service 41, 61–2, 64, 113

Valdez, Luis 180, 182, 187; *see also Bernabé* (Valdez); El Teatro Campesino (ETC)
Vasconcelos, José 186–7
Vietnam Campesino (ETC) 192
The Virginian (Wister) 132
The Virginian (Wister) 41
Vizenor, Gerald 268, 271n29

Waldau, Roy S. 121
Waldron, Nelson 45
Wallace, Henry 108
Waterheart legend (Dene peoples) 250–1, 270n14
Waters, Steve 240
Watt-Cloutier, Sheila 12, 264, 265; as inspiration for character in *Sila* 252–3,

254–5; as nominee for Nobel Peace Prize 252
Weaver, Jace 9, 122
Wells Fargo Bank logo 50
West Fever (Dippie) 133
Westling, Louise 263
"When Green Was Pink: Environmental Dissent in Cold War America" (Ellis) 140
White, Richard: on cattle industry 131; on coerced Native performers 44; on culture of hate crimes against indigenous peoples 37; on genocide/ecocide of the plains 30; on glorification of machines 92; on myth of subsistence miner/logger 70; on railroad industry 32; on risk of erasure of culture 214–15
white actors, playing characters of color 28, 49–50
white supremacy: and Darwin's theory of evolution 25–6; in frontier mythos 19; hooks on as not form of absolute power 171; lawlessness by white men concept 40–1; present day self-appointed border guards 40; *see also* Anglo-Saxonism; environmental racism; institutionalized racism
white women, as victims 39
wilderness preservation movement 61–2; *see also Girl of the Golden West* (Belasco); *The Great Divide* (Moody)
Wild West shows 21, 26–7, 42, 53n23
The Wild West: The Drama of Civilization (Cody) 8, 23, 41–7; assimilationist views in 25; Battle of Little Bighorn 45–6; dehumanization of Natives 46; induced terror as entertainment 44; Native performers seen as authentication 44; reviews of 44, 45; role in propagating frontier mythos 23, 24, 45–6
Willox, Ashlee Cunsolo 12, 266
Wilson, Lanford 188
Wilson, Woodrow 64
Winter, William 67–8
Wister, Owen 41, 132
Wobblies 135, 136–8
women conflated with nature *see* gendering of nature, as virgin/whore binary
Woods, Rufus 112
Works Progress Administration (WPA) 9, 91; *see also* Federal Theatre Project (FTP)
World Trade Organization (WTO), Seattle protests against 230, 282n5
Worster, Donald: on cattle industry as capitalist institution 131; on Dust Bowl era disaster 102, 104; on Marxist theory of machines as "second nature" 92, 93; on Progressive conservation 77, 88–9; on U.S. imperial/commercial relationship with land 96, 97–8
The Wrath of Grapes (UFW) 217–18

Yazzie, Rhiana 49–50
Yellow Hair 23–4
Yellowstone National Park 41, 62
Yosemite National Park 63

Zinn, Howard 105
Zoographies: The Question of the Animal (Calarco) 258

Printed in the United States
by Baker & Taylor Publisher Services